SPACE
FORCE
PIONEERS

Titles in the Series

The Other Space Race: Eisenhower and the Quest for Aerospace Security
An Untaken Road: Strategy, Technology, and the Mobile Intercontinental Ballistic Missile
Strategy: Context and Adaptation from Archidamus to Airpower
Cassandra in Oz: Counterinsurgency and Future War
Cyberspace in Peace and War
Limiting Risk in America's Wars: Airpower, Asymmetrics, and a New Strategic Paradigm
Always at War: Organizational Culture in Strategic Air Command, 1946–62
How the Few Became the Proud: Crafting the Marine Corps Mystique, 1874–1918
Assured Destruction: Building the Ballistic Missile Culture of the U.S. Air Force
Mars Adapting: Military Change during War
Cyberspace in Peace and War, Second Edition
Rise of the Mavericks: The U.S. Air Force Security Service and the Cold War
Standing Up Space Force: The Road to the Nation's Sixth Armed Service
From Yeomanettes to Fighter Jets: A Century of Women in the U.S. Navy

Transforming War
Paul J. Springer, editor

To ensure success, the conduct of war requires rapid and effective adaptation to changing circumstances. While every conflict involves a degree of flexibility and innovation, there are certain changes that have occurred throughout history that stand out because they fundamentally altered the conduct of warfare. The most prominent of these changes have been labeled revolutions in military affairs (RMAs). These so-called revolutions include technological innovations as well as entirely new approaches to strategy. Revolutionary ideas in military theory, doctrine, and operations have also permanently changed the methods, means, and objectives of warfare.

This series examines fundamental transformations that have occurred in warfare. It places particular emphasis upon RMAs to examine how the development of a new idea or device can alter not only the conduct of wars but also their effect upon participants, supporters, and uninvolved parties. The unifying concept of the series is not geographical or temporal; rather, it is the notion of change in conflict and its subsequent impact. This has allowed the incorporation of a wide variety of scholars, approaches, disciplines, and conclusions to be brought under the umbrella of the series. The works include biographies, examinations of transformative events, and analyses of key technological innovations that provide a greater understanding of how and why modern conflict is carried out and how it may change the battlefields of the future.

SPACE FORCE PIONEERS

TRAILBLAZERS OF THE SIXTH BRANCH

EDITED BY **DAVID CHRISTOPHER ARNOLD**

Naval Institute Press
Annapolis, Maryland

To the On-Duty Crew

Naval Institute Press
291 Wood Road
Annapolis, MD 21402

© 2024 by the U.S. Naval Institute
All rights reserved. No part of this book may be reproduced or utilized in any form or by any means, electronic or mechanical, including photocopying and recording, or by any information storage and retrieval system, without permission in writing from the publisher.

Library of Congress Cataloging-in-Publication Data

Names: Arnold, David Christopher, editor.
Title: Space Force pioneers : trailblazers of the sixth branch / edited by
 David Christopher Arnold.
Other titles: Trailblazers of the sixth branch
Description: Annapolis, Maryland : Naval Institute Press, [2024] | Series:
 Transforming war | Includes bibliographical references and index.
Identifiers: LCCN 2024019364 (print) | LCCN 2024019365 (ebook) | ISBN
 9781682471449 (hardback) | ISBN 9781682471739 (ebook)
Subjects: LCSH: United States. Space Force—History. | Astronautics—United States—History. | Astronautics—United States—Biography. | Astronautics and state—United States—History. | Outer space—Exploration—United States—History. | BISAC: HISTORY / Military / Aviation & Space | HISTORY / United States / 20th Century
Classification: LCC UG1523 .S6334 2024 (print) | LCC UG1523 (ebook) | DDC 358/.800922—dc23/eng/20240516
LC record available at https://lccn.loc.gov/2024019364
LC ebook record available at https://lccn.loc.gov/2024019365

♾ Print editions meet the requirements of ANSI/NISO z39.48–1992 (Permanence of Paper).
Printed in the United States of America.

32 31 30 29 28 27 26 25 24   9 8 7 6 5 4 3 2 1
First printing

The opinions expressed in this book are those of the authors and should not be construed as carrying the official sanction of the Department of Defense, Department of the Air Force, or other agencies and departments of the U.S. government.

CONTENTS

Preface ix

    Introduction 1
    DAVID CHRISTOPHER ARNOLD

1 Toward the United States Space Force 13
    RICK STURDEVANT

2 The First Guardian: Bernard A. Schriever 45
    JOHN G. TERINO

3 The Space Force's Godfather: John B. "Bruce" Medaris 67
    LISA BECKENBAUGH

4 Problem Solver: Osmond J. Ritland 93
    DAVID CHRISTOPHER ARNOLD

5 The Space Force's Revolutionary Commander: Thomas S. Power 118
    BRENT D. ZIARNICK

6 Father of the Space Command: James V. Hartinger 144
    MARGARET C. MARTIN

7 Father of the Space Force: Thomas S. Moorman Jr. 166
    WILLIAM D. SANDERS

8 Stars among the Stars: Charles F. Bolden, Kevin P. Chilton, and Susan J. Helms 194
    STEPHEN J. GARBER AND JENNIFER M. ROSS-NAZZAL

9 Maturing Leadership of Space: Lance W. Lord, C. Robert Kehler, and William L. Shelton 224
    HEATHER P. VENABLE

10 The Blind Kid from Alabama: John E. Hyten 254
    GREGORY W. BALL AND WADE A. SCROGHAM

11 Present at the Creation: John W. "Jay" Raymond 278
    LANCE JANDA

    Conclusion: The Space Force Needs Visionary Leaders 298
    DAVID CHRISTOPHER ARNOLD

Contributors 303
Index 307

PREFACE

After decades of discussion, the U.S. military added a sixth branch of the armed forces in December 2019, the United States Space Force. The military services have been developing space capabilities since the end of World War II, but it wasn't until the 1957 launch of Sputnik that the space race really began. All the services leaped at the chance to lead the military into space. President Dwight D. Eisenhower chose the Navy to launch the first satellite, which failed, and then the Army launched the first American satellite in space. But after losing that short race, the Air Force coined the word "aerospace" and surged ahead in the American military's own space race. Over time, space became critical to strategic leaders as well as to operational and tactical warfighters. By the 1991 Gulf War, space so was important to the fight that Air Force Chief of Staff Gen. Merrill A. McPeak referred to the conflict as "the first space war." However, control of space programs was disorganized and inefficient. After several evaluations of space management and organization and lots of criticism of the government's handling of space programs, and with congressional and White House support, the Department of Defense stood up a separate military space organization whose mission is to defend the space domain, coequal with the services that defend the other domains.

The U.S. Air Force has often been accused of ignoring its history because it was the youngest military service and therefore had so little history to focus on. None of these statements are true, of course, but it is entirely fitting as we look at over sixty years of the space age in the rearview mirror, with the newest military service just a handful of years into its existence, that we examine just how much space history the U.S. Space Force is built on. There are many more examples out there of the giants on whose shoulders we stand, to paraphrase Sir Isaac Newton, but the long climb to the top begins here.

—David Christopher Arnold
Washington, DC

INTRODUCTION
DAVID CHRISTOPHER ARNOLD

> Events of recent years have made it clear to the man in the street that he is indeed living in the Space Age. This realization marks not only the achievement of incredible scientific advances, but also a change in national awareness. Space for all too long, has been popularly regarded as a mysterious, forbidding, and unapproachable realm, peopled only by cartoon characters and the creations of science fiction writers. Now it has become evident the future of our Nation—indeed the future of free men everywhere—may depend on what we can achieve in space.[1]
>
> —Gen. Bernard A. Schriever, 1961
> Commander, Air Force Systems Command

Just as World War I's airmen derived the missions of military airpower (observation, bombardment, pursuit, and logistics), so the Cold War's leaders derived the missions of spacepower (reconnaissance, surveillance, weather, communications, navigation, and launch). U.S. Air Force Gen. Bernard A. Schriever was a space thought leader for many years in and out of uniform, but when asked by a young major working on his dissertation in 2001 if Schriever thought of himself as the father of military space, Schriever responded, "I know I'm not," although he did allow that he was the father of the intercontinental ballistic missile (ICBM).[2] Schriever's denial of his place in military space derived from the many changes that occurred after he left the day-to-day management and leadership of military space programs in 1961 to lead all of the Air Force's research and development efforts. Yet in 1998, in recognition of his tremendous influence on military space, the Air Force renamed the nation's premier space operations base in his honor.

As Carl von Clausewitz put it, "Historical examples always have the advantage of being more realistic and of bringing ideas they are illustrating to life."[3] In these chapters, you will find events in space history that are "carefully detailed" to bring them into clearer focus. Here you will find many important events in military space history "thoroughly detailed," which "is more instructive than ten that are only touched on," as Clausewitz argued. Indeed, Clausewitz's protagonist, Napoleon Bonaparte himself, studied great military leaders because although "the duties and knowledge of an engineer or artillery officer, may be learned in treatises . . . the science of strategy is only to be acquired by experience, and by studying the campaigns of all the great captains. . . . Your own genius will be enlightened and improved by this study, and you will learn to reject all maxims foreign to the principles of these great commanders."[4] Gen. George Washington, when asked as early as 1775 for his thoughts on which military books to read, recommended to his officers who wished to become better leaders books including Humphrey Bland's book on military discipline and another on the art of war by Count Turpin de Crissé.[5] Thus, if one wishes to be a great captain, one must study the great captains. This is a book about some great captains of military spacepower.

Synopsis

This volume aims to enhance our understanding of leadership in military space through the experiences of past space leaders. Unlike leadership of the U.S. Air Force, where there is a sizeable literature both in comprehensive histories of the service and biographies of senior leaders, leadership in military space continues to remain a mystery to many, with relatively little in-depth research because of the public face of the civil space program represented by the National Aeronautics and Space Administration (NASA) compared to the more secretive military space community, the relative dominance of air leaders of space programs, and the influence of civilian leaders on these programs. In short, this book will help pave the way for discussions about leadership in military space, particularly important at the dawn of the new military space service.

These general officers have had a much greater influence than simply what happened while they were at the helm of a major program—they helped to create and shape the culture of military space organizations. Psychologist Edgar Schein, who spent his life studying organizational culture and leadership, argued that "culture is ultimately a characteristic of a group, just as personality and character are ultimately characteristics of an individual." In short, "leadership creates changes; if those changes produce success for a group and the leader's vision and values are adopted, a culture evolves and survives." If they fail, we "usually never hear about it . . . [but] When leaders produce a whole new organization . . . we hold them up as 'models' of great leadership. . . . Leadership is necessary, but it succeeds only when the new way fits what was needed."[6] United States Space Force (USSF) Lt. Col. William D. Sanders digs deeper into military space culture in a 2022 article by arguing that the heritage of space in the Air Force, and in the greater military space culture, shows us "several competing traditions: engineers, operators, integrators, and warfighters."[7] It is in the deep roots of these cultures, and the wider military culture, that these leaders worked. These four competing traditions form the outline of the organization of this book but, as the reader will see, at best these are loose categories.

Historian Thomas Parke Hughes points out in his book *Rescuing Prometheus* when writing about the leadership of the ICBM programs in the 1950s that these officers were "motivated by the conviction that they were responding to a national emergency, [so] they single-mindedly and rationally dedicated the enormous funds at their disposal to providing national defense."[8] Schriever tried to develop deterrence through strength and capability, not the threat of mutually assured destruction.[9] As he put it later, "I am looking for ways to avoid killing people. . . . We need to do something other than find ways to kill people better."[10] The leaders in this book created changes and success, but the context and the culture changed over the decades.

The people in this volume, and others, used spacepower to change the character of war in the twentieth century. A concept known to Clausewitz scholars, the character of war consists of who fights, why they fight, and how they fight. These characteristics of war are important in the analysis of both why wars happen and how they might proceed.[11] Space systems changed who fights air campaigns. By 1945 tens of thousands of airmen had been lost in the skies over Germany. But with space systems, a single bomber today can hit as many targets as a fleet of B-17 bombers and with greater accuracy and less loss of life, for both aviators and those on the ground. Space systems also changed how the American military fights not only by enabling more accurate targeting but also through more accurate coordination among the military services. Route packages in the skies over Vietnam kept Air Force and Navy aircraft from running into each other. But with satellite communications, the services can coordinate their actions, and the joint force has become a truly integrated operation. And of course, none of these satellites are crewed by fragile humans in space, but all of them are remotely operated, circling the globe, providing their data to computers and people on the earth. Space systems also provide intelligence data that military people and national decisionmakers never had before 1960.

Space systems have accelerated the pace of change in the character of war and simultaneously reduced the time needed to make decisions about war. War is no longer what Clausewitz called "solely the concern of the government" because war has become "deprived of its most dangerous feature—its tendency toward the extreme" as people, governments, and indeed the military put more limits on war using space systems.[12] Why they fight, of course, is because nation-states still fight for their national interests, and soldiers still fight for their teammates, because the "paradoxical trinity" remains relevant today.[13] But space systems have made war both less violent, by not involving the entire nation-state, and more deadly, by enabling quick decisions and lightning-fast actions.

Finally, the concept of "revolutions in military affairs" is one that has been around for a very long time, suggested in the 1950s by Michael Roberts, who wrote about the changes that occur during a military revolution. As Roberts and, later, Geoffrey Parker pointed out, the requirements for a revolution to occur (or to have occurred) range from technological and organizational changes to social, doctrinal, and political-economic changes.[14] The invention and employment of space-based capabilities changed the ways political and military leaders employed the military instrument of power. The generals in this book implemented huge technological programs that changed the way that leaders think about their enemies' capabilities, that aided in nuclear deterrence, and that have become essential to the world economy. The Air Force itself endured huge organizational changes: from being a service that got involved in space because of interservice rivalry, to one that invented the word "aerospace" to assert the "operationally indivisible medium consisting of the total expanse beyond the earth's surface" of air and space, then arguing in a 1996 white paper the Air Force was "on an evolutionary path to a *space and air* force," to eventually turning over the space mission to an independent military service.[15] Today,

United States Space Command argues its area of responsibility surrounds the earth above one hundred kilometers to, one could assume, infinity.[16] Space is a place, a domain, and a mission.[17]

But people execute the space mission. And those people must be led by—and are often inspired by—other people. In the military, many—although not all—of these leaders and inspirations are general officers. We have chosen several to cover in this volume as the makers of the United States Space Force. We picked this group of general officers for a couple of reasons. First, this volume stands on the shoulders of giants. This book aims to modernize and complement an earlier work by Edgar Puryear Jr., *Stars in Flight: A Study in Air Force Character and Leadership* (Presidio Press, 1981). In biographical chapters of Henry "Hap" Arnold, Carl Spaatz, Hoyt Vandenberg, Nathan Twining, and Thomas White, Puryear examines how men devoted to country, flying, and an independent air service built the U.S. Air Force. In this volume, we examine people devoted to country, space capabilities, and, eventually, an independent space service.

Second, this book expands on examinations of the role of leadership in the history of the U.S. Air Force. The most read is DeWitt S. Copp's *A Few Great Captains: The Men and Events that Shaped the Development of U.S. Air Power*, which Doubleday first published in 1980. Copp's book used Hap Arnold, Frank Andrews, Carl Spaatz, and Ira Eaker to trace the evolution of American aviation to 1939 in a narrative format. Shirley Thomas' multivolume *Men of Space* series first appeared in 1960. Volume 1 featured prominent people like Robert Goddard, Bernard Schriever, John Paul Stapp, Konstantin Tsiolkovsky, James Van Allen, Wernher von Braun, Theodore von Kármán, John von Neumann, and Chuck Yeager, only three of whom were in the U.S. military. Other volumes followed that featured engineers, scientists, astronauts, and leaders in the space business in twenty-five- to thirty-page chapters focused on a particular theme. Her eight-volume series was a decade-long project to document the "completely distinctive and individualistic" space figures of Thomas' period in history.[18] John L. Frisbee's 1987 edited volume *Makers of the United States Air Force* is closer to the concept for the volume now in your hands than other works described previously. Frisbee looked at American aviators who were not the most famous names of World War II but who nonetheless had an enormous impact on the U.S. Air Force. They include people like Benjamin Davis,

Benjamin Foulois, Harold George, George Kenney, and Elwood Quesada. Others like Hoyt Vandenberg, Nathan Twining, and Schriever also appear.

As should be obvious by now, it has been forty years or more since many of those works were published, and a great deal has changed in the military during that time, including the creation of a new military service directly connected to space. This volume offers an examination of modern space leaders but traces the value of leadership in military space as well. In short, there is not another book that offers a comprehensive and contemporary examination of leadership in military space and how these leaders led—and changed—the space business. Additionally, the contributing authors are leaders themselves. As such, their chapters bring invaluable insight based on firsthand experience as well as deriving insights from personal interviews with some of these space leaders.

In the opening chapter, longtime military space historian Rick Sturdevant provides an overview of technological and organizational evolution over the course of seventy-five years. Focusing primarily on the U.S. Air Force, his narrative explains how emerging technologies and changing international threats "pushed" the U.S. military to create new organizations. The need to satisfy new requirements, as defined by those organizations, simultaneously "pulled" technology forward. Over the span of decades, military space capabilities originally fielded for strategic purposes became increasingly useful, indeed critical, for satisfying theater and tactical warfighting needs and, later, even economic ones.

What follows are biographical chapters in which space leaders' careers are examined, an approach that proves illuminating because of the varied nature of individual leadership styles. Each leadership case study is qualitatively researched using information derived from the personal writings of their subjects and, in some cases, oral histories and interviews with the subjects, along with other historical sources. Each chapter places its subject in context by addressing the space environment in which the respective leader served, his/her leadership style, his/her performance, and any lasting contribution he/she made to military space while serving in or out of uniform. Collectively, the chapters offer insight and analysis into how different leaders have operated, as well as chronicling changes in military space over the past four decades.

John G. Terino, Air Command and Staff College professor, writes about U.S. Air Force Gen. Bernard A. Schriever. A German immigrant to the United States in 1917, Schriever learned to fly in the 1930s after earning an engineering degree at Texas A&M University in 1931. Although he left the military and flew for the airlines in the late 1930s, Schriever returned to the Army Air Forces for World War II, serving in the Pacific theater and rising to the rank of colonel. While working in a variety of technical assignments on the air staff before and after the independence of the air service, Schriever came to be a leading advocate for ICBMs and eventually took command of the Air Force's ICBM developmental organization in California. From there, he furthered the development of the Atlas ICBM and introduced concurrent development of the Titan missile. Even before the beginning of the space race in October 1957, Schriever had added military space programs to his portfolio. He introduced systems engineering and project management to these programs, producing new weapons systems at a faster rate than before. Although Schriever left California for greater responsibilities within the air service, he remained involved in the military's space and missile programs until his retirement in 1966 and well beyond. In 1998 the service renamed Falcon Air Force Base in Colorado in his honor.

Air Command and Staff College's Lisa Beckenbaugh discusses U.S. Army Maj. Gen. John B. "Bruce" Medaris. A career Army ordnance officer with experience in World War II including the Normandy invasion, Medaris began investigating the use of guided missiles for future warfighting while on the staff at the Pentagon in 1953. In February 1956 he took command of Redstone Arsenal in Huntsville, Alabama, where he worked and became friends with Wernher von Braun. Medaris was involved in preparing ballistic missiles for the Army and working on a satellite program that eventually became the Explorer I satellite in 1958, the first U.S. satellite system to reach orbit. Medaris was opposed to the decision to take the von Braun team from the Army in the run-up to the creation of NASA but eventually relented, setting the stage for the moon landings.

David Christopher Arnold, National War College professor, writes about Air Force Maj. Gen. Osmond J. Ritland. Born in 1909 and raised in California, Ritland learned to fly in the 1930s following three years in the engineering program at San Diego State College. Ritland left the Army to fly for United Airlines on the West Coast until World War II, when he returned to the military and became a developmental test pilot at Wright Field in Ohio. Ritland flew over two hundred different types of airplanes during the period, including one of the first two XP-51 fighters and early jet aircraft like the XP-59. After the war he rejoined the test community and worked on developing new capabilities like the ejection seat until his assignment to the Pentagon in 1954, where he worked on Air Force nuclear programs and then became the service's program lead for the U-2 program. In 1956 Schriever assigned Ritland as his principal deputy in California and gave him responsibility for military space programs. When Schriever left California for Washington, Ritland became commander of the ballistic missile division, with responsibility, in addition to ICBM programs, for the Midas, Samos, and Discoverer satellite programs, the last of which was the Corona reconnaissance satellite program. In 1962 he became chief of manned space flight for the Air Force, retiring in 1965.

Brent D. Ziarnick, a retired Air Force Reserve officer and a civilian Air Command and Staff College faculty member, tackles Gen. Thomas S. Power. A B-24 and B-29 pilot during World War II, Power was one of the few officers without any college to achieve general officer rank in the Army Air Forces. After the war, he served in a variety of nonflying assignments, including as commander of Air Research and Development Command when Schriever was building the Los Angeles team into a space and missile development organization. Taking over Strategic Air Command in 1957 following the tenure of Curtis LeMay, Power integrated space missions into the organization. Often overlooked by Air Force historians, he advocated that nuclear propulsion programs, military human spaceflight programs, and other systems should be controlled by Strategic Air Command, while also developing doctrine and advocating for equipment and organizations for space systems.

Margaret Martin, head of the history department at the United States Air Force Academy, writes about Air Force Gen. James V. Hartinger. Joining the Army at age eighteen for World War II, Hartinger attended West Point after the war and joined the Air Force in 1949, becoming a fighter pilot and eventually serving in Korea in 1953. Hartinger had a variety of staff assignments and leadership roles in the fighter community, including flying over one hundred air combat missions in Vietnam. In 1968 he was the test director of the F-111 program. In 1980 he became commander of the North American Aerospace Defense Command (NORAD) and, following a series of internal service discussions, the first commander of Air Force Space Command (AFSPC) in 1982. During this major internal reorganization of the Air Force, Hartinger managed the transfer of assets and systems to the new command from other major commands and made a point of educating his staff on space, including assigning readings and then conducting pop quizzes at staff meetings on space topics. The headquarters building for Space Operations Command, AFSPC's modern descendant, is named for him.

William D. Sanders, U.S. Space Force, writes about Air Force Gen. Thomas S. Moorman Jr. Graduating from Dartmouth with a degree in history and political science in 1962, Moorman served in intelligence and reconnaissance roles throughout the 1960s, including a tour in Thailand during the Vietnam War. In the early 1970s he served on the air staff in a variety of roles and helped draft the national space policies of both President Gerald Ford and President Jimmy Carter. At NORAD, Moorman was responsible for the space catalog and also served as speechwriter for Hartinger during the establishment of AFSPC. From 1990 to 1994 he served as AFSPC commander and then, following a reorganization, as AFSPC vice commander. He completed his military career as vice chief of staff of the Air Force, the first nonpilot to hold that position. After retirement, he continued to serve in the space community, including on the 2001 space commission.

NASA historians Stephen J. Garber and Jennifer Ross-Nazzal write about three general officers who were also astronauts: Charles F. Bolden, Kevin P. Chilton, and Susan J. Helms. Astronauts hold a special place in the pantheon of military space, beginning with the Mercury astronauts who were all military test pilots, but rarely do military astronauts rise to the senior officer ranks in the military. A U.S. Marine with a technical background who served as an astronaut, Bolden both achieved general officer rank and after retirement served as NASA administrator. Only one astronaut, Chilton, has achieved four-star rank; as a test pilot he became a shuttle astronaut, flying three times in space. After a storied spaceflight career, in 1998 Chilton left NASA to "rejoin" the Air Force, achieving four stars and serving as AFSPC commander and then commander of United States Strategic Command (USSTRATCOM). Susan Helms was in the first Air Force Academy class to include women and spent much of her early career in aeronautical engineering assignments, including attending Air Force test pilot school. She flew on four shuttle missions, and when she returned to the Air Force, quickly earned accolades and promotions, culminating in her assignment as commander of Fourteenth Air Force in California.

Heather P. Venable of Air Command and Staff College writes about three transitional space leaders: Lance W. Lord, Robert C. Kehler, and William L. Shelton. Lord's ascent to the position of the first four-star nonpilot to lead AFSPC represented a change in the service's philosophy about who should run its space missions. Selected in the wake of United States Space Command's disestablishment after the September 11 attacks in favor of the creation of USSTRATCOM, Lord embarked on a tenure of stability and little change for the Air Force. Although Lord himself was not a career space officer but an officer who had come up through the ICBM ranks, he pushed for the creation of the space professional community, including the introduction of a new badge for space operators. Kehler, also a nonpilot, nontechnical officer like Lord, came up through the ICBM ranks and transitioned into the space business. Kehler's transition to space, which began as a colonel, represented a new approach. Following the 2001 space commission report, Kehler became director of the national security space integration office for the undersecretary of the Air Force, which some saw as a staff-in-waiting for a future Space Force. The next major change occurred with the tenure of William Shelton, who held multiple degrees in astronautical engineering and achieved command of AFSPC as the first officer who served his career primarily in space, to lead the Air Force's space command.

Department of the Air Force space historians Gregory Ball and Wade Scrogham write about Air Force Gen. John E. Hyten. Hyten was the son of a Saturn V engineer who grew up in Huntsville, Alabama; his relief of Shelton as AFSPC commander represented continuity in the most important space leadership position in the Air Force. Hyten's career, beginning with his first engineering assignments, was full of technical responsibilities, including important military space programs. As AFSPC commander, he introduced functional and organizational changes that fundamentally altered space operations in the command. After his tenure as AFSPC boss, Hyten took over USSTRATCOM from Kehler and then became the first space officer to be vice chairman of the joint chiefs of staff.

Lance Janda, Cameron University historian, writes about Air Force and Space Force Gen. John C. "Jay" Raymond who, in many ways, represents both continuity and change. Following a first assignment in ICBMs, Raymond spent the next three decades as a space officer in the Air Force, rising to become the last AFSPC commander until he was appointed the first chief of space operations for the USSF in 2019.

The other military branches know, understand, and even celebrate their history. The Air Force is the least historical of the military branches—perhaps because it was the youngest until 2019, when the U.S. Space Force stood up, or perhaps because it was the most technological service until then. Whatever the reason, learning from our past is vitally important. The USSF needs to know its history because space history began well before 2019 when the service gained its independence from a service that only six decades before had gained its independence from the Army. With this common ground in hand, we can share our stories of how we contributed to the creation of the USSF. In essence, we are all "trailblazers of the United States Space Force," and the stories here are our stories, too.

Notes

1. B. A. Schriever, "Foreword," in Shirley Thomas, *Men of Space: Profiles of the Leaders in Space Research, Development, and Exploration*, 8 vols. (Philadelphia: Chilton, 1961), 3:vii.
2. Gen. Bernard A. Schriever, interview with the author, Washington, DC, June 27, 2001.
3. Carl von Clausewitz, *On War*, ed. Michael Howard and Peter Paret (Princeton: Princeton University Press, 1976), 171–73.
4. Maxims LXXVII and LXXVIII in "Military Maxims of Napoleon," n.d., sourced to "Napoleon's Maxims of War. With notes by General Burnod. Translated from French by Lieut. General Sir G. C. D'Aguilar, C. B., and published by David McKay of Philadelphia in 1902," http://www.military-info.com/freebies/maximsn.htm. Maxim LXXVIII: "Peruse again and again the campaigns of Alexander, Hannibal, Caesar, Gustavus Adolphus, Turenne, Engene, and Frederick. Model yourself upon them. This is the only means of becoming a great captain, and of acquiring the secret of the art of war."
5. Oliver L. Spaulding Jr., "The Military Studies of George Washington," *The American Historical Review* 29, no. 4 (July 1924): 678.
6. Edgar H. Schein with Peter Schein, *Organizational Culture and Leadership*, 5th ed. (Hoboken, NJ: Wiley, 2007), 127–31.
7. Lt. Col. William D. Sanders, "Space Force Culture: A Dialogue of Competing Traditions," *Air and Space Operations Review* 1, no. 2 (Summer 2022): 27–41.
8. Thomas Parke Hughes, *Rescuing Prometheus* (New York: Pantheon Books, 1998), 10.
9. Maj. Gen. Bernard A. Schriever, "ICBM—A Step toward Space Conquest," address presented to the Space Flight Symposium, San Diego, February 19, 1957, cited in David Christopher Arnold, *Spying from Space: Constructing America's Satellite Command and Control Systems* (College Station: Texas A&M University Press, 2005), 7–8 (174n16).

10. Schriever, interview by author, cited in Arnold, 8 (174n17).
11. Clausewitz, 585–94.
12. Clausewitz, 89, 589.
13. For a discussion of the trinity, see Clausewitz, 75–89.
14. See especially Geoffrey Parker, *The Military Revolution: Military Innovation and the Rise of the West 1500–1800*, 2nd ed. (Cambridge: Cambridge University Press, 1996).
15. Air Force Manual 1–2, *Air Doctrine, United States Air Force Basic Doctrine*, December 1, 1959, quoted in Robert Frank Futrell, *Basic Thinking in the United States Air Force, 1961–1984*, vol. 2 of *Ideas, Concepts, Doctrine* (Maxwell Air Force Base, AL: Air University Press, 1989), 714–15; Gen. Ronald R. Fogleman and Secretary Sheila E. Widnall, "Global Engagement: A Vision for the 21st Century Air Force," November 1996, 9, https://www.airforcemag.com/PDF/DocumentFile/Documents/2008/GlobEngage_112296.pdf.
16. "Frequently Asked Questions" (n.d.), https://www.spacecom.mil/About/Frequently-Asked-Questions/. One hundred kilometers is the generally accepted von Kármán line where "space" begins and above which aerodynamic principles used for flight are ineffective and Kepler's laws have precedence. The "line" is not codified in international law, however.
17. United States Space Force, *Spacepower: Doctrine for Space Forces*, June 2020, 4.
18. Shirley Thomas, "Author's Introduction," in *Men of Space*, 1:xiii; Valerie J. Nelson, "Shirley Thomas Perkins, 85; USC Teacher Quit Career in Radio, TV to Write on Exploration of Space," *Los Angeles Times*, August 6, 2005, http://articles.latimes.com/2005/aug/06/local/me-perkins6.

1 TOWARD THE UNITED STATES SPACE FORCE

RICK STURDEVANT

The establishment of the United States Space Force (USSF) on December 20, 2019, marked the birth of the first U.S. military service since the U.S. Air Force itself in September 1947 and signaled the emergence of a guardian culture different from the cultures of U.S. airmen, soldiers, sailors, or marines. This remarkable event occurred, however, only after roughly three-quarters of a century punctuated by significant technological advances, shifting organizational structures, perceived threats to U.S. national security in space, and the timely presence of a few gifted military and civilian leaders. Some naive observers of the ceremonial celebration in December 2019 might have thought it historically inevitable, but U.S. military space historians reflected instead on the significant challenges between 1943 and 2019 and all the technological, organizational, and political twists and turns, advances, setbacks, triumphs, and disappointments along the path toward the USSF becoming a reality.

World War II and the Early Cold War

In midsummer 1943 Professor Theodore von Kármán, director of the Guggenheim Aeronautical Laboratory California Institute of Technology (GALCIT), received from the Army Air Forces (AAF) top-secret photographs suggesting Nazi Germany's development of the V-2 ballistic missile. Col. William Howard Joiner of the AAF, a senior liaison officer at the institute, asked the GALCIT chief rocket project engineer if one of their engines could propel a long-range missile. On November 20, 1943, von Kármán proposed a four-phase expansion of his group's rocket engine research. That report, bearing the designation JPL-1, would subsequently change GALCIT's official name to the Jet Propulsion Laboratory (JPL). When the AAF showed little interest in funding expansion of JPL's rocket engine research and development, Joiner's fellow liaison officer, Capt. Robert Staver of Army Ordnance, pressed his organization's rocket development division, which had been formed in September 1943, for support. This resulted in an Army Ordnance–California Institute of Technology project contract on July 1, 1944, for development of a guided missile capable of delivering an explosive payload of one thousand pounds over 150 miles with an accuracy within 3 miles of the target.[1]

Two months later, Commanding General of the U.S. Army Air Forces Gen. Henry H. "Hap" Arnold asked von Kármán to lead an assembly of renowned scientists and engineers, calling itself the Scientific Advisory Group, to assess the current state of airpower research based in large measure on Nazi Germany's technological advances and to guide long-range research and development after World War II ended. The Scientific Advisory Group issued a preliminary report, *Where We Stand*, in August 1945 and delivered its final study, *Toward New Horizons*, in mid-December 1945. In the first of thirty-three volumes, von Kármán elaborated on "Science: The Key to Air Supremacy" and predicted the operational success of intercontinental ballistic missiles (ICBMs). Realizing that such rockets might also serve as space launch vehicles, he declared in one simple sentence, "The satellite is a definite possibility."[2]

Meanwhile, in September 1945, General Arnold and Franklin Collbohm of Douglas Aircraft Company secured Donald Douglas' support to organize a research and development (RAND) group of civilian scientists and engineers at Santa Monica, California, to function independently from the company's existing research and engineering division. In a November 1945 report to Secretary of War Robert Patterson, Arnold became one of the first senior military leaders to describe the future of warfare in terms of potential uses for ICBMs and satellites. He appointed Maj. Gen. Curtis E. LeMay in December 1945 as the Pentagon's first deputy chief of air staff for research and development to oversee Project RAND and other research.[3]

Intrigued after reviewing a May 1945 space study by Wernher von Braun, in early October 1945 Cdr. Harvey Hall from the U.S. Navy Bureau of

Aeronautics presided over the first meeting of a seven-member committee to establish the feasibility of space rocketry. Since he knew it was unrealistic to consider funding an earth satellite vehicle program from the Navy budget alone, Hall engaged AAF officers in the Pentagon and, in March 1946, proposed to LeMay a joint Navy-AAF venture. Furious that the Navy was encroaching on what LeMay considered an exclusive AAF area of responsibility, he refused to consider Hall's proposal and directed Project RAND to report as soon as possible on the feasibility of the AAF developing and launching a satellite. This resulted on May 2, 1946, in RAND's first report, a 250-page engineering analysis titled *Preliminary Design of an Experimental World-Circling Spaceship*.[4]

Organizing Military Space

By the early 1950s a reduction in the size and weight of nuclear weapons rendered possible their delivery by long-range ballistic missiles. It also became increasingly evident that rockets carrying warheads through space could also carry satellites into orbit. RAND issued a Project Feed Back Summary Report on March 1, 1954. In two volumes, editors James E. Lipp and Robert M. Salter explained, in much greater detail than the 1946 RAND report, the feasibility of a military satellite system.[5]

Meanwhile, classified intelligence about Soviet advances in rocket and nuclear technology caused some Pentagon senior leaders to seek accelerated U.S. efforts in those fields. For that purpose and with Secretary of the Air Force Harold Talbott's approval, Trevor Gardner, his special assistant for research and development, assembled the scientific advisory strategic missiles evaluation committee—more popularly known as the Teapot Committee—chaired by Dr. John von Neumann in October 1953. Based largely on the committee's succinct ten-page report in February 1954, the newly established Western Development Division (WDD) stood up on July 1, 1954 in El Segundo, California. Brig. Gen. Bernard Schriever became the first WDD commander; his mission, development of the Atlas ICBM, was the highest national priority. Although WDD aligned under Air Research and Development Command (ARDC), Schriever had permission to report directly to President Dwight D. Eisenhower. He also personally selected most of his original team members, later referring to them as his "old timers."[6]

While WDD contracted with industry for rocket development and procurement, the U.S. Army relied on its in-house von Braun team. On February 1, 1956, it established the Army Ballistic Missile Agency at Redstone Arsenal in Huntsville, Alabama, and placed von Braun's team under the new agency and its first commander, Maj. Gen. John B. Medaris. With unparalleled authority from the secretary of the Army to hasten the pace of both missile development and production, Medaris and von Braun soon proposed using Army rockets to launch satellites.[7]

The Soviet Union's launch of Sputnik, the world's first manmade satellite, on October 4, 1957, added urgency to launching a U.S. spacecraft. Although the U.S. Naval Research Laboratory already had received authorization to proceed with its Vanguard project as a contribution to space exploration during the International Geophysical Year, a rapidly emerging U.S. national security requirement for space-based reconnaissance gained more attention from senior U.S. defense leaders. Based on ARDC commander Lt. Gen. Thomas Power's assignment of the Air Force military satellite project to WDD two years earlier, on October 10, 1955, a handful of engineering officers working on satellite subsystems at Wright Air Development Center in Ohio had moved secretly in civilian attire to WDD in January and February 1956 to work on a satellite program designated Weapon System 117L.[8]

Primarily to forestall uncontrolled, costly interservice rivalry in the ongoing missile and emerging military space programs, Secretary of Defense Neil H. McElroy created the Advanced Research Projects Agency (ARPA) on February 8, 1958. Roy W. Johnson, its first director, quickly established it as an oversight agency that allocated funds and missile and space projects to the military services. Until passage of the National Aeronautics and Space Administration (NASA) Act later in 1958, ARPA oversaw both U.S. civil and military space programs. At the end of 1959, however, ARPA's responsibility as the central Department of Defense (DoD) organization for military space projects also ceased.[9]

The 1960s witnessed other noteworthy organizational changes related to military space. Successful launch and return of the first reconnaissance satellite photos from space of Soviet military sites led to the establishment of the National Reconnaissance Office. With appointment of Robert McNamara as secretary of defense, General Schriever managed to convince the John F.

Kennedy administration to disestablish ARDC and create Air Force Systems Command (AFSC) in April 1961. In Colorado Springs, Colorado, Air Defense Command—redesignated as Aerospace Defense Command (ADCOM) in 1968—began accumulating space-related operational responsibilities, such as space surveillance and the ground-launched antisatellite system called Program 437. After the launch of Defense Support Program early warning satellites in the 1970s, ADCOM became responsible for their on-orbit control and data processing.[10]

Toward Air Force Space Command

As military space systems matured in the early 1970s, the Air Force assigned operational responsibility for satellites to different commands, depending on which organization had the greatest functional need for a particular space-based capability. The dispersal of systems made it difficult to coordinate either requirements or operational concepts from a "total system" perspective, and it forced the air staff to perform programmatic tasks that belonged more properly at a major command. Furthermore, as military space systems became more sophisticated, some possessed multiple capabilities that made it difficult to assign them to a specific command based on function.[11]

Both the technically oriented New Horizons II study chaired by Brig. Gen. David D. Bradburn in 1975 and the comprehensive Future Air Force Space Policy and Objectives Pentagon study by Headquarters (HQ) U.S. Air Force (USAF)/deputy chief of staff for plans and operations in early 1977 blamed the service's inefficient utilization of space assets on an inadequate understanding of capabilities and lack of clearly articulated goals for the operational use of space. The latter study included a large matrix listing systems across the top and functions down the left side, with color coding for each organization involved in each of the functions for each of the systems. Because of the many colors associated with the different functions and systems, the matrix chart became known as the "Navajo blanket." Summarizing the situation in a 1977 *Air University Review* article, Col. Morgan Sanborn wrote, "The point is that space has become an amalgam of systems and users.... The need for a separate space command within the Air Force... seems obvious."[12]

Acting on that recommendation, Chief of Staff of the Air Force (CSAF) Gen. David C. Jones formed a committee with less than a dozen members who were, according to one participant, "ensconced" in a room next to General Jones' office in the Pentagon. Headed by Assistant Vice Chief of Staff Lt. Gen. Wilbur L. "Bill" Creech, with Brig. Gen. Robert T. Herres as his very active deputy and Col. Harold W. "Pete" Todd as the primary action officer, the group worked "diligently five days a week . . . for several months" to identify the maximum number of ADCOM functions that could be retained and accomplished with the least resources. Known familiarly as the "Green Book study" and more formally as the "Proposal for a Reorganization of USAF Air Defense and Space Surveillance/Warning Resources," the committee's final report recommended eliminating ADCOM, with air defense assets going to Tactical Air Command (TAC) and space surveillance and early warning to Strategic Air Command (SAC).[13]

Almost simultaneously with General Creech's study, Congress debated shifting ADCOM functions to TAC and SAC. Testifying before the House Committee on Appropriations in early February 1977, HQ USAF director of programs Maj. Gen. Abbott C. Greenleaf had cautioned against such action on the grounds it "would cause fragmentation of a specialized area of responsibility, would not provide any manpower/force savings and possibly cause a major reorganization of the NORAD [North American Aerospace Defense Command] defense structure." By early summer 1977, however, the appropriations committee "directed its Surveys and Investigations Staff to conduct a review of need to support a separate ADC[OM] headquarters and command structure and what savings and other program improvements might result from transferring the responsibilities and assets of ADC[OM] to SAC and TAC."[14]

Despite the diehards' best efforts and a three-month delay due to a lawsuit brought by a group of ADCOM civilians, HQ USAF proceeded to dismantle the command. Air defense resources transferred to TAC on October 1, 1979, with missile warning and space surveillance resources going to SAC on December 1, 1979. Meanwhile, an aerospace defense center had been activated from ADCOM remnants in Colorado Springs to train and equip people to support the space surveillance and missile warning missions.[15]

At that point, consolation for Undersecretary of the Air Force Hans Mark, retired Gen. James Hill, and others who supported creation of a space command lay in the possibility that one might rise, phoenixlike, from the aerospace defense center. They had reason to remain cautiously optimistic. When Secretary John Stetson had approved the "reorganization" of ADCOM in late summer 1978, he had suggested that USAF Chief of Staff Gen. Lew Allen Jr. appoint a special high-level group of general officers to advise how the service should organize itself to perform space operations. In response, General Allen authorized an executive committee led by Lt. Gen. Andrew B. Anderson Jr., the HQ USAF deputy chief of staff for plans and operations, to examine all aspects of space mission management and to propose organizational alternatives. The committee, in turn, created a working group that included several lower-ranking officers who later would occupy important positions in an operational space command: Col. Gaylord W. "Wes" Clark, Lt. Col. Thomas S. Moorman Jr., and Maj. Robert S. Dickman. Their product, the Space Mission Organization Planning Study, was a top-secret report released on February 5, 1979, that set the parameters for discussions over the next three years.[16]

The study identified five organizational options for the space mission. As General Hill pointed out, the first—to retain the current organizational structure—had been preempted by the decision to disestablish ADCOM. That left four alternatives: give the space mission to SAC, give it to AFSC, create a separate space operations service under AFSC, or establish a new major command for space operations. General Hill prodded General Allen to support establishing a space command using "as its cadre the personnel presently assigned to ADCOM for space management." On the eve of his retirement in December 1979, General Hill again addressed the chief of staff of the Air Force on this issue:

> Unless we make an explicit organizational decision which assigns to a single organization the Air Force responsibilities in space operations once and for all, we will be faced with negative long-term impacts on resource management and planning. In my judgment, we can no longer afford the luxury of so many groups and diversified interests sharing responsibility for the space activities that have progressed beyond development and are (now) operational.[17]

A 1980 Air Force Scientific Advisory Board "summer study on space," under the leadership of former secretary of the Air Force John L. McLucas, reinforced General Hill's opinion by concluding that although the service had done well during the preceding fifteen years to turn experimental systems into reliable, operational ones, it was organized inadequately for operational exploitation of space and placed insufficient emphasis on the inclusion of space systems as essential elements in an integrated force structure. That CSAF Allen remained aloof might have reflected his preference for compromise and his inclination to proceed cautiously until a strong servicewide consensus outweighed the more parochial interests of individual commands. His personal doubts about the need for change almost certainly were matched by his awareness that precipitous action might generate unwanted opposition from the joint chiefs of staff (JCS), the other services, and DoD agencies.[18]

Nonetheless, high-level policy and doctrine statements ensured the continuation of spirited discussions about how the Air Force ought to organize space operations. Presidential directive number 37 on May 11, 1978, had asserted the nation's right to free passage and unhampered operation of its property in space and, consequently, its right to defend that property against hostile threats. Publication of Air Force Manual 1–1, *Functions and Basic Doctrine of the United States Air Force*, on February 14, 1979, for the first time officially identified space operations as one of the service's basic missions. That recharged the ongoing debate about whether space was primarily an operating medium or a mission area.[19]

By autumn 1981 several organizational changes related to space operations put the Air Force on a path to the creation of a space command. In October 1979 a defense space operations committee had become the primary advisory body to the secretary of defense on issues related to national security space. On the air staff, in the "F ring" in the Pentagon basement, a space division was spun off from the reconnaissance division with a congressional mandate to focus on tactical employment of national capabilities. Secretary of the Air Force Hans Mark said in September 1980 that the Air Force, to strengthen its forces, must improve its ability to conduct operations in space and must develop proper organizational arrangements to deal with that new role, which meant creating a deputy commander for space operations position within the AFSC space division at Los Angeles. In September 1981 HQ USAF created a directorate of space operations under the deputy chief of staff for plans and operations (XO).[20]

In late summer 1980 a pair of relatively unsung heroes in Air Force space history—Lt. Gen. Jerome F. O'Malley and Maj. Gen. John T. Chain Jr.—had become XO and director of operations, respectively. At his first staff meeting,

General O'Malley asked his staff to prioritize what they believed were the most significant issues facing the service. Afterward, when General Chain mentioned to his staff that properly organizing for space operations came in second only to maintaining strategic deterrent posture, Col. Earl S. Van Inwegen, then chief of General Chain's space division, remarked, "Well, we've got ten 'Billy Mitchells' of space here ready to work the problem." General Chain directed Colonel Van Inwegen's team—including Roger DeKok, Bill Savage, John Hungerford, Vito Pagano, and John Angel—to generate a "fester briefing" that outlined for General O'Malley an operational space organization and that explained why it should be a separate Air Force major command. When Colonel Van Inwegen asked why the general called it a "fester briefing," Chain replied, "I'm going to put it on General O'Malley's desk, and it's going to sit there and fester until he does something about it." The best rationale Colonel Van Inwegen's team could provide for an organizational change was that "space operations, such as they were in the Air Force, were primarily almost totally research and development in nature, run by Space Division/SAMSO [Space and Missile Systems Organization] and the Air Force Systems Command or . . . the NRO [National Reconnaissance Office]." When they finally took the briefing to General O'Malley, "he came unglued," because he knew such reasoning would not impress General Allen, then chief of staff and a former AFSC commander. General O'Malley, who feared it would harm their cause significantly if General Allen reacted negatively to the briefing, said he would not allow it to go forward until the team came up with a better rationale.[21]

Two more high-level analyses related to Air Force space organization appeared in late 1980 and early 1981. First, using the *Space Mission Organization Planning Study* as a point of departure, the space organization study by the Deputy Assistant Secretary of the Air Force for Space Plans and Policy favored a new operational command for space. Five months later, in May 1981, General O'Malley approved a space policy and requirements study that focused operationally on the service's space posture and the best means of providing required space capabilities.[22]

Meanwhile, the issue of whether DoD required a shuttle operations and planning complex separate from NASA's Johnson Space Center and, if needed, where the complex would be located slowly made its way toward resolution through the Pentagon bureaucracy. Although an October 1978 program management directive said DoD shuttle operations would be conducted in a controlled mode at Johnson Space Center, a May 1979 secretary of the Air Force memorandum asserted the service stood a better chance of obtaining funds from Congress if it combined its own shuttle and satellite control centers to form a new consolidated space operations center (CSOC). A June 1979 site survey briefing introduced the concept of a "hypothetical site east of Peterson Air Force Base" near Colorado Springs for the CSOC, and a December 1979 final report identified the Colorado Springs site as the preferred location for the new center. By the end of September 1980, the Air Force had completed the environmental impact statement and land acquisition for situating CSOC facilities, but a request by the New Mexico congressional delegation for review of the site selection process and criteria delayed finalization of the site decision until March 1981. Building the CSOC ultimately influenced the decision to establish a separate space command in Colorado Springs.[23]

By summer 1981 a unique cast of characters occupied key senior leadership positions from which to influence creation of a space command. On the air staff, General O'Malley remained XO, with General Chain having become his assistant in July 1981 and Colonel Van Inwegen assigned as the first director of space operations in September. Lt. Gen. James V. Hartinger, who had been NORAD commander since January 1980, won a personal campaign to receive his fourth star on October 1, 1981. One of Hartinger's West Point classmates, Lt. Gen. Richard C. Henry, continued to serve as commander of space division in Los Angeles, while another, Gen. Robert T. "Tom" Marsh, became AFSC commander in February 1981.

Testifying before the Senate Armed Services Committee (SASC) in November 1981, Air Force Undersecretary Edward C. "Pete" Aldridge Jr. had acknowledged the need for a more coordinated, integrated approach to military space operations and suggested the establishment of "some form of a 'space command'" as the right answer. That same month, he told an audience at the American Astronautical Society national conference, "I believe the right answer may be to form a space command to operate our satellites and launch services."[24]

In the U.S. House of Representatives, on December 8, 1981, Rep. Ken Kramer of Colorado introduced a bill requiring the Air Force to report on the desirability of creating a space command and renaming the service itself the "United States Aerospace Force." Secretary of the Air Force Verne Orr and

General Allen opposed the name change but acknowledged they were seriously considering a new command. Two months later, on February 9, 1982, Representative Kramer addressed the military reform caucus, saying the United States should reorient its approach conceptually and organizationally to protect more directly its presence on the "high seas of space."[25]

Pursuant to General Allen's directive, Hartinger's and Marsh's respective staffs formed working groups to develop the air staff briefing. Although the groups met periodically to review each other's work, they did not exactly find common ground. Considering the fragmented management of Air Force space activities among twenty-six different organizations, the absence of a clearly defined operational advocate for space systems, and the lack of concrete plans for using space systems in wartime, Hartinger's staff pushed vigorously for immediate, revolutionary action to create a separate major command for space operations. Marsh's staff, on the other hand, favored a slower, evolutionary approach. When General O'Malley, who had been visiting the Air Force Academy, dropped by the Chidlaw Building in Colorado Springs on April 15, 1982, Hartinger's staff showed him an extra briefing chart they had prepared to depict how a space command might be formed at once. Liking what he saw, O'Malley told Hartinger to bring the chart when he came to Washington to brief General Allen.[26]

That all-important briefing session occurred two days later, April 17, 1982, which happened to be General Hartinger's fifty-seventh birthday. After hearing AFSC's formal presentation on the "Space Organizational Issue," General O'Malley objected to its vagueness about when an operational space command might be formed. As the discussion subsided, General Hartinger revealed his more specific slide showing how the Air Force might create, without delay, a space command on par with SAC, TAC, and Military Airlift Command. General O'Malley exclaimed, "That's it! That's what we need!" CSAF Allen calmly concurred, "Okay, Jimmy, you go back out to Colorado Springs, and let's get a space command started." The air staff space operations steering committee, chaired by General Chain, who had succeeded General O'Malley as the XO when the latter became Air Force vice chief of staff in June 1982, subsequently worked to refine the organizational concept and plan the transition. On June 21, 1982, the service officially announced its decision to form Space Command effective September 1, 1982.[27]

Establishing and activating Space Command did not mean immediate consolidation of responsibilities for space-related operations. It took until May 1983 for SAC to transfer control of all ground-based early warning and space surveillance sensors, plus Defense Support Program and Defense Meteorological Satellite Program resources. Other programs followed slowly: management of global positioning system (GPS) resources came in 1984, followed by the Air Force Satellite Control Network common-user element in 1987, space launch in 1990, Air Force astronauts in 1991, and the Air Force satellite communications system in 1992.[28]

United States Space Command

Air Force Space Command's first commander, Gen. James V. Hartinger, and his staff began developing procedures and a rationale for a unified space command soon after September 1982, and President Ronald Reagan's announcement of his Strategic Defense Initiative (SDI) on March 23, 1983, provided a crucial boost. A unified space command seemed the most sensible organization to become the operational focus for SDI planning and system operations. When in April 1983 the joint chiefs of staff requested organizational suggestions on how best to support SDI, General Hartinger forwarded his proposal for a unified space command. Secretary of Defense Caspar Weinberger responded to Hartinger's report on May 9, saying he thought Space Command would provide the "necessary structure" for a cohesive ballistic missile defense program that involved space-based elements. He surprised Hartinger, however, by suggesting the issue of whether to create a unified or specified command remained open, because the latter would preserve Air Force management and command prerogatives but still permit representation from the other services. Since the Navy had made a unified rather than a specified command its price for supporting an Air Force major command for space, Air Force Chief of Staff Gen. Charles Gabriel opted to side with General Hartinger and presented a rationale for establishment of a unified space command to the joint chiefs on June 7, 1983. The JCS, in turn, sent Secretary Weinberger a memorandum with the rationale for a unified space command and, in November 1983, he recommended that President Reagan approve establishing such a command to help ensure SDI success.[29]

Although in February 1984 the JCS created a joint planning staff for space to make necessary preparations for a unified space command, President Reagan did not officially approve the formation of U.S. Space Command until November 20, 1984. Nine months later, on August 30, 1985, the president accepted JCS recommendations for U.S. Space Command and named Gen. Robert T. Herres—already multi-hatted as binational NORAD commander in chief, specified Aerospace Defense Command commander in chief, and Space Command commander—as the first commander in chief of U.S. Space Command (USSPACECOM). The command's activation ceremony occurred on the parade grounds at Peterson Air Force Base (AFB), Colorado, and the new unified command's headquarters was in Colorado Springs. In November 1987 the headquarters relocated to the Ent Building on Peterson, where it remained pending construction of a new building just northeast of the Air Force Space Command (AFSPC) headquarters.[30]

During Operations Desert Shield and Desert Storm and for the first time during actual warfare, USSPACECOM utilized the full range of space-based capabilities, which led some authorities to call Desert Storm the "first space war." Over the next decade, USSPACECOM "operationalized" and "normalized" the medium of space in support of U.S. domination of the battlespace during coalition operations in the Balkans, Southwest Asia, and Afghanistan. The employment of military satellite systems, particularly GPS and military satellite communications, in those conflicts drove a significant transformation in U.S. warfighting and in warfare generally. Recognizing that transformation, USSPACECOM published its *Long Range Plan* in April 1998, which examined the strategic environment of the future and presented a vision for 2020 that embraced control of space, global engagement, full-force integration, and global partnerships. In February 2000 Gen. Howell M. Estes III approved vision statements for "dominating the space dimension of military operations to protect U.S. interests and investments" and to "integrate space forces into warfighting capabilities across the full spectrum of conflict."[31]

The September 11, 2001, terrorist attacks on the World Trade Center and the Pentagon starkly manifested a need for more concerted homeland defense, which compelled senior leaders to support creating United States Northern Command. But not wanting a proliferation of unified commands beyond the ten identified in the unified command plan, the existence of another command had to cease. Consequently, Secretary of Defense Donald Rumsfeld announced USSPACECOM would be disestablished and its combatant command responsibilities merged with those of U.S. Strategic Command (USSTRATCOM). The disestablishment occurred on October 1, 2002, when operational space responsibilities went to USSTRATCOM.[32]

The Space Commission

While international turmoil worked to unravel the existence of the first USSPACECOM, Secretary Rumsfeld took steps to strengthen U.S. military space posture based on recommendations from the congressionally mandated "Report of the Commission to Assess National Security Space Management and Organization," which was completed on January 11, 2001, prior to President George W. Bush's appointing Rumsfeld as secretary of defense. Labeling the United States an attractive candidate for a "space Pearl Harbor," the report provided an in-depth analysis of ways to reduce the perceived vulnerability across all relevant levels and sectors of the U.S. national security space enterprise. To answer its specific charter, the commission also assessed four organizational approaches: a new military department for space, a space corps, an assistant secretary of defense for space, and a separate major force program for space. Its report concluded that "to deter and defend against hostile actions directed at the interests of the United States," a "rechartered Air Force" was "best suited to organize, train and equip space forces" for the "nearer term," but "in the mid term, a Space Corps within the Air Force may be appropriate to meet this requirement; in the longer term it may be met by a military department for space."[33]

Between May 2001 and July 2003, Secretary Rumsfeld implemented most of the recommendations from the Commission to Assess National Security Space Management and Organization. These included creating a policy coordinating committee for space within the National Security Council structure, designating the Department of the Air Force as the DoD executive agent for space, and creating a "virtual" major force program for space "to increase visibility into the resources allocated for space activities." He provided guidance to ensure space education at all levels, including professional military education, would ensure a substantial cadre of both military and civilian space-qualified professionals within each of the military departments. Among steps to "support the adoption of a 'cradle-to-grave' approach for space," Rumsfeld moved Space and Missile Systems Center from Air Force Materiel Command, where it had been assigned after the disestablishment of Air Force Systems Command on July 1, 1992, to AFSPC. He also directed the secretary of the Air Force to assign a single-hatted, four-star officer to command AFSPC, thereby enabling that individual to focus exclusively on organizing, training, and equipping space forces and relieving future AFSPC commanders from being encumbered by triple-hatted command responsibilities for USSPACECOM and NORAD.[34]

During the first decades of the twenty-first century, an almost entirely new host of space launch vehicles and military satellites advanced traditional U.S. on-orbit capabilities and even extended them in novel ways. In 2002 Delta IV and Atlas V evolved expendable launch vehicles, both in multiple design configurations, which provided assured access to space. The availability of SpaceX's Falcon 9 rocket—with its recoverable, reusable first stage—in 2010 reduced launch costs, and demonstration of that company's Falcon Heavy in 2018 promised the same for heavier payloads going into higher orbits. For the early warning mission, the first Space-Based Infrared System payloads went into highly elliptical orbit in 2006, and the system's satellites began replacing older Defense Support Program spacecraft in geosynchronous orbit in 2011. For satellite communications (SATCOM), the narrowband Mobile User Objective System began supplanting the Ultra-High-Frequency Follow-on in 2012; the Wideband Global SATCOM system did the same with respect to the Defense Satellite Communications System III in 2007; and Advanced Extremely High-Frequency satellites following Milstar in 2010 for protected SATCOM. GPS had achieved full operational capability in 1995; new models further enhanced positioning, navigation, and timing. Although proposed years earlier, the Space-Based Space Surveillance pathfinder satellite introduced on-orbit space domain awareness in 2010, and the first pair of Geosynchronous Space Situational Awareness Program craft enhanced space domain awareness capability in 2014. An extraordinary experimental addition to the U.S. military space presence had come in 2010 with the first 224-day mission of the X-37B orbital test vehicle, a reusable, multipurpose, robotic spaceplane.[35]

Reorganizing Military Space

Meanwhile, it became apparent to U.S. military leaders, not to mention U.S. adversaries, that the strategic environment—given U.S. reliance on space-based capabilities in warfighting—demanded further organizational adjustment. As one of its initial steps, USSTRATCOM established a joint functional component command (JFCC) for space and global strike in January 2005 as a unified component under the dual-hatted Eighth Air Force commander. Its charter called for optimizing operational-level planning, execution, and force management for USSTRATCOM's nuclear deterrence mission. The USSTRATCOM commander reorganized the JFCC on July 19, 2006, into two separate components: the joint functional component command for global strike and integration, headquartered at Barksdale AFB, Louisiana, and the joint functional component command for space (JFCC Space), headquartered at Vandenberg AFB, California. Air Force Space Command's Fourteenth Air Force commander was dual-hatted as JFCC Space commander.[36]

Over time, as the character of space activities and the threat in that domain continued to change, military leaders at the Pentagon, USSTRATCOM, and AFSPC perceived a need for further organizational adjustments. The joint space operations center (JSpOC), which opened at Vandenberg on May 18, 2005, reflected some of those adjustments. On June 1, 2015, JFCC Space began a six-month pilot program to incorporate commercial operators within its JSpOC; the success of that experiment resulted in a decision in June 2016 to formalize a commercial integration cell within the JSpOC. Meanwhile, USSTRATCOM partnering agreements to share space situational awareness data with a growing number of nations led to plans for placing allied personnel within the JSpOC. Together, those changes resulted in the JSpOC being redesignated, on July 18, 2018, as the combined space operations center.[37]

At a higher level, two organizational changes drew the defense and intelligence communities closer to one another. In January 2015, following direction from the undersecretary of defense, USSTRATCOM commander Adm. Cecil D. Haney established the joint space doctrine and tactics forum. Cochaired by the USSTRATCOM commander and the National Reconnaissance Office director, the forum included the commanders of JFCC Space, AFSPC, U.S. Army Space and Missile Defense Command/Army Forces Strategic Command, U.S. Fleet Cyber Command, and U.S. Marine Corps Forces Strategic Command. The forum members, who met a few times each year, emphasized the foundational role played by intelligence as they sought collectively to improve operational capabilities in space. On June 23, 2015, Deputy Secretary of Defense Robert O. Work announced the development of a joint interagency combined space operations center (JICSpOC) within existing facilities at Schriever AFB, Colorado, to "improve processes and procedures, ensuring data fusion among DoD, intelligence community, interagency, allied and commercial space entities" and to provide backup to the JSpOC. The new JICSpOC opened in October 2016. On April 1, 2017, it formally received a new name to reflect its structure and purpose more accurately: the National Space Defense Center.[38]

In mid-December 2016, less than two months after Gen. John W. "Jay" Raymond assumed command of AFSPC, he sent the CSAF a position paper titled "Improving DoD Execution of Space Responsibilities" to "inform and scope CSAF's decision space regarding DoD Space Organization." Composed by the AFSPC commander's action group, that paper summarized how key reports, over the course of more than two decades, noted the absence of a coherent management structure and caused growing congressional concern. With respect to space operations, the paper suggested advocating for "a normalized service component alignment" that would relieve the Fourteenth Air Force commander of JFCC Space responsibilities and assign those to the AFSPC commander, thereby making him joint forces space component commander (JFSCC) for USSTRATCOM command and control purposes. The decision space included the possibility, however, of reestablishing a "separate U.S. Space Command."[39]

Since both General Raymond and Gen. John E. Hyten, former AFPSC commander and now USTRATCOM commander, agreed on a JFSCC organizational construct, progress toward fleshing it out gained momentum early in 2017. By March the reorganization effort had become "Transforming AFSPC into a Warfighting HQ," with its commander having both the traditional role of organizing, training, and equipping Air Force space forces and also commanding and controlling multiservice space forces for the USSTRATCOM commander. On June 16, 2017, USSTRATCOM issued an order to execute the "Restructure Implementation Plan."[40] General Hyten presided over a ceremony formally inactivating JFCC Space and activating JFSCC on December 1, 2017, at Vandenberg AFB, where the new JFSCC flag was unfurled. In a memorandum expressing his initial guidance and intent to all JFSCC personnel, General Raymond emphasized becoming "an intel-driven organization, with integrated and synchronized indications and warning and operational intelligence across the enterprise." He intended, furthermore, to strengthen current capabilities "to deliver timely/tailored space effects to Combatant Commanders and to defend the space JOA [Joint Operational Area]."[41]

Reestablishing U.S. Space Command
Months prior to JFSCC's activation, planners at AFSPC headquarters in Colorado Springs had begun contemplating JFSCC's transition from a component command into a subordinate unified command under USSTRATCOM. Undoubtedly, that planning effort had gained impetus from language in section 1602 of the House Armed Services Committee version of the fiscal year (FY) 2018 National Defense Authorization Act (NDAA). The House version directed the establishment, by January 1, 2019, of "United States Space Command as a subordinate unified command under United States Strategic Command." The final FY18 NDAA, approved by Congress and signed by President Donald Trump on December 12, 2017, however, contained no such mandate, calling instead for a DoD review of national security space organization and structure.[42] Nonetheless, key USSTRATCOM, JFSCC, and AFSPC staff members were emboldened by the news in late April 2018 that a "mark" inserted by the House Armed Services Committee strategic forces subcommittee into the draft FY19 NDAA called for a subunified space command. Shortly after, the White House labeled the bill's language "premature," asserting that the White House would consult with Congress once DoD delivered its congressionally mandated review.[43]

As the FY19 NDAA draft, which retained the language supporting a subunified space command, made its way through the House of Representatives and to the Senate, officials in the White House and DoD expressed alarm. From the president on down, they called for larger, more immediate organizational and managerial changes to meet what they characterized as rapidly growing threats from adversaries, especially China and Russia. In his remarks at a National Space Council meeting on June 18, 2018, President Trump announced, "Very importantly, I'm hereby directing the Department of Defense and Pentagon to immediately begin the process necessary to establish a space force as the sixth branch of the armed forces." The president continued, "General Dunford [Chairman of the Joint Chiefs of Staff Gen. Joseph Dunford], if you would carry that assignment out, I would be very greatly honored, also." The next day, a letter signed jointly by Secretary of the Air Force Heather Wilson, USAF Chief of Staff Gen. David L. Goldfein, and Command Master Sergeant of the Air Force Kaleth O. Wright informed "Fellow Airmen" that they looked forward "to begin planning to establish a space force as a new military service within the department."[44]

By early July, with President Trump directing congressional and media attention toward the idea of establishing a U.S. military service for space, officials in the Pentagon and at USSTRATCOM in Omaha and AFSPC headquarters in Colorado Springs quietly discussed replacing the notion of a subunified command for space with planning for a separate unified combatant command for space. On August 9, 2018, when DoD submitted its "Final Report on Organizational and Management Structure for the National Security Space Components," it specifically stated, "To further accelerate warfighting capability, the Department recommends creating a new U.S. Space Command to become a unified combatant command."[45] At that point, General Raymond harbored a heightened sense of urgency about reestablishing USSPACECOM. He stood up the USSPACECOM joint planning group—more familiarly known as Task Force Sierra—at AFSPC headquarters, admitting his immediate focus was trying to normalize JFSCC into a component part of the new USSPACECOM, even as officials in the Pentagon struggled to ensure the "subunified" language in the FY19 NDAA was changed to "unified."[46]

Secretary of Defense James Mattis sent President Trump a memorandum, dated October 5, 2018, recommending direct establishment of a separate unified combatant command for space. If the president gave that direction, Mattis pledged to work with Congress to have the statutory requirement for a subunified command amended or repealed. Responding on December 18 to Mattis' recommendation, the president directed reestablishing USSPACECOM "as a functional Unified Combatant Command." It soon became apparent, however, that his directive did not equate to USSPACECOM's immediate creation. Numerous criteria or conditions had to be satisfied, essential steps taken, questions answered, and issues resolved. These included presidential nomination and Senate confirmation of a commander, unified command plan updates, modification of the FY19 NDAA language regarding a subunified command, assignment of mission responsibilities, organization and force structure, acquisition authorities, USSTRATCOM divestiture and USSPACECOM implementation planning, a strategic basing decision, and military construction funding for new facilities.[47]

Not until August 29, 2019, was a second U.S. Space Command officially established during a ceremony at the White House. It had two subordinate commands: Combined Force Space Component Command and Joint Task Force Space Defense. The Combined Force Space Component Command planned, integrated, conducted, and assessed global space operations for delivering space capabilities. The Joint Task Force Space Defense conducted space superiority operations—in unified action with mission partners—to deter aggression, defend U.S. and allied interests, and defeat adversaries throughout the continuum of conflict.[48]

The United States Space Force

The phoenixlike reappearance of USSPACECOM set the stage for even more dramatic organizational change, which had been gestating somewhat chronologically in parallel with it but whose roots stretched back decades deeper in history. For example, during the 2001–2 academic year at the U.S. Army School of Advanced Military Studies at the Army Command and General Staff College in Fort Leavenworth, Kansas, Maj. Jeffrey R. Swegel, USAF, had submitted a study titled "A Fork in the Path to the Heavens: The Emergence of an Independent Space Force." The author perceived that the "transition of space from a war-enabling medium to a warfighting medium" placed the United States "at a defining moment in history with regard to U.S. aerospace dominance." He concluded, "If the U.S. is to maintain its lead, it should soon move to separate space forces along the same model as that of the creation of the USAF in 1947."[49]

It took more than a decade and half, however, for that proposition to receive legislative support. One of the earliest salvos that laid the foundation for the culminating push for a separate space service, or at least a corps within the Department of the Air Force, came at a House Armed Services Committee subcommittee on strategic forces hearing on September 27, 2016. Titled "National Security Space: 21st Century Challenges, 20th Century Organization," that hearing included two legislators who played significant roles in establishing a separate space service: Alabama Republican Mike Rogers and Tennessee Democrat Jim Cooper. Rogers set the tone, stating, "There is a fundamental question before us today. Is the Department of Defense strategically postured to effectively respond to these threats and to prioritize the changed space domain over the long term? It is all too clear that we are not." Testifying at that hearing, former deputy secretary of defense John J. Hamre offered several options to improve the space enterprise, one of which was "a separate service that is worrying about space" under the Department of the Air Force, "parallel to the . . . Marine Corps, where it is in the Department of the Navy."[50]

Another major step forward came at the Space Foundation's annual symposium in Colorado Springs in April 2017, where Congressman Rogers spoke about past "studies" of possible space organizations since the 2001 Rumsfeld report. He announced, "I've made space organization and management my number one priority for the rest of this Congress." Rogers concluded, "My vision for the future is a separate Space Force within the Department of Defense, just like the Air Force, which had to be separated from the Army in order to be prioritized and become a world-class military service."[51]

With the tone established for the rest of the legislative year, Rogers and Cooper added language into the draft FY18 NDAA. It included creating a Space Corps within the Department of the Air Force. Over the ensuing few months, however, senior DoD officials argued against the House proposal. Air Force Chief of Staff Gen. David Goldfein argued that "it would actually move us in the wrong direction and slow us down from where we need to go." Air Force Secretary Heather Wilson believed it "would create additional seams between the Services, disrupt ongoing efforts to establish a warfighting culture and new capabilities, and require costly duplication of personnel and resources." Secretary Mattis strongly urged that Congress "reconsider the proposal of a separate service Space Corps." The opinions of senior military leaders, combined with Senate opposition to the House version, resulted in the proposal for a Space Corps dematerializing in 2017.[52]

The FY18 NDAA included several other provisions that were expected to strengthen the military space enterprise. For example, it called for disestablishing the principal DoD space advisor position, which had been filled by the secretary of the Air Force. It also eliminated the defense space council (to be replaced by a reinvigorated national space council), established the AFSPC commander as JFCC Space under USSTRATCOM, and redesignated the operationally responsive space office as the space rapid capabilities office. The

FY18 NDAA also halted the Air Force's attempt to implement something that Cooper and Rogers labeled "a hastily developed half-measure initiated by the Secretary of the Air Force, which at best only added a box on the organization chart." The two congressional advocates for a Space Corps asserted they would not "allow the United States national security space enterprise to continue to drift towards a Space Pearl Harbor."[53]

In mid-March 2018, roughly two weeks after DoD had delivered its congressionally mandated "Interim Report on Organizational and Management Structure for the National Security Space Components of the Department of Defense," President Trump floated the idea of a "space force" during a speech at Miramar Naval Air Station in San Diego, arguing that "my new national strategy for space recognizes that space is a warfighting domain, just like the land, air, and sea. We have the Air Force. We'll have the space force." While the national media paid little attention on that occasion, reporters certainly took notice on June 18 when the president spoke at a meeting of the national space council, chaired by Vice President Mike Pence. The president declared, "Very importantly, I'm hereby directing the Department of Defense and Pentagon to immediately begin the process necessary to establish a space force as the sixth branch of the armed forces. That's a big statement. We are going to have the Air Force and we are going to have the Space Force."[54]

When DoD submitted its "Final Report on Organizational and Management Structure for the National Security Space Components of the Department of Defense" on August 9, it highlighted five immediate steps the department planned to take to lay the groundwork for a separate service while Congress mulled creation of the Space Force. The report called for establishing "several of the component parts of the Space Force": a space development agency, a space operations force to support combatant commands, a legislative proposal for the FY20 budget cycle regarding a Space Force, a space governance council to vet all decisions prior to sending the legislative proposal to the White House, and additional information about the new space-focused unified command. Even if DoD leaders had fallen into line behind the president, not all members of Congress were enthusiastically supportive. Senator Jack Reed, for example, cautioned, "I think we have to reorganize our space forces because our threats are now in multiple dimensions. But I think creating a separate service with all the infrastructure and bureaucracy is not the way to go."[55]

On February 19, 2019, more than six months after President Trump vocally directed a sixth military service dedicated to space, the White House released Space Policy Directive 4, "Establishment of the United States Space Force," which provided specific guidance on the missions and organization of the proposed new service. The Space Force would include all the uniformed and civilian personnel "conducting and directly supporting space operations from all Department of Defense Armed Forces" and would assume responsibility for all "major military space acquisition programs." Finally, the directive directed that the Space Force develop career tracks for military and civilian personnel across all "relevant specialties," which included space operations, intelligence, engineering, science, acquisitions, and cyber.[56]

Two days later, acting secretary of defense Patrick Michael Shanahan signed a memorandum directing the secretary of the Air Force to "organize and lead a team of civilian and military subject matter experts from across the DoD to conduct the detailed planning necessary to establish the U.S. Space Force." On February 22, in response to Shanahan's memorandum, Secretary Wilson directed a space force planning task force (SFPTF) be created for that purpose. She appointed Maj. Gen. Clinton E. Crosier, a career space officer, as the director, expecting him to deliver a plan that included transition phases from pre-establishment through full operational capability.[57]

USSTRATCOM commander Gen. John E. Hyten testified before the SASC on February 26. Saying that Congress soon would receive the DoD legislative proposal for a Space Force, the general, who had been chief of the space control division in the HQ USAF space operations and integration directorate at the Pentagon seventeen years earlier, remarked, "I was around when we transitioned the old U.S. Space Command to U.S. Strategic Command in 2002. Now we're kind of going back the other way." He recalled, "I watched us almost break the space mission when we did that, because . . . we slapped billets and said, 'These 500-plus billets are going to move from Colorado to Omaha.' And I'll just say the people didn't come with them automatically." Hyten pledged, "I just want you to know that I support the concept of the space force inside the Air Force that the president is now pushing," because a separate military department for space in the near term risked creating bureaucratic excess.[58]

Two days later, Maj. Gen. John E. Shaw, AFSPC deputy commander, convened a meeting at Peterson AFB to lay the groundwork for AFSPC support to General Crosier's SFPTF. The AFSPC team adopted Task Force Tango as its name and began wrestling with developing a field-level organizational construct for the Space Force. Throughout the planning process, Task Force Tango met frequently with the SFPTF and steadfastly supported some variant of an organizational and mission structure with three field commands: one for operations, a second for acquisitions, and a third for training or support.[59]

While the SFPFT and Task Force Tango members continued working behind closed doors through late spring 2019, congressional activity ratcheted up. On April 4 the Center for Strategic and International Studies released its *Space Threat Assessment 2019*, which contained a foreword by Representative Cooper: "This is the year of decision. . . . The House of Representatives has overwhelmingly and on a bipartisan basis supported a new 'Space Corps' for several years, and the president has recently demanded a 'Space Force.' The Pentagon has responded with a proposal that assembles a Space Force that resembles the House's Space Corps proposal. This year's National Defense Authorization Act (NDAA) will decide the outcome."[60]

A week later, the SASC held a hearing on the proposed Space Force, with testimony from acting Secretary Shanahan, General Dunford, Secretary Wilson, and USSTRATCOM leader General Hyten. The senators, many of whom remained skeptical about the need for a new military service, asked the senior defense leaders probing questions about how a space force would fix or at least ameliorate problems ranging from acquiring space systems to bureaucratic redundancies. What would a new service cost the American taxpayer? How would such a new service improve national security? Would a separate space force really reduce U.S. vulnerability in the space domain?[61]

On May 14, 2019, the House Appropriations Committee released its markup of the FY20 NDAA, but it did not authorize a Space Force and committed only $15 million to study and refine plans for potentially establishing one. An urgent attempt to sway congressional opinion came a few days later when a substantial number of former senior leaders, including a former defense secretary and multiple retired general officers, submitted an open letter strongly encouraging Congress "to establish the U.S. Space Force, to realize the full potential of space power and space capabilities in order to protect and advance U.S. vital national interests." Representative Cooper and Representative Rogers forced the issue legislatively by introducing an amendment to the FY20 NDAA bill that authorized a space corps rather than a space force, thereby requiring reconciliation between the Senate and House versions of the bill. The White House responded on July 9 with a statement of administration policy that requested several modifications to the House version of the FY20 NDAA, which lacked the provision for a Space Force. Three days later, the House of Representatives passed H.R. 2500, the FY20 NDAA, by a vote of 220 to 197. After the Senate passed its version, ironing out the final NDAA language took several months.[62]

Finally, late in the afternoon on December 9, 2019, Congress released its NDAA conference report. The report contained the language that, when signed by the president, established the United States Space Force as a sixth branch of the armed forces under the Department of the Air Force. Congress gave the secretary of the Air Force eighteen months to establish the new service. More specifically, and somewhat surprisingly, the NDAA redesignated Headquarters, Air Force Space Command as Headquarters, United States Space Force. At the time, members of Air Force Space Command were "assigned" immediately to the USSF, and the NDAA gave the secretary of the Air Force authority to transfer Air Force personnel into the new service at an appropriate time. Furthermore, the final draft of the legislation created a chief of space operations (CSO) to oversee the new service. That general officer, chosen from the U.S. Air Force, would report directly to the secretary of the Air Force and would become a member of the joint chiefs of staff after one year. The legislation noted that the individual selected as CSO could also serve as the commander of USSPACECOM for one year. Finally, the legislation stipulated that the individual who was the AFSPC commander on the day before the legislation was signed into law could become the CSO without Senate confirmation. This obviously paved the way for President Trump to appoint General Raymond—already serving as AFSPC and USSPACECOM commander—as the first CSO.[63]

Conclusions

Two days after the conference report was released, on December 11, 2019, the House of Representatives voted to approve the FY2020 NDAA by a vote of 377 to 48. Just under a week later, the Senate followed suit and approved the bill with a vote of 86 to 8. All that remained was for President Trump to sign the bill. The presidential signing ceremony for the NDAA was scheduled for Friday, December 20, 2019, at Joint Base Andrews, Maryland.[64]

Over the ensuing twenty-month period, the organizational structures and activations for the three USSF field commands occurred. On October 21, 2020, HQ USSF was redesignated as HQ Space Operations Command, followed by redesignation of Space and Missile Systems Center as Space Systems Command on August 13, 2021, and, finally, the creation of Space Training and Readiness Command—the most unique of the three—on August 23, 2021.[65] Operational space units, with their respective mission responsibilities, were assigned under each field command. By 2022 building a unique service culture was well under way but remained ongoing into the foreseeable future.

Notes

1. Clayton Koppes, *JPL and the American Space Program: A History of the Jet Propulsion Laboratory* (New Haven: Yale University Press, 1982), 18–21.
2. Theodore von Kármán, *Toward New Horizons: Science, the Key to Air Supremacy. Commemorative Edition, 1950–1992* (Washington, DC: HQ Air Force Systems Command, 1992); Michael H. Gorn, *The Universal Man: Theodore von Kármán's Life in Aeronautics* (Washington, DC: Smithsonian Institution Press, 1992), 108–17.
3. David N. Spires, *Beyond Horizons: A Half Century of Air Force Space Leadership* (Peterson AFB, CO: HQ Air Force Space Command, 1998), 14–15.
4. Harvey Hall, *History of a Physicist: Joys and Woes* (Fort Belvoir, VA: self-published, 2000), 128–42; R. Cargill Hall, "Early U.S. Satellite Proposals," in *The History of Rocket Technology: Essays on Research, Development, and Utility*, ed. Eugene M. Emme (Detroit: Wayne State University Press, 1964), 69–74; Spires, *Beyond Horizons*, 14–16; Project RAND, *Preliminary Design of an Experimental World-Circling Spaceship* (Santa Monica, CA: Douglas Aircraft Company, May 1946).
5. J. E. Lipp and R. M. Salter, eds., *Project Feed Back Summary Report* (Santa Monica, CA: RAND, March 1, 1954), https://www.rand.org/content/dam/rand/pubs/reports/2015/R262z1.pdf and https://www.rand.org/pubs/reports/R262z2.html.
6. Jacob Neufeld, *The Development of Ballistic Missiles in the United States Air Force, 1945–1960* (Washington, DC: Office of Air Force History, 1990); Space and Missile Systems Center (SMC), "Historical Overview of the Space and Missile Systems Center, 1954–2003" (Los Angeles AFB, CA: SMC/HO, 2003).
7. Paul H. Satterfield and David S. Akens, "Historical Monograph—Army Ordnance Satellite Program," Army Ballistic Missile Agency History Office, November 1, 1958, https://apps.dtic.mil/sti/pdfs/ADA434326.pdf; John B. Medaris with Arthur Gordon, *Countdown for Decision* (New York: G. P. Putnam's Sons, 1960), 98–112.
8. Brent D. Ziarnick, *Tough Tommy's Space Force: General Thomas S. Power and the Air Force Space Program* (Maxwell AFB, AL: Air University Press, 2019), 34–44; Curtis Peebles, *High Frontier: The U.S. Air Force and the Military Space Program* (Washington, DC: Air

Force History and Museums Program, 1997), 7–10; Dwayne Day, "Bill King and the Space Cadets," *The Space Review*, July 6, 2009, https://www.thespacereview.com/article/1408/1; Rick W. Sturdevant, "Retrospective on a Rocket Pioneer: Robert C. Truax and American Rocket Development," paper presented at the AIAA SPACE 2008 Conference and Exposition, September 10, 2008, San Diego, CA.

9. Stephen M. Rothstein, *Dead on Arrival? The Development of the Aerospace Concept, 1944–1958* (Maxwell AFB, AL: Air University Press, 2000), 60–62; Richard J. Barber Associates, "The Advanced Research Projects Agency, 1958–1974," report, December 1975, https://apps.dtic.mil/sti/pdfs/ADA154363.pdf; David N. Spires, "The Air Force and Military Space Missions: The Critical Years, 1957–1961," in *The U.S. Air Force in Space, 1945 to the Twenty-First Century*, ed. R. Cargill Hall and Jacob Neufeld (Washington, DC: USAF History and Museums Program, 1998), 33–45.
10. Rick W. Sturdevant, "The United States Air Force Organizes for Space: The Operational Quest," in *Organizing for the Use of Space: Historical Perspectives on a Persistent Issue*, ed. Roger D. Launius (San Diego, CA: Univelt, 1995), 170–74.
11. Spires, *Beyond Horizons*, 176–77.
12. Gen. Russell E. Dougherty, CINCSAC, to Gen. David C. Jones, HQ USAF/CC, April 14, 1977, in David N. Spires, *Orbital Futures: Selected Documents in Air Force Space History*, 2 vols. (Peterson AFB, CO: Air Force Space Command, 2004), 1:558–60; Brig. Gen. Earl S. Van Inwegen III, USAF (ret.), "The Air Force Develops an Operational Organization for Space," in Hall and Neufeld, 136; Sturdevant, "The United States Air Force Organizes for Space," 176; quoted in Robert Frank Futrell, *Ideas, Concepts, Doctrine: Basic Thinking in the United States Air Force, 1961–1984*, 2 vols. (Maxwell AFB, AL: Air University Press, 1989), 2:686.
13. Van Inwegen, "The Air Force Develops," 136–37; Brig. Gen. Earl S. Van Inwegen, USAF (ret.), interview with the author, November 1, 1995; USAF Special Study Group, "Proposal For: A Reorganization of USAF Air Defense and Surveillance/Warning Resources," January 1978.
14. "Congressional Testimony on Missions of TAC and ADCOM: Questions and Answers for the Record, House of Representatives Subcommittee of the Committee on Appropriations," February 8, 1977; House of Representatives, 95th Congress, 1st sess., "Department of Defense Appropriation Bill, 1978," report no. 95-451, 71; and "House Unit Studies Shifting ADC[OM] Functions to TAC, SAC," *Aerospace Daily*, June 30, 1977, 341.
15. Robert Kipp, "The Reorganization of 1979 and the Space Organization Issue," Air Force Space Command History Office, March 8, 1988; Futrell, 2:689; Spires, *Beyond Horizons*, 193–96.
16. John C. Stetson, SecAF, to Gen. Lew Allen, HQ USAF/CC, "ADCOM Reorganization," September 11, 1978, in Spires, *Orbital Futures*, 1:563–64.
17. Capt. T. Tony Tunyavongs, "A Political History of the Establishment of Space Command," *Quest: The History of Spaceflight Quarterly* 9, no. 1 (2001): 37.
18. Gen. James E. Hill, CINCAD, to Gen. Lew Allen, HQ USAF/CC, February 5, 1979, and HQ UAF/XO, Space Mission Organization Planning Study, February 5, 1979, in

Spires, *Orbital Futures*, 1:565–602; Sturdevant, "The United States Air Force Organizes for Space," 177; Spires, *Beyond Horizons*, 196–99.

19. Sturdevant, "The United States Air Force Organizes for Space," 177–78; R. Cargill Hall, "National Space Policy and Its Interaction with the U.S. Military Space Program," in *Military Space and National Policy: Record and Interpretation* (Washington, DC: George C. Marshall Institute, 2006), 11–13.
20. Robert Kipp, "Background Paper on Formation of Space Command," Air Force Space Command History Office, March 8, 1988; Van Inwegen, "The Air Force Develops," 138–40; Van Inwegen interview.
21. Van Inwegen interview.
22. Spires, *Orbital Futures*, 1:610–61.
23. "Consolidated Space Operations Center Lacks Adequate DOD Planning," General Accounting Office Report MASAD-82-14, January 29, 1982, http://archive.gao.gov/d41t14/117451.pdf; Hans Mark, *An Anxious Peace: A Cold War Memoir* (College Station: Texas A&M Press, 2019), 367–72.
24. Sturdevant, "The United States Air Force Organizes for Space," 178; Van Inwegen, "The Air Force Develops," 141.
25. Verne Orr to Ken Kramer, December 11, 1981, and Ken Kramer to Melvin Price, February 2, 1982 in Spires, *Orbital Futures*, 1:662–70; Sturdevant, "The United States Air Force Organizes for Space," 178; Tunyavongs, 37–38.
26. Gen. James V. Hartinger, USAF (ret.), interview with the author, August 20, 1992; Van Inwegen interview; Van Inwegen, "The Air Force Develops,"142; Maj. Gen. Thomas S. Moorman Jr., special assistant for SDI to AFSC/CV, interview transcript, by Robert Kipp and Thomas Fuller, AFSPC/HO, July 27, 1988, 15–18.
27. Spires, *Beyond Horizons*, 202–7; Sturdevant, "The United States Air Force Organizes for Space," 180; Moorman interview, 18–20; Hartinger interview; James V. Hartinger, *General Jim Hartinger: From One Stripe to Four Stars* (Colorado Springs, CO: Phantom Press, 1997), 245–48.
28. Spires, *Beyond Horizons*, 214, 216, 296–97; AFSPC/PA, "Air Force Space Command: Two Decades of Space," news release, August 23, 2022.
29. Spires, *Orbital Futures*, 480–81, 696–701.
30. Dr. Thomas Fuller, USSPACECOM/HO, chronology, "Formation of U.S. Space Command," ca. 1986; Dr. Thomas Fuller, USSPACECOM/HO, paper, "Formation of U.S. Space Command," August 1987; Memo, Ronald Reagan to Caspar W. Weinberger, "Unified Command for Space," November 20, 1984; ASD/PA, "Formation of United States Space Command (USSPACECOM)," news release, November 30, 1984; USSPACECOM/PA, "United States Space Command," fact sheet, November 1985.
31. Rick W. Sturdevant, AFSPC/HO, "United States Space Command, 1985–2002," background paper, August 16, 2018; Gen. Howell M. Estes III, USCINCSPACE, *Long Range Plan*, March 1998.
32. Memo with attachment, Lt. Gen. George W. Casey Jr., Director, Joint Staff, to Chief of Staff, U.S. Army et al., "Promulgation of Unified Command Plan 2002 (with Change-1

and Change-2 incorporated)," February 4, 2003; attachment, Unified Command Plan, April 30, 2002 (with Change-1 dated 30 July 2002 and Change-2 dated 10 January 2003 incorporated).

33. Donald H. Rumsfeld et al., "Report of the Commission to Assess National Security Space Management and Organization—Executive Summary," Pursuant to Public Law 106–65, January 11, 2001.

34. See Spires, *Orbital Futures*, 1244–65, for copies of memo, Donald Rumsfeld, SECDEF, to John Warner, SASC Chair, May 8, 2001; memo with attachment, Donald Rumsfeld, SECDEF, to Secretaries of the Military Departments et al., "National Security Space Management and Organization," October 18, 2001; DOD Directive 5101.2, "DoD Executive Agent for Space," June 3, 2003; and memo, SECAF to USECAF, "Delegation of DoD Executive Agent for Space Responsibilities," July 7, 2003.

35. Rick W. Sturdevant, HQ SpOC/HO, "Selected U.S. Military Space Systems," fact sheet, October 5, 2021.

36. Gen. James E. Cartwright, USMC, USSTRATCOM/CC, for USSTRATCOM Joint Functional Component Commander for Space and Global Strike, "Joint Functional Component for Space and Global Strike—Implementation Directive," January 5, 2018; JFCC SGS, "Joint Functional Component for Space and Global Strike—Concept of Operations," May 6, 2005; USSTRATCOM, "JFCC for Space and Global Strike Achieves Initial Operational Capability," news release, December 1, 2005, SD 2212; Report, Government Accountability Office, "Military Transformation: Additional Actions Needed by U.S. Strategic Command to Strengthen Implementation of Its Many Missions and New Organization," September 8, 2006, 2213; Fact Sheet, JFCC Space, "Joint Functional Component Command for Space (JFCC-SPACE)," January 2008; Briefing, Lt. Gen. William Shelton, CDR JFCC Space, "JFCC Space Perspective," June 10, 2008.

37. News Release, 1st Lt. Lucas Ritter, 30 SW/PA, "Joint Space Operations Center Opens at Vandenberg," May 26, 2005; Briefing, Brig. Gen. Timothy Coffin, DCDR JFCC Space, "Joint Functional Component Command for Space," April 23, 2014; Capt. Nicholas Mercurio, "USAF Commercial Integration Cell Pilot Program Underway," news release, June 9, 2015; Caleb Henry, "Air Force Formalizing Long-Term Commercial Integration Cell at JSpOC," *Defense Daily*, June 13, 2016; USSTRATCOM, "USSTRATCOM Space Control and Space Surveillance," fact sheet, January 2014; JFSCC, "Combined Space Operations Center Established at Vandenberg AFB," news release, July 19, 2018.

38. USSTRATCOM, "Defense Intelligence Communities Collaborate for Space Resiliency," news release, September 2, 2015; USSTRATCOM, "USSTRATCOM Welcomes Defense and Intelligence Communities to Strengthen Space Resilience," news release, August 29, 2016; DoD, "New Joint Interagency Combined Space Operations Center to Be Established," news release, September 11, 2015; 50 SW/PA, "Yardley Named September's Innovator of the Month," news release, October 21, 2015; 50 SW/PA, "Lieutenants Keep JICSpOC Standup on Schedule," news release, October 21, 2015; Stephen K. Hunter, "Standup of the Joint Interagency Combined Space Operations Center," November 13, 2015; National Reconnaissance Office, "Space Operations Center Gets

New Name, USSTRATCOM Begins Expanded Multinational Space Effort," news release, April 5, 2017.
39. Point Paper, AFSPC/CCX, "Improving DoD Execution of Space Responsibilities," December 14, 2016.
40. Background Paper, AFSPC/A2/3/6OP, "Alternate Command Relationships," January 5, 2017; Briefing, AFSPC, "USSTRAT ReOrg Options," January 24, 2017; AFSPC, "AFSTRAT Force Development Concept," March 8, 2017; AFSPC, "Transforming AFSPC into a Warfighting HQ [V6_CC final]," March 10, 2017; Background Paper, AFSPC/CCX, "Opportunities Offered by Proposed Space Re-Org," April 12, 2017; Background Paper, AFSPC/A2/3/6OP, "Joint Force Space Component Commander Organization Way Ahead," April 24, 2017; Briefing, AFSPC/CC, "Joint Force Space Component Commander," April 28, 2017; Background Paper, AFSPC/A2/3/6W, "Joint Force Space Component Command Implementation," June 19, 2017; Memo, Gen. John W. Raymond, AFSPC/CC, to All AFSPC and JFCC Space Personnel, "Commander's Initial Guidance and Intent for JFSC Implementation," July 21, 2017.
41. Willis Jacobson, "Space Forces Restructured at Vandenberg Air Force Base as New Leaders Take Over," *Santa Maria Times*, December 1, 2017; News Release, Maj. Cody Chiles, 30 SW/PA, "Maj. Gen. Whiting Takes Command of 14th Air Force," December 2, 2017; News Release, AFSPC/PA, "Air Force Command Becomes JFSCC, Joint Space Forces Restructure," December 3, 2017; Memo, Gen. John W. Raymond, AFSPC/CC, to All JFSCC Personnel, "Commander's Initial Guidance and Intent," December 1, 2017.
42. HASC, "National Defense Authorization Act for Fiscal Year 2018—Report of the Committee on Armed Services, House of Representatives, on H.R. 2810 together with Additional Views," July 6, 2017; 115th Congress, H.R. 2810, July 14, 2017; Report, SAF/LL, "FY18 National Defense Authorization Act (NDAA) Summary," November 17, 2017.
43. Steve Hirsch, "House Subcommittee Proposes Creation of New Space Command in Mark of Fiscal 2019 Policy Bill," *Air Force Magazine*, April 25, 2018; Position Paper, AFSPC/CCX, "Subordinate Unified Command vs Joint Force Space Component Command," May 2, 2018, SD 2296; Ellen Mitchell, "Amendment to Slow Space Force Shot Down in House Defense Markup," *The Hill*, May 9, 2018; Paul McLeary, "Fights Over DoD Bureaucracy, Space Force Make Tensions Flare on HASC," *Breaking Defense*, May 10, 2018; Office of Management and Budget, "Statement of Administration Policy: H.R. 5515—National Defense Authorization Act for Fiscal Year 2019," May 22, 2018; Marcia Smith, "White House Objects to HASC's Call for U.S. Space Command," *Space Policy Online*, May 23, 2018.
44. News Release, White House Press Office, "Remarks by President Trump at a Meeting with the National Space Council and Signing of Space Policy Directive-3," June 18, 2018; Letter, SecAF, CSAF, and CMSgtAF to Fellow Airmen, "Establishing a Space Force," June 19, 2018.
45. Background Paper, AFSPC/CCX, "U.S. Space Command History Documents," August 6, 2018; Report, DoD to Congressional Defense Committees, "Final Report on

Organizational and Management Structure for the National Security Space Component of the Department of Defense," August 9, 2018.
46. Memo, Gen. Joseph L. Dunford Jr., CJCS, et al., to SecDef, "Military Advice for Space Warfighting," August 29, 2018; Rick W. Sturdevant, AFSPC/HO, "USSPACECOM Joint Planning Group Msn/Structure Discussion," notes, August 29, 2018.
47. Memo, SecDef to POTUS, "Establishment of U.S. Space Command as a Unified Combatant Command," October 5, 2018; Memo, POTUS to SecDef, "Establishment of United States Space Command as a Unified Combatant Command," December 18, 2018; Elizabeth McLaughlin, "Trump Signs Memo Establishing U.S. Space Command, a New Combatant Command but Not a Military Service," ABC News, December 18, 2018; Tom Roeder, "Trump Revives U.S. Space Command, Likely Again Based in Colorado Springs," *The Gazette*, December 18, 2018.
48. Memo, SecDef, "Establishment of United States Space Command," August 29, 2019; Press Release, Maj. Cody Chiles, CFSCC/PA, "Combined Force Space Component Command Established at Vandenberg AFB," August 30, 2019.
49. Jeffrey R. Swegel, "A Fork in the Path to the Heavens: The Emergence of an Independent Space Force," School of Advanced Military Studies, U.S. Army Command and General Staff College, Fort Leavenworth, Kansas, 2002, https://apps.dtic.mil/sti/pdfs/ADA403851.pdf.
50. Report, SASC, "National Security Space: 21st Century Challenges, 20th Century Organization," September 27, 2016.
51. Mike Rogers, "Remarks at 2017 Space Symposium," *Strategic Studies Quarterly* 11, no. 2 (Summer 2017), https://www.airuniversity.af.edu/Portals/10/SSQ/documents/Volume-11_Issue-2/Rogers.pdf.
52. Wilson Brissett, "Separate Space Corps Included in Strategic Forces Markup," *Air and Space Forces Magazine*, June 21, 2017; Travis J. Tritten, "Jim Mattis Urges House to Abandon Space Corps Proposal," *Washington Examiner*, July 12, 2017; SECDEF, "Letter to the Honorable Michael R. Turner," July 11, 2017; SECAF, "Letter to the Honorable Michael Turner," July 11, 2017; Sydney J. Freedberg Jr., "Space Corps, What Is It Good For? Not Much: Air Force Leaders," *Breaking Defense*, June 21, 2017.
53. Mike Rogers and Jim Cooper, "Chairman Rogers and Ranking Member Cooper Joint Statement on Fundamental Space Reform," November 8, 2017, https://mikerogers.house.gov/news/documentsingle.aspx?DocumentID=600.
54. DoD, "Report to Congressional Defense Committees—Interim Report on Organizational and Management Structure for the National Security Space Components of the Department of Defense," March 1, 18; Sarah Kaplan, "Trump Floats Idea of 'Space Force,'" *Washington Post*, March 13, 2018; White House, "Remarks by President Trump at a Meeting with the National Space Council and Signing of Space Policy Directive 3," June 18, 2018, https://trumpwhitehouse.archives.gov/briefings-statements/remarks-president-trump-meeting-national-space-council-signing-space-policy-directive-3/.
55. DoD, "Final Report on Organizational and Management Structure for the National Security Space Components of the Department of Defense," August 9, 2018; Joe

Gould, "Space Force 'Not the Way to Go,' Says Key Democrat," *Defense News*, August 13, 2018.
56. Donald J. Trump, "Space Policy Directive 4," February 19, 2019, https://trumpwhitehouse.archives.gov/presidential-actions/text-space-policy-directive-4-establishment-united-states-space-force/.
57. Memo for Record, Acting SECDEF, "U.S. Space Force Planning Team," February 21, 2019.
58. Vivienne Machi, "Hyten Approves Putting Space Force under Air Force, But Notes Need for Full Department Eventually," *Defense Daily*, February 26, 2019.
59. Briefing, SFPTF, "DRAFT US Space Force Macro Organizational Framework Team," February 28, 2019; Transcript, USSF/HO, "Interview with Colonel Jack Fischer," January 3, 2020.
60. Center for Strategic and International Studies, "Space Threat Assessment 2019," April 4, 2019, https://www.csis.org/analysis/space-threat-assessment-2019.
61. SASC press release, "Inhofe, Reed Announce Hearing on U.S. Space Force," April 4, 2019; Transcript, SASC, "Hearing to Receive Testimony on the Proposal to Establish a United States Space Force," April 11, 2019.
62. House Appropriations Committee, "HAC NDAA FY2020 Markup," May 13, 2019; Sandra Erwin, "House Appropriators Deny Space Force Funding, Call on DOD to Study Alternatives," *Space News*, May 19, 2019; Senate Armed Services Committee, "FY2020 NDAA Executive Summary," May 22, 2019; Sandra Erwin, "Senate Armed Services OKs Space Force with Conditions," *Space News*, May 23, 2019.
63. House of Representatives, "National Defense Authorization Act for Fiscal Year 2020–Conference Report to Accompany S. 1790," December 9, 2019, https://www.congress.gov/116/crpt/hrpt333/CRPT-116hrpt333.pdf; Sandra Erwin, "NDAA Conference Agreement Establishes U.S. Space Force, Directs Major Overhaul of Space Acquisitions," *Space News*, December 9, 2019.
64. Patricia Zengerle, "U.S. House Approves Space Force, Family Leave in $738 Billion Defense Bill," Reuters, December 11, 2019; Connor O'Brien, "Senate Approves Bill to Establish Space Force, Federal Paid Parental Leave," *Politico*, December 17, 2019; Donald J. Trump, "Statement on Signing the National Defense Authorization Act for Fiscal Year 2020," December 20, 2019.
65. Fact Sheet, AFHRA, "Space Operations Command," October 27, 2021, https://www.afhra.af.mil/About-Us/Fact-Sheets/Display/Article/2886917/space-operations-command-ussf/; Fact Sheet, AFHRA, "Space Systems Command," November 29, 2021, https://www.afhra.af.mil/About-Us/Fact-Sheets/Display/Article/2886931/space-systems-command-ussf/; Fact Sheet, AFHRA, "Space Training and Readiness Command," November 10, 2021, https://www.afhra.af.mil/About-Us/Fact-Sheets/Display/Article/2886936/space-training-and-readiness-command-ussf/#:~:text=Redesignated%20as%20Space%20Training%20and,Oct%201993%2D1%20Apr%202013.

2 THE FIRST GUARDIAN
Bernard A. Schriever
JOHN G. TERINO

Gen. Bernard A. Schriever's career spanned the rise of industrial America into full global power status, the maturation of the air age, and the dawn of the space age. Schriever represented archetypes for every element of these manifestations of military and civilian technology during his long time in uniform. His classical engineering degree gave him credibility and useful expertise that set him apart from most of the Air Corps, combat missions and logistics leadership during World War II imparted real-world experience in achieving results under pressure, and his managerial acumen fundamentally altered research and acquisition practice in the American military for decades when he led Air Force acquisition during the early Cold War.[1] Schriever had a large, lasting, and enduring impact on the Air Force, the aerospace industry, and national efforts in space exploration and exploitation. He successfully transitioned from a classic military leader of the industrial age into a model technocratic system builder of the electronic and space age. In his many accomplishments, he transcended the lofty cultural and technical expectations for aviators during the early twentieth century and embodied the ideals expected of airmen, and he became the exemplar for what the U.S. Space Force seeks in its personnel. In simple terms, Bernard Schriever was the first guardian.[2]

As aviation captured the imagination of the world and heralded a new age of progressive optimism in the first four decades of powered flight, fliers symbolized the spirit of the new age as the vanguard of the future.[3] Later, those who piloted the rockets that left earth also assumed representation of the future and pushed the vision of tomorrow beyond the atmosphere. The scientists and engineers who developed rocketry and space travel, in spite of attempts to the contrary, were never accorded the reputation as vanguards of the future that the astronaut-pilots earned.[4] Nevertheless, they were vital to building the resources and capacity to turn high-technology ideas into reality.[5] Today, as the utility of space has become commonplace and new spacefaring journeys and missions beyond earth orbit are imminent, the nascent U.S. Space Force seeks paradigmatic leaders as icons for the future of the service as well.[6]

Bernard Schriever exemplifies all three of the ideals presented in the previous paragraph. He was an archetypal airman in the founding generation of American military aviators, he embodied the ideal technical manager-operator Air Force leaders sought as the service gained independence and established itself as a high-tech military entity, and he represents precisely the type of guardian the Space Force seeks as it builds its independent culture. Schriever is often described as the "the father of Air Force missiles and space."[7] Tall, athletic, and handsome, Schriever epitomized the popular image of a pilot, was a model officer leader in the U.S. Air Force, and, most significantly, embodied a new breed of professional, the military system builder. Several military system builders emerged in the middle decades of the twentieth century to operationalize challenging technologies for the American military, and Schriever was the most influential of them all.[8] He was the paradigmatic military and technological system builder, with a deep influence on the Air Force, the Space Force, the National Aeronautics and Space Administration (NASA) and the Apollo program, and the aerospace industry.

The Air Force was not unique in the expectation that leaders of aviation would both define the future and lead the nation into it through the mastery of flight and technology.[9] Young, athletic, and technically competent men like

Bernard Schriever embodied this vision. But his legacy far surpassed those of his peers and generations of Air Force professionals. From humble beginnings as the child of a single mother newly emigrated to the United States and struggling to make ends meet, he became a college graduate, a commissioned officer, a pilot, an engineer, a military system builder, a four-star general, and, ultimately, a legendary figure who influences aspects of the American defense community to this day. Schriever is not just the "man who built the missiles"; he is the man who built the system that built the missiles, mastered the management and leadership required to build those weapons, and ensured that the nation's aerospace industry could support American ambitions in space and for high-tech acquisition for decades after his retirement and death.[10] His story represents a journey of accomplishment that saw him rise from athletic overachiever to qualified engineer by the time he was twenty years old, overcome long odds to earn Air Corps wings, further his engineering education, survive combat in World War II, lead the most important defense acquisition effort of the 1950s, and develop technological systems for his own service and for the joint force.

Aviation profoundly influenced society and culture in the early twentieth century, especially in America. After a century of unparalleled mechanical progress that transformed agriculture, industry, transportation, warfare, and expectations for the future, the powered aircraft inspired a completely different sort of vision for humanity because it transcended the bounds of earth.[11] A sense rapidly took hold that airpower could usher in a new age of peace, cooperation, and realization of higher goals for all. In the minds of many, it stimulated a drive for technological solutions to prevent a repeat of the industrial slaughter of World War I and to advance society as a whole.[12] For some, the promise of aviation could either avert conflict by bringing the world closer together or deter war through the threat of massive destruction from the air.[13] Aviation represented the future, and pilots had a distinct and unique role to play as leaders of that future in the understanding of many Americans as government support for both military and civil aviation increased in the late 1920s.[14]

The Air Corps tapped into these cultural ideals as part of its recruiting efforts to encourage the right sort of young men to join as pilots.[15] While military service, especially in the Army, was not popular during the Roaring Twenties, the prospect of being a pilot had tremendous social, cultural, and economic cachet, and the Air Corps benefited as applicants sought wings in droves.[16] Even as all air services struggled during World War I to determine the ideal attributes for aircrew, a broad consensus regarding the physical, mental, and emotional aptitudes required for pilots emerged in the decades after the conflict.[17] During the war, there were intense debates in the United States regarding conscripting pilots from all normal pools of military-eligible men or seeking only elites from the best colleges.[18]

In the early postwar era, air service leaders routinely defined the high standards required for pilots in both general and specific terms in public messaging. Brig. Gen. William "Billy" Mitchell wrote that the air service wanted "able-bodied young men coming up to the physical requirements and having what is termed 'a college education.'"[19] The attributes of "young men who go to our colleges and not only are proficient in their studies, but in athletics such as football, baseball, tennis, polo" were prized and available in abundance, according to Mitchell.[20] Beyond education and athleticism, he believed, the sense of teamwork and cooperation from playing football and baseball made them superior combat pilots.[21] Mitchell argued that sports such as American football, baseball, basketball, hockey, and tennis inclined participants toward the type of "thinking discipline" required of good pilots.[22] Mitchell's writings not only reflected the prevailing sentiments and culture of military aviators in the United States and abroad but also mirrored the official standards published by the Army Air Corps. With budgets and resources tight and numbers of applicants exceeding available slots, the Air Corps could select the cream of the crop for most of the interwar period.[23] Since some college education, preferably in engineering, was required, it was axiomatic that pilots would come from an elite group because white males in the United States with college experience were a small percentage of the population.[24]

Future generals Henry "Hap" Arnold and Ira C. Eaker promulgated the same message in the books they coauthored to encourage young men to join the Air Corps. They detailed both the official standards and their own perceptions regarding attributes of successful aviation cadets. Arnold and Eaker, like Mitchell, reinforced the idea that fine coordination of mind and muscle, developed from sports, was desired.[25] In their book *Winged Warfare*, after repeating the standard requirements—being between twenty and twenty-seven years old, with perfect physique complete with normality in eyes, heart, lungs, and neuromuscular coordination, and a college degree or at least two years' training at a university—Arnold and Eaker asserted that because "the American boy has a mechanical background," America had "the raw pilot material for the largest and most superb air force in the world."[26] The dominant perception, and the actual reality, for many years was that the Air Corps only took the best it could get.[27] Many wanted to be pilots, but only a select few would actually qualify. Even fewer would earn their wings.

Early Years

Among the thousands who had the "right stuff" when standards were high was Bernard Schriever. Born in Bremen, Germany, on September 14, 1910, he arrived, with his mother and younger brother, in the United States at the age of six in February 1917. His father had been in the United States for years because the German ocean liner on which he worked had been interred in New York harbor since the start of World War I. By April 1917 the United States was officially at war with Germany, and the Schriever family was reunited in New Jersey. To escape virulent anti-German sentiment, they moved to New Braunfels, a suburb of San Antonio, Texas, a region settled heavily by German immigrants since the 1840s.[28] After the untimely death of his father in an industrial accident and some hard times adjusting while his mother found gainful employment herself, the family ended up living adjacent to the historic Brackenridge Park golf course in San Antonio.[29] In addition to her duties managing the household of wealthy banker Edward Chandler and his family, his mother Elizabeth ran a concession stand near the twelfth hole, and "Bennie" Schriever and his brother, Gerhard, helped make ends meet by caddying on the course.[30]

Hard work, good grades, and excellence in sports were also important as Schriever graduated high school at sixteen and enrolled in Texas A&M University to study engineering. He captained the varsity golf team for two years, played baseball, and was a Reserve Officers Training Corps cadet as well. He had grown up with the Army since San Antonio was then, and still is, a major military town. The future was literally in the air around Schriever as the skies of San Antonio filled daily with dozens of aircraft.[31] The Alamo City was the center of Army Air Corps flight training as home to a number of military airfields and Fort Sam Houston, giving San Antonio one of the largest concentrations of soldiers in the chronically understrength interwar Army.[32] Through his teenage years, Schriever excelled at golf and met many local Air Corps pilots who played rounds at Brackenridge.[33] Before graduating from college, Bennie had established basic character traits, social acumen, and the strong drive to complete tasks that formed the core of his professional identity. When he earned his wings on June 29, 1933, he had already overcome long odds by actually attending and graduating from college. He beat even longer odds when he became a pilot in the Air Corps because the washout rate was extraordinarily high.[34] Army fliers truly represented an elite group because of the robust educational and physical standards demanded. Schriever was one of thousands of young men who flew for the Air Corps before World War II. However, over the next fifteen years, he took significant steps to establish himself as much more than an average military flier as he transitioned from an archetypal airman in the Army to a technical leader of the new Air Force.

In the summer of 1933, Lieutenant Schriever began his active military career by reporting for duty with the 9th Bombardment Squadron at March Field in Riverside, California. As an officer with a reserve commission, he could look forward to flying for about a year or two on active duty unless he earned one of the very few regular commissions available to junior officers. Regardless, his Air Corps training, certification, and experience qualified him for jobs in the rapidly expanding civilian airlines.[35] Assignment to March Field was significant in Schriever's development for several reasons. His commander, Lt. Col. Henry H. "Hap" Arnold, was not a typical Air Corps leader. Arnold learned to fly with Orville and Wilbur Wright and was one of the first three qualified military aviators.[36] More importantly, he had a deep and enduring appreciation for the role of technology as a basis for advancing the potential of airpower.[37] Unlike his military contemporaries, Arnold cultivated relationships with scientists, engineers, and industrialists who advanced emerging science and technology. As far back as 1913, working for chief signal officer Brig. Gen. George Scriven, Arnold oversaw aircraft production and specification. He personally supervised engine tests at various locations, which included frequent visits to the National Bureau of Standards test facility.[38] As a lieutenant serving in an infantry unit in the Philippines, Arnold petitioned his former boss in the signal corps to study aeronautics and engineering at either the Massachusetts Institute of Technology or Cornell University, a request denied by General

Scriven.[39] But his predilection for finding and advancing technically savvy officers manifested itself in more than a few Air Force careers.

By the time Schriever reported for duty at March Field, Arnold had renewed and expanded his relationship with Nobel laureate and California Institute of Technology (Caltech) president, physicist Robert Millikan. Millikan recruited renowned aerospace engineer Theodore von Kármán to lead the Guggenheim Aeronautical Laboratory at Caltech (GALCIT) in 1930, and Arnold made his acquaintance when he led Air Corps research and development as commander of the matériel division at Wright Field in Ohio before his California assignment. Von Kármán's presence further enhanced the GALCIT reputation for leading-edge aeronautical research. A few years later, von Kármán helped found the Aerojet Corporation, one of the first commercial rocket propulsion companies.[40] Schriever later became very familiar with research emerging from Caltech, many graduates of the program, and the products of Aerojet, but before all that happened, he enhanced his reputation as an officer, a pilot, and a golfer while he and his family became especially close to the Arnolds.

Bennie went west with his whole family, and since he was unmarried, his mother filled in at his side at many social events. Mrs. Bee Arnold and Elizabeth Schriever were roughly the same age and both spoke German, and they became fast friends. Schriever also dated Arnold's daughter Lois and became close friends with his son Bruce.[41] Winning local tournaments and setting club records at the nearby Victoria Country Club in Riverside also elevated the young officer's profile at March Field.[42] Finally, Schriever gained entry into another significant group as a regular in Maj. Carl "Tooey" Spaatz's late-night poker games with other commanders at March Field.[43] While these social connections were good, none would have mattered if Schriever had not shone as a competent pilot and officer. Despite his successes, Bennie did not earn a regular commission and was back home in San Antonio by March 1935.

The stint out of uniform was short, as he was back on duty by the summer in charge of a Civilian Conservation Corps camp on the Arizona–New Mexico border for the next year.[44] For Schriever and many officers, this experience was both a leadership challenge of the highest order and great experience for the future mobilization and expansion of the service for World War II. Leading civilians without the legal authority of the articles of war was quite a challenge for young men barely older than their charges. Unifying disparate cadres of people motivated by different goals into successful teams provided valuable lessons for Lieutenant Schriever. Simple leadership lessons such as thoroughly studying the challenge at hand before acting, picking the right people to accomplish it, backing people who are loyal and tell you the truth about their efforts, and giving them the latitude to accomplish their task became the bedrock of his managerial approach for the rest of his career.[45]

Expansion and improvement of the Air Corps starting in 1936 brought many reserve officers back to active duty.[46] Schriever returned and was assigned to fly fighters in the Panama Canal Zone. Brig. Gen. George H. Brett, the Air Corps commander of the zone, learned of Schriever's golf prowess and assigned him to his staff as an aide to help in the improvement of his own game.[47] While in that position, Bennie met General Brett's daughter, Dora, and began a courtship leading to marriage two years later. The ceremony took place in Washington, and since travel from Panama was too arduous, the assistant chief of the Air Corps, Brig. Gen. Arnold, and his wife Bee stood in for the Bretts.

After failing once more to earn a regular commission in 1937, Schriever signed on with Northwest Airlines. While he was based in Seattle in late 1937, Schriever's flying skills improved as he regularly flew at night and in bad weather, but he did not have the job for long.[48] In March 1938 General Arnold flew out to Seattle for an inspection tour at Boeing. Although he rarely played the game, Arnold specifically arranged for a golf outing with Schriever. Arnold encouraged Schriever to apply for a regular commission again since openings would be filled by competitive examination. Schriever responded favorably to Arnold's entreaty and gave up the lucrative airline job. Once again, Schriever's work ethic, talent, and skills set him apart from the crowd. He caught the attention of senior mentors, and on October 1, 1938, he finally became a regular officer in the U.S. Army Air Corps.[49]

The Schriever family relocated to Hamilton Field just north of San Francisco, where Bennie was again assigned to a bomber squadron. Their first child, Brett Arnold, was born there in March 1939. A few months later, they were on the move again to Wright Field near Dayton, Ohio, where Bennie became a test pilot and enrolled in July 1940 in the Air Corps engineering school.[50] His maturity, analytical abilities, and flying skill were certainly factors in this change of station. However, the fact that General Brett was now chief of the

Air Corps matériel division and commandant of the engineering school certainly did not count against him.[51]

World War II

With the onset of hostilities in Europe, American military expansion was rapid, and experienced officers assumed new, larger responsibilities. By the time Schriever graduated from the engineering school, Brig. Gen. George Kenney was the commandant. Among the most respected and educated officers in the Air Corps, Kenney was an expert in attack aviation, logistics, and education. Kenney attended the Massachusetts Institute of Technology as an undergraduate, worked several years in different engineering jobs, flew combat missions during World War I, and also graduated from the school he now commanded in 1921. In terms of professional technical education, Kenney was among the best in the upper reaches of the Army Air Corps in 1941.[52] Kenney rated Schriever as "superior" academically upon graduation. Bennie also secured a slot at Stanford University as a full-time graduate student in aeronautical engineering. Just halfway through his courses when Pearl Harbor was attacked, Schriever remained in school through graduation in June 1942 before he was assigned to fly bombers in the Pacific theater.[53] Flying combat required more than piloting an aircraft, and Captain Schriever got the chance to showcase his skills as a logistician, staff officer, and commander as well, but his educational tours helped establish his credentials as a system builder after the war.

The Fifth Air Force was in Australia when Schriever reported for what would be a three-year combat tour. Waging war on a global scale was challenging, and fighting in the Pacific, more so than many other theaters, required teamwork, creative problem-solving, and significant logistical and mechanical acuity to generate effective combat power. Schriever's former commandant Kenney led the Fifth Air Force, bringing his considerable personal, technical, and professional skills to bear under difficult circumstances. Kenney's right-hand man was Ennis Whitehead, also an educated engineer, a World War I pilot, and a top graduate of the Air Corps engineering school.[54] Schriever served under senior leaders with backgrounds and understandings similar to his own in a theater where creative problem-solving regarding tactics, logistics, doctrine, and aircraft modifications compensated for a lack of resources.[55] After Schriever had spent a few months flying B-17s in combat, Kenney put Bennie's skills to work as the chief of maintenance and engineering for the Fifth Air Force service command. Schriever quickly proved his bona fides in terms of results, and both his rank and responsibilities elevated before he went home after V-J day as a full colonel.[56]

Scientific Advisory Board

The postwar Air Force, even after demobilization, was much larger than the Air Corps had ever been. Large numbers of World War II veterans were offered regular commissions as the service built upon the combat experience earned during the conflict. Schriever was no average officer; instead, his advanced education and expertise elevated him to the elite of the Air Force in terms of qualifications to fulfill Hap Arnold's vision for a highly technical military organization.[57] Arnold wanted to ensure the Air Force maintained a close connection with leading scientists and engineers. The relationships that he had pioneered had paid off during World War II, and before he retired, he took steps to link the service even more closely with those who could advance America's technological edge.[58] Schriever repeated these steps to prepare the Air Force for space and missile capabilities.

First, Arnold met with von Kármán to mobilize his talents forecasting important technologies for the Air Force for decades to come. This established the Scientific Advisory Group (later called the Scientific Advisory Board) for the Army Air Forces and established the need for technically savvy officers.[59] Second, in January 1946 Arnold summoned his protégé, Colonel Schriever, to his office and passed him the proverbial torch regarding managing and cultivating science and technology for the Air Force.[60] For the better part of the next twenty years, Schriever would change the relationship between the Air Force and technology acquisition, management, and development as the paradigmatic technical manager-operator for other Air Force officers to emulate through staff work on technology evaluation and forecasting, by shepherding America's first intercontinental ballistic missiles (ICBMs) into reality, by initiating the first operational reconnaissance satellite programs, and, eventually, by establishing Air Force Systems Command to both develop and acquire new technology for the service.

Arnold ensured that scientists had access to the top ranks of the Air Force by elevating research and development to the air staff. Maj. Gen. Curtis LeMay, a young, combat-hardened officer who had worked his way into the inner circle of top Air Force leaders through operational success, was the first deputy chief of staff for research and development. Schriever held numerous positions in development and liaison with the Scientific Advisory Board in the years immediately after World War II before and after he attended the National War College. In these positions, he networked with the top technical talent of the services, the scientists and engineers who worked with the Air Force, and civilian leaders of the military, and he dealt with the red tape that characterizes the upper echelons of the senior staff. His intelligence, good looks, athletic build, and articulate presentation skills served him well, even when he ran into formidable opposition while briefing ideas that were hard to sell to entrenched operational commanders or bureaucratic satraps resistant to the need to adjust to the coming revolutions in electronics, propulsion, and materials that heralded the dawn of both the jet age and the space age.

All the military services were interested in rockets after witnessing the impact, literally and figuratively, of the German V-2 program in the latter years of World War II. The technology transfer efforts of the United States gave the Army both actual rockets and support equipment as well as many of the personnel associated with Wernher von Braun's rocket team. The Air Force already had a relationship with the "suicide squad" and the Jet Propulsion Laboratory at Caltech as well as some emerging efforts with other traditional aviation companies such as Convair. However, a combination of complacency (because of the expected long-term advantage provided by the atomic bomb) and budget cuts impacting the military in the postwar drawdown reduced the pace of missile programs significantly. After the test of the Soviet atomic bomb in 1949, a dawning realization that America's communist rival might develop an intercontinental rocket capable of striking the American heartland fueled action to mobilize U.S. resources to counter this emerging threat. This perception also provided an opportunity to develop new methods to marshal talent, accelerate development, and pioneer new management theories to deliver results more quickly. Bernard Schriever was ideally positioned to be the face of this new effort, but it was not a foregone conclusion that he would get the job.

Concurrency
Air Force leaders had to be convinced that ballistic missiles could be developed more rapidly than the air-breathing missiles already under development.

There is no doubt that the bomber barons who emerged as the vanguard of leaders of the postwar Air Force had a certain affinity and comfort with long-range aircraft. Their preference for air-breathing missiles such as the Snark and Navaho that were already in the works was not simply because they looked like airplanes. Many believed that the technical challenge to create air-breathing systems would be easier to solve than the issues with weight, propulsion, and guidance associated with ICBMs. Schriever and his allies in the scientific community, on the air staff, and on the Defense Department staff did their homework and believed that with the right prioritization, sufficient funding, and the right technical managers, ICBMs could be operational within about six years. Moreover, they recognized that allowing the Soviets to gain operational nuclear ballistic missiles before the United States would give the adversary the ability to launch a surprise first strike that could not be countered since manned bombers could not launch fast enough.

Schriever also recognized that he could not rely on standard development and acquisition practices to make the ICBM a reality in a compressed timeline. First, he knew that the Air Force did not have enough technical expertise in the areas that missile development required. Second, he also knew that there were no large aircraft corporations that had the right talent. Third, he also recognized that the infrastructure to design, build, test, and employ these weapons did not really exist. The challenge before him was incredibly daunting and wide-ranging and required a radically different approach to succeed, even before he was charged with developing multiple types of missiles and the capabilities to use them for space launch in addition to weapons delivery.

Schriever championed a new form of acquisition and development management for the Air Force, the military, and even NASA that is still employed in many respects today. To gain the necessary expertise, Schriever essentially contracted with established corporations to employ the emerging discipline of systems engineering to integrate cutting-edge technologies into a final product.[61] The process was christened "concurrency" because all subsystems were developed concurrently, that is, at the same time, under the watchful eyes of an engineering management staff that orchestrated the schedule to ensure all parts were integrated before they were fitted into the final assembly. Simon Ramo and Dean Wooldridge, friends and graduates of Caltech, had reunited at Hughes Aircraft in 1946 and assembled a team of exceptionally talented young engineers whose products were research and development of cutting-edge avionics, fire control systems, and ultimately missiles as well. They also branched out into manufacturing so their designs could move beyond prototyping. Ultimately, they left Hughes in 1953 and established their own company, the Ramo-Wooldridge Corporation, to provide specialized expertise in systems engineering. This entity became the Air Force's partner in developing ICBMs and other technologies when Schriever became the first commander of the Western Development Division (WDD) of the Air Research and Development Command in April 1954. Brigadier General Schriever moved his family to Santa Monica, and WDD was established on an old Catholic school campus in Inglewood, California.

Under Schriever's leadership, WDD charted a new course in technology acquisition and management. Instead of relying on the Air Force's tried and true practice of working with a single prime contractor like a well-established aircraft manufacturer to produce a complete airplane, Schriever's team served as the prime contractor with Ramo-Wooldridge responsible for systems engineering and technical direction. To save time and ensure delivery before the Soviet Union could field a fully operational fleet of nuclear missiles, Schriever and the Air Force assistant secretary for research and development, Trevor Gardner, maneuvered to brief President Dwight Eisenhower on the need to prioritize the ICBM project in order to eliminate bureaucratic impediments. Assigned the highest priority for weapons systems development in 1955, Schriever and Ramo developed their new managerial approach, concurrency. Additionally, because of the importance of the project and the need to develop the industrial base to advance missiles and rockets, there was a commitment to redundancy as well. WDD acquired not one ICBM, but two. Both the Atlas and Titan missiles were developed simultaneously with different contractors working on each. Later, the intermediate-range Thor missile was also included in the effort.[62]

Schriever spent the better part of the next five years shuttling around the country in an exhausting schedule as he simultaneously assembled his team of officers, procured test ranges and launch sites, built infrastructure to support all facets of the programs, and met with politicians, generals, and contractors and played golf with many of them. It was all part of conducting the business of completing the crash development of multiple missiles, including initiating the development of the solid-fuel Minuteman ICBM and the first spy satellites.

In April 1959 he became the commander of Air Research and Development Command and pinned on his third star. In 1961 his overhaul of acquisition practice for the Air Force was complete when he earned his fourth star as the commander of the newly established Air Force Systems Command, where he spent the final five years of his career until his retirement in August 1966.

Schriever did not complete all these programs by himself, instead relying on dozens of people he handpicked to lead the programs and the hundreds of people working on them. For example, Schriever brought Osmond Ritland from the air staff, where he had been working on the U-2 program with the Central Intelligence Agency, to be his deputy in Los Angeles and to run space programs. Schriever brought over nuclear engineer Otto Glasser to run the Atlas missile program. He hired Charles Terhune away from Kirtland Air Force Base, New Mexico, to be the deputy director of technical operations. Schriever also hired from Kirtland Maj. Forrest McCartney to develop the Corona reconnaissance satellite and Lt. Col. Charles "Moose" Mathison to recover the film buckets. He brought in Ed Hall to develop the Minuteman solid-fuel ICBM program, eventually replacing Hall with Sam Phillips. Ritland retired as a two-star general, Glasser, Terhune, and McCartney as three-stars, and Phillips as a four-star, all of them leading important space and missile programs in and out of the Air Force.

Schriever ensured systems engineering traveled beyond the Air Force into other space and missile programs through the people he mentored. The Navy applied the systems engineering approach developed to field the Atlas system in the development of its Polaris missile, which also relied on solid-fuel rockets like in Hall's Minuteman program. While Wernher von Braun and his rocket team in Huntsville, Alabama, were successful in developing ballistic missiles for the Army and rockets that powered the Apollo program for NASA by relying on the arsenal system of management, they received a boost from Schriever's efforts that made the moon landing possible before the goal established by President John Kennedy. In 1963 NASA administrator James Webb requested General Schriever release Brig. Gen. Sam Phillips, his Minuteman missile program director, to NASA. Schriever only agreed to do so if Phillips was put in charge of the Apollo program. Webb agreed, and Phillips introduced configuration control and other Air Force acquisition management techniques in addition to almost two hundred Air Force officers into NASA to get Apollo back on schedule. The ultimate success of the moon landing can be attributed to Phillips' ability to apply the lessons and procedures learned while working with Schriever on a complex, technologically dependent project.[63]

Conclusions

After retirement in 1966, General Schriever lived in Washington, served on corporate boards, consulted on research and development, and played golf. Ultimately, his marriage to Dora fell apart, and he remarried. Gen. Lance Lord, the commander of Space Command in 2005, honored General Schriever with the first set of space operator "wings," calling Schriever, shortly before his death, America's first space operator.[64]

In a 1999 interview with NASA, Schriever recognized the changes his systems, especially space systems, had brought:

> Well, space overall has had a tremendous impact on national security. . . . It's going to be a while yet, I think, before we restructure and rethink some of the ways in which we are going to have to arm ourselves, because we're in the space business now, but there's still that interaction between ground, sea, air, and space, and they have to be integrated, and they are being integrated now, but they weren't really integrated. They did a great job in the Gulf War. That's the first war that I would put in the category of Arnold's brains, and brains are going to play a more and more important role, because the sophistication of precision weapons, the speed of light that relates to information. . . . So we have a challenge of optimizing our capability in a completely new environment. Space has intruded, you might say, in many ways, and in other ways it can bring about what I consider a spread in our deterrent overall capability. . . . And our first job is to have a military force that deters. The military is there really to prevent wars, and to prevent wars, you have to be able to fight them.[65]

Almost two decades later, Congress created the independent U.S. Space Force. In a foundational document, the service notes that it seeks people of character, connection, commitment, and courage to advance mastery of space in service to the nation.[66] More than a century before the Space Force was established, Bennie Schriever demonstrated those qualities in his quest to be an Air Corps pilot. He marshalled his connections, talent, and commitment to excellence to gain the education and skills for an ideal technical leader in the new Air Force. Finally, when the nation needed someone with the courage to develop the missiles that serve as both the ultimate deterrent and as the reliable gateway to space for decades, he was ready to lead that organization. There can be no finer example than Bernard Schriever, the first guardian.

Notes

1. On the United States as the industrial ideal, see Thomas P. Hughes, *American Genesis: A Century of Invention and Technological Enthusiasm* (New York: Viking Penguin, 1989).
2. As I developed this chapter, I relied on discussions and conversations with my colleagues at the Air Command and Staff College, Air University, and the Air Force Academy to help refine and strengthen my views and arguments regarding Schriever. In particular, I owe the maturation of the idea of Schriever as the first guardian to a lunch conversation with Lt. Col. Joseph Ladymon.
3. For examples of air-mindedness and its impact on culture at this time, particularly in the United States, see Robert Wohl, *A Passion for Wings: Aviation and the Western Imagination, 1908–1918* (New Haven: Yale University Press, 1994); Robert Wohl, *The Spectacle of Flight: Aviation and the Western Imagination, 1920–1950* (New Haven: Yale University Press, 2005); Joseph J. Corn, *The Winged Gospel: America's Romance with Aviation* (Baltimore: Johns Hopkins University Press, 2001); and A. Bowdoin Van Riper, *Imagining Flight: Aviation and Popular Culture* (College Station: Texas A&M University Press, 2004). For a more general examination of human fascination with flight, see Bayla Singer, *Like Sex with Gods: An Unorthodox History of Flying* (College Station: Texas A&M University Press, 2003).
4. The idea of pilots as spacefaring leaders is explored in a number of works, most popularly in Tom Wolfe, *The Right Stuff* (New York: Bantam Books, 1980). Matthew H. Hersch, "'Capsules Are Swallowed': The Mythology of the Pilot in American Spaceflight," in *Spacefarers: Images of Astronauts and Cosmonauts in the Heroic Era of Spaceflight*, ed. Michael J. Neufeld (Washington, DC: Smithsonian Institution Scholarly Press, 2013), 35–55, discusses both elements of this issue. Pilots as the ideal transitioning from the atmosphere to space were perceived differently from the scientists and engineers who designed and built the craft that launched the astronauts in public perception in spite of attempts to change that idea. Schriever has a foot in both camps since he was originally a military pilot and is the man who built the missiles. Hersch further explores the meaning and role of astronauts in the American space program in *Inventing the American Astronaut* (New York: Palgrave Macmillan, 2012).
5. Public perception of the role of pilots and their links to the future, especially that of General Schriever himself, is articulated in the *TIME* magazine cover story from April 1, 1957, "ARMED FORCES: The Bird and the Watcher," 2, 5.

6. For examples illustrating this effort, see Sandra Irwin, "Space Force Talent Strategy Is a Departure from the Norm," *Space News*, September 21, 2021 (particularly about the publication of *The Guardian Ideal*); Reagan Mullin, "The Space Force's Critical Lesson for the Rest of the Military," *War on the Rocks*, December 15, 2021; and Center for Strategic and International Studies, transcript, "Discussing Two Years of the Space Force with General Raymond," January 26, 2022.
7. Even though the introduction of this volume includes a passage detailing how Schriever once denied applying this term to himself, in retrospect, it is well deserved. Among the many works or sites where this term or a similarly worded accolade is applied to General Schriever are Randolph J. Sanders, *Schriever Air Force Base: A History in Pictures* (Washington, DC: Government Publishing Office, 2017), iv; Jacob Neufeld, "General Bernard A. Schriever: Technological Visionary," *Air Power History* 51, no. 1 (Spring 2004): 43; Karl P. Mueller, "General Bernard Schriever: Guiding the Air Force to New Frontiers," in *Airpower Pioneers: From Billy Mitchell to Dave Deptula*, ed. John Andreas Olsen (Annapolis, MD: Naval Institute Press, 2023), 177; "In Memoriam: General Bernard Adolph Schriever USAF (Ret.), 1910–2005," *Air Power History* 52, no. 3 (Fall 2005): 67; and even the official U.S. Air Force biography entry, "General Bernard Adolph Schriever," https://www.af.mil/About-Us/Biographies/Display/Article/104877/general-bernard-adolph-schriever/. In *Makers of the United States Air Force* (1996; repr., Washington, DC: Office of Air Force History, 1987), 281, Air Force historian John L. Frisbee characterizes him "as the officer most closely associated with the development of ballistic missiles." Finally, the inscription on Schriever's headstone at Arlington National Cemetery states, "Father of the Air Force's Ballistic Missile and Space Programs."
8. I am employing the term system builder as developed in the scholarship of Thomas P. Hughes in a number of his works, but particularly in "The Evolution of Large Technological Systems" in *The Social Construction of Technological Systems: New Directions in the Sociology and History of Technology*, eds. Wiebe E. Bijker, Thomas P. Hughes, and Trevor Pinch (Cambridge, MA: MIT Press, 1987). He develops the term more fully in both *American Genesis* and *Rescuing Prometheus* (New York: Pantheon Books, 1998). I assert the cadre of military system builders includes Gen. Leslie Groves, Gen. Curtis LeMay, Adm. Hyman Rickover, Gen. Bernard Schriever, Adm. William Raborn, Adm. Levering Smith, Brig. Gen. Robert F. McDermott, and Gen. Wilbur "Bill" Creech. Each of them held leadership positions in organizations for an extended period, exercised personnel, budgetary, and command authority beyond the parameters of normal assignments, and developed a military-technical system that fundamentally changed normal operations of the American military.
9. For ease of expression, I will regularly employ the term Air Force in this essay even when the term Air Service, Air Corps, or Army Air Forces may be more historically accurate for the period discussed.
10. Walter Boyne, "The Man Who Built the Missiles," *Air Force*, October 2000, 81. See also Frisbee, 281, where he writes that Schriever "would be responsible for research, development, and acquisition of all new weapons used by the United States Air Force."

11. For a general outline on the impact of various technologies on life at the turn of the twentieth century, see Stephen Kern, *The Culture of Time and Space, 1890–1918* (Cambridge, MA: Harvard University Press, 2003); regarding the airplane specifically, see 242–47. On the impact of mechanization on transportation in particular, see Wolfgang Schivelbusch, *The Railway Journey: The Industrialization of Time and Space in the 19th Century* (Berkeley: University of California Press, 1986). A broader contemporary perspective regarding the implications of the new materials and technologies of the early twentieth century can be found in Lewis Mumford, *Technics and Civilization* (New York: Harcourt, Brace, 1934), especially chap. 5.
12. A succinct summary on prescient science fiction for aviation, with particular attention to the writings of H. G. Wells, before powered flight in addition to ideas on the peril and promise of airpower before, during, and after World War I can be found in the opening chapters of Michael S. Sherry, *The Rise of American Air Power: The Creation of Armageddon* (New Haven: Yale University Press, 1987).
13. See Mumford, 266; William "Billy" Mitchell, *Winged Defense: The Development and Possibilities of Modern Air Power–Economic and Military* (1925; repr., Tuscaloosa: University of Alabama Press, 2009), ix, 19; Corn, chap. 2; Kern, 242–47; and Sherry, 38, 43. Another examination of the impact of flight and other technologies that created the modern world and accelerated national aviation networks is David T. Courtwright, *Sky as Frontier: Adventure, Aviation, and Empire* (College Station: Texas A&M University Press, 2005), 12–14, and elsewhere in the book in general.
14. The Air Corps Act stimulated some growth and capability for Army aviation, often at the expense of the rest of the service. Edward M. Coffman, *The Regulars: The American Army, 1898–1941* (Cambridge, MA: Belknap Press of Harvard University Press, 2004), 277. For more details, see James P. Tate, *The Army and Its Air Corps: Army Policy toward Aviation, 1919–1941* (Maxwell Air Force Base, AL: Air University Press, 1998), 27–48. For more on civil aviation, see, for example, Roger E. Bilstein, *Flight in America: From the Wrights to the Astronauts* (Baltimore: Johns Hopkins University Press, 1984, 1994), especially chap. 3.
15. Rebecca Hancock Cameron, *Training to Fly: Military Flight Training, 1907–1945* (Washington, DC: Air Force History and Museums Program, 1999), 216–17.
16. Regarding typical American ambivalence about a professional standing Army, see Coffman, 292–94. For the appeal of military aviation, see Beirne Lay Jr., *I Wanted Wings* (New York: Grosset & Dunlap, 1937). On selectivity, see *A History of Military Aviation in San Antonio* (San Antonio, TX: U.S. Government, September 1996), 18.
17. Lee Kennett, *The First Air War: 1914–1918* (New York: Free Press, 1991), chap. 7. Also, for broad generalities regarding the physical and mental requirements established during World War I and continuing through the next two decades, see Mark K. Wells, *Courage and Air Warfare: The Allied Aircrew Experience in the Second World War* (1995; repr., Portland, OR: Frank Cass, 2000), chap. 1.
18. Linda R. Robertson, *The Dream of Civilized Warfare: World War I Flying Aces and the American Imagination* (Minneapolis: University of Minnesota Press, 2003), 53–60. For

an account of the lasting influence of elite college airmen in World War I and beyond, see Marc Wortman, *The Millionaires' Unit: The Aristocratic Flyboys Who Fought the Great War and Invented American Air Power* (New York: Public Affairs, 2006).

19. William Mitchell, *Our Air Force: The Keystone of National Defense* (New York: E. P. Dutton, 1921), 115.
20. Mitchell, *Winged Defense*, 24.
21. Mitchell, *Winged Defense*, 172.
22. Gen. William Mitchell, *Skyways: A Book on Modern Aeronautics* (Philadelphia: J. B. Lippincott, 1930), 63–66.
23. *History of Military Aviation in San Antonio*, 18.
24. A quick glance at reliable statistics would put the percentage at somewhere between 5 and 25 at most; see National Center for Education Statistics, "Percentage of Persons 25 to 29 Years Old with Selected Levels of Educational Attainment, by Race/Ethnicity and Sex: Selected years, 1920 through 2013," October 2013, and U.S. Census, "A Half-Century of Learning: Historical Statistics on Educational Attainment in the United States, 1940 to 2000," April 6, 2006.
25. H. H. Arnold and Ira C. Eaker, *This Flying Game* (New York: Funk and Wagnalls, 1936), 108–9. The desire for athletic ability was a long-standing ideal among the leaders of aviation in the Army (see Cameron, 272).
26. Maj. Gen. H. H. Arnold and Col. Ira C. Eaker, *Winged Warfare* (New York: Harper & Brothers, 1941), 29–30.
27. One of the many sources that detail the impact of popular attitudes toward flight, educational standards and availability, and the impact of the Great Depression on the militaries of World War II—in particular, the air services—with direct application to this chapter is Eric M. Bergerud, *Fire in the Sky: The Air War in the South Pacific* (Boulder, CO: Westview Press, 2001), 310–13, 316–19.
28. The best book-length biography of Schriever is Neil Sheehan's *A Fiery Peace in a Cold War: Bernard Schriever and the Ultimate Weapon* (New York: Random House, 2009).
29. Alamo City Golf Trail, "About Brackenridge Park Golf Course" (n.d.). It is the oldest public course in Texas and the long-time home of the Texas Open championship, and it was designed by noted golf architect A. W. Tillinghast. Schriever twice won both the state junior championship and the San Antonio city championship.
30. Sheehan (6–13, 19, 24–26, 125, 253) chronicles the formative impact of golf—financially, socially, athletically, professionally, emotionally, and intellectually—on Schriever. His passion for golf followed him throughout his life and often influenced his Air Force career. Schriever even installed a putting green at his long-time official residence at Andrews Air Force Base while he was the commander of Air Research and Development Command and then Air Force Systems Command, as chronicled by Perry Jamieson and Mary Lee Jefferson in *Belle Chance: A Commander's Haven* (Washington, DC: Air Force History and Museums Program, 2001), 28.
31. Gen. Bernard A. Schriever, oral history interview by Dr. Edgar F. Puryear Jr., June 29, 1977, transcript, 42, Air Force Historical Research Agency, K239.0512–1492. After

Chapter 2

mid-1922, for the next two decades, San Antonio was the Army's flying training center, and every Army aviator finished flight training at Kelly Field; *A History of Military Aviation in San Antonio*, 17.

32. Coffman, 234. Another indication of the size of the garrison and amount of aerial activity can be found in the 1927 special feature *Wings: Grandeur in the Sky* in *Wings*, directed by William A. Wellman (Hollywood, CA: Paramount, 2012). After Randolph Field became operational in 1931, there were at least five military flying fields within fifteen miles of Schriever's house in San Antonio (Brooks, Dodd, Duncan, Kelly, and Randolph fields). See Arnold and Eaker, *This Flying Game*, photo page between 88 and 89 depicting an aerial view of San Antonio with four military airfields annotated (Duncan Field was part of Kelly Field). Duncan Field's relation to Kelly Field is explained more fully in *A History of Military Aviation in San Antonio*, 28, 113.

33. His mother even dated one of his future instructor pilots (Sheehan, 14). Additionally, when he won the 1932 Texas State amateur golf championship, he defeated Capt. Ken Rogers, one of his flight instructors at Kelly Field (Neufeld, "General Bernard A. Schriever: Technological Visionary," 38).

34. Wells details that only 3,505 rated pilots were trained by the Army between 1923 and 1939 (6n13). For examples regarding the high selectivity, risk, and long odds of completing training, see Cameron, 253, 272. With a washout rate of over 50 percent, Schriever truly beat long odds to become a bomber pilot.

35. For details on the regular versus reserve status and its impact on aviation cadets and their assignments, status, and future, see Cameron, 242, 245, 260–61. Arnold and Eaker, *This Flying Game*, 87, assert that the Army Air Corps has trained more fliers than any other school, including 90 percent of airline pilots.

36. Richard G. Davis, *HAP: Henry H. Arnold, Military Aviator* (Washington, DC: Air Force History and Museums Program, 1997), 3.

37. The best biographical account of this side of General Arnold can be found in Dik Alan Daso, *Hap Arnold and the Evolution of American Airpower* (Washington, DC: Smithsonian Books, 2001). Also Roger Bilstein, "Donald Douglas: From Aeronautics to Aerospace," in *Realizing the Dream of Flight: Biographical Essays in Honor of the Centennial of Flight, 1903–2003*, ed. Virginia P. Dawson and Mark D. Bowles (Washington, DC: NASA, 2003), 101.

38. Daso, 63–65. See also Wayne Biddle, *Barons of the Sky* (New York: Simon and Schuster, 1991), 184–85.

39. Cameron, 62.

40. Daso, 127–28, 141–42, and Fraser MacDonald, *Escape from Earth: A Secret History of the Space Rocket* (New York: Public Affairs, 2019), 87–98. GALCIT later became NASA's Jet Propulsion Laboratory.

41. John C. Lonnquest, "The Face of Atlas: General Bernard Schriever and the Development of the Atlas Intercontinental Ballistic Missile, 1953–1960" (PhD diss., Duke University, 1996), 49.

42. Sheehan, 19.
43. Spaatz rose to prominence as a bomber commander in World War II and became the first chief of staff of the U.S. Air Force. Schriever, oral history transcript, 5; DeWitt S. Copp, *A Few Great Captains: The Men and Events that Shaped the Development of U.S. Air Power* (McLean, VA: EPM Publications, 1980), 209; Sheehan, 17.
44. The Civilian Conservation Corps was a New Deal stimulus effort to both improve stewardship of public lands and put hundreds of thousands of young men to work planting trees, fighting forest fires, and draining marshes, among other duties. It had numerous positive effects on the Army and on the youth of the nation; Charles E. Heller, "The U.S. Army, the Civilian Conservation Corps, and Leadership for World War II, 1933–1942," *Armed Forces & Society* 36, no. 3 (2010): 439–53.
45. For details on the Army experience writ large with the Civilian Conservation Corps, see Coffman, 242–45. For Schriever's experience, see Sheehan, 23–24.
46. Although the Air Corps Act of 1926 authorized a five-year buildup of the service, the onset of the Great Depression coupled with a penurious Congress stunted that growth. The Drum Board and the Baker Board, among other initiatives, provided an impetus for modest expansion until the war clouds of World War II ushered in a massive enlargement of all things military (Tate, 140–70). More details on budget pressure and personnel increases can be found in John F. Shiner, "The Heyday of the GHQ Air Force, 1935–1939," in Nalty, *Winged Shield* (Washington, DC: Air Force History and Museums Program, 1997), 135–37.
47. The definitive biography of Brett is Douglas A. Cox, *Airpower Leadership on the Front Line: Lt. Gen. George H. Brett and Combat Command* (Maxwell Air Force Base, AL: Air University Press, 2006). For more details on how Schriever became Brett's aide, see Sheehan, 24–25.
48. Jacob Neufeld, *Bernard A. Schriever: Challenging the Unknown* (Washington, DC: Office of Air Force History, 2005), 3–4.
49. Sheehan, 26–27.
50. The engineering school was the direct forerunner to the Air Force Institute of Technology, the modern scientific and technical graduate school of the Air Force. See "Air Force Institute of Technology," March 2022, https://www.af.mil/About-Us/Fact-Sheets/Display/Article/590443/air-force-institute-of-technology/.
51. Sheehan, 28–29.
52. For details on Kenney, see Thomas E. Griffith Jr., *MacArthur's Airman: General George C. Kenney and the War in the Southwest Pacific* (Lawrence: University Press of Kansas, 1998), 5–6, 7–16, 20–21, or Sheehan, 30.
53. Sheehan, 31.
54. Griffith, 58–59.
55. For insights on the innovative and creative application of doctrine and technology in the southwest Pacific, see Bergerud, 288–91, and Griffith, 63–64, 81–84, 118–19. Captain Paul "Pappy" Gunn gets much of the credit for technical innovations regarding aircraft; John R. Bruning, *Indestructible: One Man's Rescue Mission that Changed the Course of*

World War II (Boston: Hachette Books, 2017), and Jay A. Stout, *Air Apaches: The True Story of the 354th Bomb Group and Its Low, Fast, and Deadly Missions in World War II* (Guilford, CT: Stackpole Books, 2019), 26–30.
56. Sheehan, 31–48, briefly details Schriever's World War II experience.
57. According to statistics cited in Vance O. Mitchell, *Air Force Officers: Personnel Policy Development, 1944–1974* (Washington, DC: Air Force History and Museums Program, 1996), by almost any definition, Schriever, and many of the officers he worked with or for (Kenney, Whitehead, and Donald Putt, for example) were among the most educated officers in the service as only 37 percent of regular officers (50) had college degrees in 1946, and this percentage did not improve much until after Sputnik. In 1954 it was 45 percent (197), in 1966 it was 65 percent (201), and by 1974 it was 85 percent (201).
58. Daso, 209–14.
59. The meeting between Arnold and von Kármán in a car at LaGuardia Airport in New York is recounted in Daso, 192, 196. Details on the Scientific Advisory Group itself can be found in Thomas A. Sturm, *The USAF Scientific Advisory Board: Its First Twenty Years, 1944–1964* (1967; new imprint, Washington DC: Office of Air Force History, 1986). Further details on both the meeting and the influence and legacy of von Kármán can be found in Michael H. Gorn, *Harnessing the Genie: Science and Technology Forecasting for the Air Force, 1944–1986* (Washington, DC: Office of Air Force History, 1988). Michael H. Gorn, ed., *Prophecy Fulfilled: "Toward New Horizons" and Its Legacy* (Washington, DC: Air Force History and Museums Program, 1994), contains the original reports written by von Kármán and the committee he formed.
60. Sheehan, xvii–xix, 118–24.
61. Systems engineering, in the words of one of its early practitioners, is "the discipline of the design of the whole, to realize a harmonious and effective ensemble, as distinct from the design of the parts." Simon Ramo, *The Business of Science: Winning and Losing in the High-Tech Age* (New York: Hill and Wang, 1988), 41.
62. Sheehan, 233–35, 261, 318.
63. Stephen B. Johnson, *The Secret of Apollo: Systems Management in American and European Space Programs* (Baltimore: Johns Hopkins University Press, 2002), 136–37.
64. Sheehan, 475.
65. Bernard A. Schriever, oral history interview with Carol Butler, Washington, DC, April 15, 1999, NASA Johnson Space Center oral history project, 26–27.
66. U.S. Space Force, "Guardian Spirit," Space Force Handbook 1-1 (April 3, 2023), 3.

3 THE SPACE FORCE'S GODFATHER

John B. "Bruce" Medaris

LISA BECKENBAUGH

The "godfather of America's space program," Maj. Gen. John Bruce Medaris, USA, may not be as well known as other space pioneers, but his work in the early days of the space race was crucial to the later success of the National Aeronautics and Space Administration (NASA). As he grew up in near poverty as the only child of a single mother at the turn of the century, no one, including Medaris himself, could have imagined that his future would include an honored career in missile development and space exploration. But former NASA administrator James Beggs argued, "Had it not been for Medaris and his organization, a lot of things that were done in the early days [of space] couldn't have been done." So, who was Bruce Medaris, and how did he get to be a successful spaceman?[1]

Early Years

In 1906 Medaris and his mother, Jessie, moved to Springfield, Ohio, after her divorce from her philandering husband. She quickly got a job as the chief accountant and treasurer for a manufacturing firm. Medaris was left in the primary care of his grandmother LeSourd, who did not keep close tabs on his activities. When his mother enrolled him in school, she said his birth year was 1900; it was not until 1940 that his mother informed him that his real birth year was 1902. Money was always an issue for the family, so at the age of nine, Medaris bought a bicycle and a paper route with a loan of seventy-five dollars he secured from the Springfield National Bank. Years later he learned that his mom had set up the loan transaction, but he made good on the loan himself and paid it off with interest in six months. Always looking for additional income, at age eleven Medaris won a bid for a lamp lighting job from the city. By high school he traded both jobs for a more lucrative one as a mail handler for the Big Four Railroad Station. When not working, Medaris was an excellent student and studied a variety of subjects, including Latin, Spanish, and touch typing. He also joined a junior military organization and drilled one hour per day, fueling a lifelong fascination with weapons.

During his second year in high school, Medaris learned that taxi drivers made more money than mail handlers, so he convinced a friend to teach him to drive and started driving a taxi at night. He was thirteen years old. By age fourteen he again switched jobs and became a uniformed conductor on the Springfield Street Railway System. In the spring of 1918, Medaris decided he could not miss the action overseas and enlisted in the Marine Corps. He told the recruiter that he was eighteen and his mother confirmed the lie, compounding her earlier falsehood about his birth year. In reality, Medaris was only sixteen years old. After basic training he shipped out to France as part of the 6th Marine Regiment. Medaris never saw combat but did do some military police duty in Bordeaux, France. He returned home in August 1919 with an honorable discharge, muster-out pay, and a train ticket back to Springfield. While he was overseas, his mother had remarried, but Medaris and his new stepfather, Emil Opferkuck, clashed quickly. Medaris soon decided it was time to move on.[2]

Medaris' Marine Corps training qualified him for a senior Reserve Officers Training Corps (ROTC) spot at Ohio State University, where he studied mechanical and electrical engineering. By 1920 he was the ROTC company commander and was married to Gwendolyn Hunter. As was usual for Medaris, he worked a variety of jobs while at college, moving on from one to another for better pay. In his second year of college, the Army announced that it would "conduct a nationwide competition to select 110 candidates for officer commissions." There were two conditions: the first required him to join the Ohio National Guard, and the second was to master geometry. His ROTC instructor helped him with the first, and Medaris bought a textbook and taught himself enough geometry in sixty days to pass the exam. Medaris scored well on the exam, ranked in the top ten in the nation, and was awarded a commission in the Army in 1921. Medaris and his wife moved to Chillicothe, Ohio, and joined the 19th Infantry Regiment at Camp Sherman. When the rest of the regiment was moved to San Franscico, Medaris was transferred to Fort Benning, Georgia, and the 29th Regiment where he served with a mule-drawn machine gun company.[3]

At Benning Medaris worked tirelessly to be the best infantry lieutenant possible. He read up on infantry tactics, kept his uniform spotless, and spent countless hours on marches and maneuvers. He made it very clear to his men that he would not tolerate any disobedience. His fellow officers found him to be "inconsiderate and arrogant," leading to a cordial dislike of Medaris. He believed that to achieve promotion, total commitment was required, reminding his new wife frequently that "duty came first." At this time, wives were little more than glorified camp followers. Gwendolyn struggled with the spartan living accommodations, rigid separation by rank, and low pay, but her life was brightened with the birth of daughter Marilyn in July 1922. Medaris was a loving but strict father. He demanded total obedience and treated Marilyn as a small adult rather than a child. After two long years at Benning, Medaris was transferred to the 33rd Infantry at Fort Clayton in the Panama Canal Zone. An overseas tour was mandatory for promotion, and this posting was seen as a plum assignment. Medaris was excited, his wife less so.[4]

Life in the Canal Zone was not unlike life at Fort Benning. Medaris continued to have very high expectations for his men, and he disciplined them promptly and equitably. He felt that in the long run, strict discipline made for contented troops. That did not mean that Medaris was satisfied; in his second year, he transferred to the ordnance corps, and the family moved to Corozal Ordnance Depot. With the continued low pay, Medaris started taking on odd jobs to supplement his income. One of those was writing opinion articles on national and international affairs for the *Panama-American* under a pen name. Army intelligence soon unmasked his identity, and he was severely reprimanded. At this point, Medaris believed he had little chance of promotion and was still struggling financially. He was fluent in Spanish and had attracted job offers from companies doing business in South America. Eventually, General Motors made him an offer that was three times his military salary, so in 1927 Medaris resigned from the Army—a move he regretted for years—and moved to Cali, Colombia, 150 miles from the coast.[5]

Medaris enjoyed the increased income, not only from his higher salary but also from a generous bonus schedule. He and Gwendolyn adjusted their lifestyle to accommodate this new wealth with servants, parties, cars, and gifts. His newfound affluence may have gone to his head; when Gwendolyn and Marilyn returned to Ohio to visit family, Medaris enjoyed the company of other women in her absence. Upon her return, Gwendolyn found out about the indiscretions and promptly returned to Ohio with Marilyn. Medaris suffered two additional blows: first, when Congress changed reenlistment regulations and Medaris was denied a return to active duty, and second, when General Motors changed the terms of his contract and he was no longer eligible for generous bonuses. Medaris realized he was ready for a change, so when the Baker Kellogg Company, a New York investment house, offered him a position, he jumped at the chance for a fresh start. It was not a good landing; within a short time the market softened, and Baker Kellogg let go of most of its staff, including Medaris. Embarrassed, Medaris returned to Ohio and took a lower-paying job with the Curtiss-Wright Flying Service, learning the ins and outs of the machine tool industry. He reconciled with his wife and on the advice of friends began trading on the stock market, enjoying the seemingly easy money and the lifestyle it afforded him. Upon returning from a business trip, Medaris learned that Gwendolyn had undergone a minor operation at a local military hospital, but there were complications leading to an inability to bear any more children. Medaris was enraged that, without his consent, she had undergone surgery that resulted in the possibility of a son being gone. Just months later the stock market collapsed, and the family lost their entire savings and Medaris his job.[6]

The Great Depression was hard on Medaris. His wife blamed him for losing all their money, and their marriage worsened. Medaris did every odd job he could find, while Gwendolyn scrimped on everything to do with the household. One day, Medaris returned home to an empty house. Gwendolyn had left him and returned to her parents. Medaris continued to send whatever money he could to his wife and child, but eventually, Gwendolyn sued for divorce and custody of Marylin. Bruce did not contest the action. According to his biographer, Medaris was too "proud, ambitious, and self-righteous" to really look at himself critically as having any part in the marriage failing.

Looking for new opportunities, Medaris moved to Cincinnati at the urging of a friend and worked as a door-to-door salesman, a losing proposition during the Depression. Walking the streets looking for work, Medaris ran into a former Army colleague. He invited Medaris to his office and the next day offered him a job with the Kroger Baking Company. Medaris was soon immersed in corporate policy and management, working long hours, but for a steady wage.

As he was moving up the corporate ladder, Medaris met Virginia Smith while playing bridge. The two dated for a while and were married on August 29, 1931. After having two children, Marta in 1933 and John in 1936, Medaris was restless with his life. He liked the stability that selling groceries gave his family, but he wanted something more. He even appealed to the War Department hoping to return to the military, but he was denied. In 1937, when his uncle offered him a car dealership in Cincinnati, Medaris jumped at the chance for a change. Looking back, Medaris said that the move to selling cars was "one of the worst decisions of my life." He personally signed a $10,000 loan agreement to secure the dealership. When his uncle unexpectedly died, the used car market dried up, and the parent firm was thrown into bankruptcy, Medaris was on the hook for the loan. The family cut expenses, sold household furniture, and Medaris returned to door-to-door sales. It took the family eight years to pay off the balance of the loan. Luckily, a friend recommended Medaris to the Dictaphone Corporation, where he became a senior salesman earning a steady wage again.[7]

World War II

The Medaris family was finally achieving some stability after a few years of turmoil until the military aims of Japan, Germany, and Italy became readily apparent. On July 1, 1939, the Army announced that Congress had authorized the expansion of the officer corps by calling up selected reserves. Medaris immediately contacted Col. Frederick McMahon, chief of the Cincinnati ordnance district, and offered his services. Because Medaris had stayed active in the reserves, serving as the president of the Cincinnati chapter of the Reserve Officers Association, McMahon wholeheartedly supported his return to the military. Within a few weeks, Medaris had orders for one year of active duty at the rank of captain. As captain of ordnance, Medaris surveyed ammunition production in southern Ohio, southern Indiana, Kentucky, and part of Tennessee, making good use of all his civilian management experience. In 1940, as he was approaching the end of his one-year orders, Medaris was called to Washington, DC, by Brig. Gen. A. B. Quinton Jr., the chief of all ordnance districts, to accelerate the expansion of production for the whole country. With his orders extended an additional year, the Medaris family moved again.[8]

General Quinton and now-Major Medaris were directed to speed up production contracts for increased tonnage to ship to Great Britain. Medaris took his directions literally, as his boss remarked, "If being too aggressive, Major Medaris has that fault. He is impatient with stupid people but in this critical situation, that enhances his value." Medaris was often brought in to unravel problem situations, like when his knowledge of Spanish worked to gain the trust of the Mexican government to produce valuable dynamite. He also organized an ammunition regulation station that scheduled shipments across the nation connected by a teletype network. His system functioned efficiently throughout the entire war. After Pearl Harbor, production goals increased exponentially and Medaris was busier than ever, but he longed to be overseas in the action. His boss did not want to lose him, so Medaris made himself "thoroughly obnoxious" about going overseas. Turning to an old friend, Brig. Gen. Henry Aurand, Medaris was able to secure a position in Iran—not ideal, but at least he was overseas and closer to the action. Promoted to lieutenant colonel, Medaris awaited his move to Iran. When it was delayed, he returned to Washington to await new orders. In May 1942 orders came for an assignment he really wanted, ordnance officer for II Corps.[9]

Medaris joined his unit in Florida for training. Finding accommodations for his family in Jacksonville, the nearest town, was hard, because of the explosion of military personnel in the area. After a few months in a drab little house, Medaris' wife had enough. She moved the children back to Ohio to stay near family. Lonely and bored by the endless training now in North Carolina, Medaris took up with a young widow living in the area. A bout with hepatitis, which ran rampant through II Corps, ended the affair, but Medaris struggled with the guilt of his actions, and his relationship with his wife suffered. In early November 1942, Medaris was given command of the 42nd Ordnance Battalion bound for North Africa. He bid his family farewell and, after a perilous journey, docked in Oran, Algeria, as part of Operation Torch. With new orders as the assistant to Col. Urban Niblo, Medaris began unloading his supplies and preparing for war.[10]

By February 1943 Niblo had been moved up and Medaris replaced him. Now-Colonel Medaris waged a running battle with everyone to get more and more supplies for the battles. He instructed his officers to "forget the books!" They had to "improvise and invent to meet the situation" they were in. After the Battle of Kasserine Pass, where the U.S. Army had its nose bloodied badly, Maj. Gen. Dwight D. Eisenhower sent Maj. Gen. George D. Patton to take over II Corps with Maj. Gen. Omar Bradley as his deputy. By late March a reinvigorated II Corps participated in the Battle of El Guettar. After the fighting, Medaris sent Patton his list of captured and/or destroyed Axis equipment. Patton asked him to look again. After another survey of the battlefield, Medaris sent Patton's assistant the same numbers and told him he would not "misrepresent" the facts: "Something is either true or it's not. And if it's not I won't bend it." Medaris fully expected to be disciplined, but instead Patton awarded him a medal, the Legion of Merit.

When Patton moved to plan the Sicily campaign, Omar Bradley became Medaris' new boss, and his next assignment was to strike Bizerte, Tunisia. Medaris had to move men and machines two hundred miles, crossing British supply lines. Bradley asked Medaris how he was going to accomplish the herculean task without upsetting the British. Medaris replied that he proposed to move without telling the British because "crossing the British supply line is worse than dealing with two committees of Congress." Bradley again asked Medaris what he was going to do and Medaris replied, "Trust me." Bradley did. Medaris moved 30,000 vehicles and 111,000 men and pieces of equipment without disturbing the British. When the attack launched, Medaris had all his men and machines in the assigned locations ready for action. The advance was a success, and the Germans surrendered.[11]

Medaris ran into all sorts of new supply problems in Sicily. He continued to complain about shortages and realized quickly that General Patton had little interest in logistics, only in killing the enemy. With shortages and overused machines, Medaris fought to keep Patton supplied to do just that. His men were pushed as close to the front lines as possible, many within the range of small arms fire. When on August 17 Messina finally fell and the last of the Italian resistance faded away, Medaris breathed a short sigh of relief. General Bradley was slated to head to England to prepare for the invasion of France. A hasty goodbye luncheon was arranged. The rumor in the handshake line was that if Bradley said, "I'll be seeing you," you were headed to England with him. If he said, "Thank you for your service," you were staying with II Corps and heading to Italy. As Bradley shook the hand of Medaris, he said the former, and Medaris was elated at the vote of confidence from the general. After a brief rest in Tunisia, Medaris and the other handpicked officers flew to England. Bradley took over First Army spearheading the invasion of France with Medaris as his ordnance officer.[12]

The first challenge Medaris faced in England was with First Army Headquarters. The personnel chief insisted that he would select the officers in Medaris' command. Medaris disagreed, and the two ended up in front of Bradley. Bradley agreed with Medaris, which allowed him to personally select the officers under his command. Medaris let all the battalions and companies know what was expected of them in the fight: "to keep the combat units adequately supplied under all conditions." This was a challenge because the one thing Medaris was woefully short of was trucks, although he did have tracked vehicles to spare. Col. Charles Patterson, the antiaircraft officer, had many trucks, but no tracked vehicles. So the two worked out a horse trade of sorts. Patterson's 40-millimeter machine guns would be towed by tracked vehicles, and Medaris would get trucks to fill with ammunition and supplies. This was not the last time Patterson and Medaris were able to make deals. After Bradley moved up and Lt. Gen. Courtney Hodges took over First Army, Medaris, Patterson, and a few others became known as "Hodges' Forty Thieves" for their propensity for acquiring equipment and supplies in inventive ways often outside of legal channels, much to the disgust of other commanders. Medaris was willing to do anything to keep the men supplied as the battle for the European continent began.[13]

Medaris landed the day after D-Day to beaches strewn with men, machines, and equipment. He immediately began trying to bring order to the chaos. A big test came when the beach ammunition dump caught on fire after some captured German ammunition exploded. Medaris raced to the scene, found the officer in charge had fled, and watched as men were being wounded by shrapnel from the exploding ammunition. Watching soldiers dragging water hoses to the huge fire, Medaris knew that water would be of no use, so he ordered men with shovels and bulldozers to begin using sand to bury the stocks and create berms to keep the flying shrapnel to a minimum. It took eight hours to bring the blaze under control, with Medaris himself often driving a bulldozer. After it was over, Medaris found the officer who had fled and relieved him of command, shipping him back to England. Medaris had no stomach for cowards. Bradley agreed, taking it one step further in awarding Medaris the Soldier's Medal for "heroism not involving conflict with an enemy."[14]

Medaris' next big test came after June 22 when a huge storm wrecked Mulberry A, the artificial harbor created to make the unloading of supplies easier. The destruction of Mulberry A caused supplies to run dangerously low. Medaris knew a big two-pronged offensive into France was being planned, so he met with General Bradley and let him know that he did not have enough ammunition stocks to support more than a one-pronged effort. Bradley asked Medaris directly if he was recommending calling off one prong of the corps attacks. Medaris replied, "Yes, sir." After further questioning by Bradley, Medaris remained steadfast. He could not support both prongs of the attack with current ammunition stocks. Bradley trusted Medaris' opinion, knowing he would only tell him the unvarnished truth, so Bradley revised his plans. Only the planned Normandy breakout went forward, with the attack on Bordeaux waiting until supplies could be built up.[15]

That trust served Medaris well a few weeks later when the Americans encountered French hedgerows made of dirt and dense scrubs. Used as fencing, the hedgerows, which were between five and sixteen feet high, were almost impassible by regular military vehicles. Sergeant Burtis Culin Jr. came up with an ingenious solution to the hedgerow problem: welding tusklike steel prongs onto a tank so it could bust through the hedgerow without exposing its thin-skinned underbelly to the enemy. Bradley brought Medaris to the demonstration and gave him two weeks to create the tusks on tanks. Medaris needed supplies from the United States, but how many? He wrestled with the quantity until he decided that one million pounds of welding rods were needed.

Washington balked. That kind of an order required operating U.S. plants for twenty-four hours a day. Bradley called Medaris in for confirmation that he really needed one million pounds. Medaris was adamant. Bradley okayed the order and insisted Washington send the requested amount. When the welding rods arrived, Medaris called everyone with welding skills to the beach along with more than eight hundred tanks. Using the German-made Czech hedgehogs, steel L- or H-shaped underwater obstacles, to create the tusks, Medaris' teams outfitted all the tanks in less than ten days. Medaris had made the right calculations, as it took almost all the welding rods he had ordered to complete the task.[16]

While Medaris may have had Bradley's unwavering support, not all senior officers thought the same way. Medaris felt that needed supplies were too slow when coming from coastal depots, so he sent "expediters" to the coastal bases to comb through the wares and send forward what was needed. Lt. Gen. Brehon Somervell, chief of all supply services in the Army, complained to Bradley endlessly that the "Hundred Bulldogs," as the expediters were known, were unnecessary and interfered with regular operations. Bradley disagreed, especially when Medaris was able to cite case after case of missing supplies to prove Somervell wrong. His actions cost Medaris dearly. Bradley recommended him for promotion to brigadier general three times—once endorsed by Eisenhower—and all three times Somervell turned them down. Medaris' disappointment at not being promoted did not dampen his enthusiasm for his job at all, especially as American forces continued to make progress liberating France.[17]

It was obvious to Bradley that Medaris was willing to do whatever it took to make sure that First Army was supplied with the best parts and ammunition. For example, Medaris made a deal with a French automaker in Paris to rebuild truck engines and make spare parts to keep trucks rolling without having to wait for replacements shipped all the way from home. Medaris discovered that British-made sighting bars were far superior to U.S.-made ones and had them produced in his ordnance shops. A visiting general noticed the unauthorized gun sights and reported Medaris to General Hodges for punishment. Hodges listened and then dismissed the general without a second thought. Medaris never faced even a cross word about the incident from Hodges. Medaris also contracted with Belgian firms for tires and tubes, spare parts for vehicles and guns, and to rebuild mortar tubes, which again was quicker than waiting for replacement parts—anything to keep First Army supplied for the fight.[18]

Medaris' improvisation came in handy a short time later. To help move ammunition and supplies more efficiently during battles, Medaris set up an unauthorized communications network for all the ordnance units. Maj. Gen. William Kean, First Army chief of staff, found out about the network and ordered it shut down. Medaris complied—to an extent. He told his operators to stop sending but not to stop listening. Medaris was convinced the network would be needed in the future. On December 16, 1944, when the Germans counterattacked in the Ardennes forest, the Battle of the Bulge began. First Army was caught by surprise. At the command post, General Kean asked Medaris where ammunition stores were located that might fall into German hands. Medaris pointed them out. Kean ordered them to be removed as fast as possible. Medaris told him that notifying those units to remove the ammunition would take at least two days if done by the book, hinting that he might have a quicker, not-by-the-book, way. Kean told Medaris to "do it his [Medaris'] way," and within twenty minutes the depots were notified and were on the move to get the ammunition out of German hands. This unofficial network ended up being the only means of communication for ordnance units during most of the battle.[19]

Medaris had a glimpse of his future a short time later. As the Allies gained the upper hand in the air war with the Luftwaffe nearly destroyed, Germany relied on its V-1 (vengeance flying bomb) and V-2 (retaliation rocket). V-1s flew slow enough that they could be intercepted, but V-2s were bigger and faster and could not be intercepted. The V-2 had been created at the Peenemunde Rocket Center on the Baltic Sea under the direction of Dr. Wernher von Braun. Medaris witnessed von Braun's work up close one day when he was in his mobile office and heard an explosion followed by a loud, low rumble. Figuring an ammunition dump had exploded, he hurried to the scene. Medaris found that men had been injured, but not from an ammunition dump explosion. Once the wounded had been cared for, Medaris and his staff climbed down into the crater to examine the remains of the weapon. He immediately remembered intelligence reports of a new German weapon. He had the V-2 parts carefully collected and sent to Supreme Headquarters Allied Expeditionary Force. Little did he know at the time that the men who created this weapon would flee oncoming Soviet forces and surrender to the Americans. Those same men were transported to the United States to work for the American rocket program eventually under the command of Medaris.[20]

As the Germany army continued to retreat, Eisenhower decided that First Army headquarters was to return home to prepare for the final invasion of Japan. The Germans surrendered on May 6, 335 days after D-Day, and by May 10 Medaris was on a plane bound for home with General Hodges. Medaris' service to First Army across Africa and Europe did not go unnoticed. Maj. Gen. Robert Wilson, the logistics chief, penned these words when Medaris was awarded the Distinguished Service Medal: "He was an outstanding organizer and administrator whose thinking was far ahead of anyone else. He was a human dynamo who could get more things done by himself than any two men. He would carry out adverse decisions to the letter even though he disagreed. But he would always argue his position cogently and courageously until it was either accepted or overruled. He was a pioneer in modern ordnance service, there was no book for this type of war. Many, many times he wrote 'the book' and his concepts were always proved sound by actual experience."

Medaris believed that his time in Europe taught him management, organization, and leadership. These lessons increased his "maturity and self-confidence," things that were to help him in the years that followed.[21]

Postwar Years

After the war, Medaris had some decisions to make, the biggest being whether he was going to apply for integration into the regular Army or return to civilian life. By this time, he had been denied promotion to general officer four times and believed it was because he was not in the regular Army—perhaps not realizing that not everyone loved him as much as General Bradley and General Hodges. Medaris took some leave to discuss it with his family, and while at home, he had a bout of stomach trouble and exhaustion that landed him in the hospital. It was there that he realized he loved the Army and did not want to leave. When he returned to full health, Medaris applied for permanent active duty and integration into the regular Army. His wish was fulfilled, and he became the deputy to Brig. Gen. Harold Nisely, ordnance officer for Army ground forces at Fort Monroe. By August 1948 Nisely retired, and Medaris, still a colonel, was picked to succeed him.[22]

The peacetime Army Medaris entered had two missions: to prepare for the next war and to revamp doctrine, policies, and procedures learned from the last war. Medaris took on the latter with "missionary zeal." He preached reform of the ordnance service and its relationship to the combat arms to anyone who would listen and to many who were not interested. There was little enthusiasm for Medaris' ideas at the Pentagon, and many of his seniors regarded him as a rebel whom they barely tolerated. Many officers remembered that some of his exploits during the war were not by the book and caused problems. Others felt he wanted to be a field commander and was not "loyal" to ordnance. Medaris could not set his ego aside and compromise, and so he spent much of his time frustrated by other officers' lack of vision—namely, his vision. It was not only other officers that Medaris rubbed the wrong way. The Army ordnance budget had to be approved by the ground forces commander, who was the principal customer of ordnance supply. Medaris went over the budget line by line with officers and civilian officials and was less than polite in his demands for changes, which furthered his salty reputation at the Pentagon.[23]

In 1948 Medaris was tapped for a special assignment because of his Spanish language skills. The United States was looking to firm up relations, especially militarily, with the Argentine government of Juan Péron. To further that goal, Secretary of State George C. Marshall invited the Argentine minister of defense, Gen. Jose Humberto Sosa Molina, to visit American military installations. At the last minute, the Spanish interpreter fell ill. The Army checked its rosters, and the only officer with the language proficiency to serve in that role was Medaris. The visit was a rousing success. At the final dinner at General of the Army Omar Bradley's home, when the Army chief of staff asked Gen. Sosa Molina if he was willing to sign a mission agreement, the general agreed as long as Medaris was the chief. Medaris was given three weeks to assemble his team of seven soldiers, gather families, and board a ship to Buenos Aires. In Argentina, Medaris and his staff worked to update the instruction at military schools with new tactics and procedures. Medaris was highly critical of State Department officials working in Argentina, arguing that they were arrogant and insensitive.

Medaris was excited when Gen. Matthew Ridgway, in charge of all South American missions, visited Buenos Aires. He regaled Ridgway with his ideas on modernizing the supply service. Ridgway was so impressed with Medaris that he dispatched a letter to Washington urging his immediate promotion to brigadier general. The letter joined the others in Medaris' file, but promotion still eluded him. When the Korean War began, Medaris fired off a letter to Ridgway offering his services, but Ridgway told Medaris to stay put as the mission in Argentina was too important. It was not until 1952 that Medaris returned home and to the Korean War.[24]

Initially, Medaris was assigned as a deputy at the Franklin Arsenal in Philadelphia, a seeming step down for the experienced colonel. Medaris asked arsenal commander Col. Ward Becker why he had been set aside. According to Medaris' biographer, Becker told Medaris that all the ordnance officers at the Pentagon had met about assignments, and not a single one of them offered to take Medaris. Becker reportedly said, "You've acquired a reputation as a troublemaker who is hard to handle. You shouldn't be surprised considering how you treated ordnance people in Africa, Sicily, and England while you were with Bradley and Hodges." Becker reminded Medaris that he was not in charge—he was a deputy and had better act as such. Medaris chafed but saluted smartly and assured Becker he would follow orders.[25]

Twelve days later, Medaris received a phone call from Lt. Gen. Louis Ford, chief of ordnance, calling him to Washington to solve a "serious jam." In person the next day, Ford explained the jam. The fighting in Korea had gone on longer than anyone in the Pentagon anticipated, causing a serious ammunition shortage. Ford ordered Medaris to find out why there were shortages and how to fix them, telling him to be "aggressive" because there was no time for "pussyfooting around." This is exactly the type of job the hardheaded, aggressive troublemaker relished. With Col. John Zierdt as his deputy, Medaris established a task force that found many problems with ammunition production.

First, after World War II, President Harry Truman was trying to balance increased domestic spending with a drawdown in military forces. Budgets were tightened across the military, and an easy target for budget cuts was structures only really needed during wartime, like supply. Thus, "production of ammunition all but ceased, and little funds were made available to preserve the war reserves on hand, or the production base that would be required to make more." Ammunition plants took longer than the estimated 180 days to restart, so the Army was only able to order a 90-day supply of ammunition that took 60 days to deliver to U.S. depots and ship to Japan, 60 days to offload and store, and an additional 30 days to get to Korea. Thus, a 90-day order of ammunition took

150 days to arrive at the front. The whole contracting and transportation system was broken. The Army was also expending artillery and small arms at a far greater rate than during World War II, leading to even more shortages. Finally, vehicles to move ammunition were scarce in Korea because all vehicles needing repairs were sent to Japan rather than being fixed in Korea.[26]

Medaris dove in to fix the many problems. Backed by Army chief of staff Gen. J. Lawton Collins, Medaris used his brash personality to force changes to the supply system within the Army in Japan and Korea, leaving behind more officers resentful of his tough tactics. Returning to Washington, Medaris spent additional months working with civilian organizations to fix the ammunition manufacturing and transport system in the United States. The Army's ammunition problems made the front page of every newspaper, triggering a congressional investigation. Gen. James Van Fleet, commanding general of U.S. and other United Nations forces, testified before Congress that his troops lacked ammunition. Ford tasked Medaris with the ordnance corps defense but warned him not to criticize Van Fleet in any way. Medaris stuck to the facts, laying out prewar budget cuts, increased ammunition expenditures, and the complexity of ammunition production. Based on his confident testimony before Congress, Collins promoted Medaris to brigadier general in April 1953.[27]

Missiles

Medaris' first job with his new star was director of industrial operations, supervising fourteen procurement districts and eight manufacturing arsenals. Relying on his background in the civilian sector and during World War II, Medaris improved management, processes, and procedures in the two years he held the position. After the war, Medaris was considering his future when he found himself on the list of officers for promotion compiled by General Ridgway. President Dwight Eisenhower agreed, and Congress confirmed Maj. Gen. John Bruce Medaris. After such a long wait for his first star, his second arrived just three years after the first. Medaris was elated but once again considered retirement. Omar Bradley asked him to join the Bulova Company, and Medaris was seriously considering the offer, going so far as to talk to his boss. Medaris' wife was happy about the move, excited about the possibility of finally establishing permanent roots in one place. But before anything was finalized, the Army called again.

As assistant chief of staff of the Army for logistics, Lt. Gen. Carter B. Magruder held the highest logistics position in the service.[28] General Magruder asked Medaris to become involved in the highly contentious missile race. While strategic bombers were the only workable delivery system for a nuclear weapon, missiles were being considered as an alternative delivery system as early as 1950. The German scientists who came to the United States with Wernher von Braun had been building and testing V-2 rockets and V-2 variants at White Sands Missile Range in New Mexico for years for the Army. When the Army jumped into the ballistic missile race with the Air Force, von Braun and associates were sent to Redstone Army Arsenal in Alabama. The redesignated guided missile development division, under the direction of von Braun, was tasked with building the two-hundred-mile-range Redstone missile. Medaris entered the fight as the Army was proposing the Jupiter rocket, an intermediate-range ballistic missile (IRBM) with a range of fifteen hundred miles, and a satellite proposal known as Project Orbiter, which was shelved while the Army and Air Force argued over the IRBM. The Army, with Navy backing, argued its Jupiter rocket was just longer-range artillery, while the Air Force argued it was a substitute for manned aircraft in the strategic nuclear mission. The Defense Department chose not to choose, and the race was on. The Army's development of the Jupiter IRBM continued as the Air Force developed its Thor IRBM.[29]

On February 1, 1956, the U.S. Army Ballistic Missile Agency (ABMA) was established at Huntsville, Alabama, with Medaris in charge. He was tasked with the research, development, production, and deployment of a totally new weapons system. He was given special authority to cut the red tape of traditional acquisition bureaucracy, not unlike what Gen. Bernard Schriever had with the Air Force. Initially, von Braun was wary of the blunt, hard-charging Medaris, who offended many with his "spit and polish" attitude, but eventually von Braun grew to like serving with the decisive Medaris. The motto Medaris extolled to von Braun and his staff was simple: "When you are told to do a job and given the resources you need, you have lost all excuses for failure. The buck stops with you."

Others saw Medaris as eccentric with a flair for showmanship, but Schriever understood him to be highly intelligent with initiative and a talent for organization—a strong adversary in the missile race. Medaris was the perfect man for the challenging missile and space job. He used his new authority to handpick officers from all over the Army, expanding von Braun's division from sixteen hundred to three thousand people. To create esprit de corps with the expanded group, Medaris created a special flag, distinctive shoulder patches, and other extra flourishes to impress visitors to Alabama. Medaris not only imposed his will on the Army organizations at Redstone Arsenal, he also convinced Huntsville city leaders to expand the city boundaries, extend city water and sewer systems, expand and improve the airport, build new roads, and set up a commission to figure out housing for the influx of thousands of new civilian workers.[30]

As more and more civilian and military personnel streamed into Huntsville, Medaris and von Braun got to work on plans for the future. Within two weeks of his arrival, Medaris had formally proposed a launch schedule of seven satellites in 1957 and 1958. Von Braun thought the plan was ambitious but doable; the Department of Defense (DoD) disagreed and scrapped the schedule. Medaris and his team continued to launch shorter-range missiles, achieving a record-setting Jupiter-C rocket launch in September 1956. The missile reached an altitude of 682 miles, going into outer space, and a flight length of 3,355 miles. Additional launches of both Jupiter and Thor rockets continued at Cape Canaveral, Florida—some successful, some not. In August 1957 a revolutionary Jupiter-C rocket, carrying a new heat shield design, successfully launched and was recovered. This rocket nose cone had an ablative material covering made of ceramic, designed to protect it from the tremendous temperatures of reentry. Called the "greatest single technological contribution made by von Braun's group to U.S. missile programs," it was another "win" for the ABMA team.[31]

The elation over these successful missile flights was dampened in November 1956, when Secretary of Defense Charles Wilson announced a new "roles and missions" statement. Medaris had been fighting to keep the Army in the missile and space race. He fought for engines, launch facilities, funding, and personnel, but all that seemed wasted when Secretary Wilson's memo announced that the Air Force would oversee all intercontinental missiles, such as Atlas, Titan, and Minuteman, capable of transporting a nuclear bomb five thousand to ten thousand miles. The Air Force would also get control of all 1,750-mile missiles, such as Thor and Jupiter. The Army was to be limited to two-hundred-mile-range missiles such as the Redstone. The Navy finally decided to go with the solid-propellent Polaris, deemed safer for ship storage, and ended its partnership with the Army on Jupiter. One bright spot for the team at Huntsville was that Secretary Wilson agreed that Thor and Jupiter could continue to compete, with a decision on which would win the 1,750-mile-range war to be made later. Medaris tried to put on a happy face with his team, telling them that the Army may have lost the long-range missile fight, but they still had a missile in competition; thus, the work and test launches of Jupiter missiles continued.[32]

Space

Everything changed significantly on October 4, 1957, when the Soviet Union successfully launched the first artificial earth satellite into an elliptical low earth orbit. According to Medaris, Sputnik was the best thing to happen for ABMA. The new secretary of defense, Neil McElroy, happened to be visiting Huntsville that same day for a briefing on the status of the Jupiter program. The briefing quickly shifted from the entire Jupiter program to one focused on the capabilities of Jupiter-C. Von Braun was so confident in this missile that he told McElroy he could have a satellite in orbit in sixty days. Medaris corrected him, saying ninety days. The problem was that the Naval Research Laboratory's Project Vanguard had the lead for the American earth satellite program. Medaris and his team were officially on hold. Medaris believed in the ability of the Jupiter program to launch, so he told his team to keep working to get two rockets ready. He was going out on a limb by authorizing work he had no budgetary authority to do.

When Sputnik 2 orbited in November of that same year, with a dog named Laika aboard, McElroy gave ABMA the green light to prepare to launch but withheld an actual "go" order. Medaris and von Braun continued working, but both men were frustrated by the missing order—Medaris enough so that he even considered retiring. It took many angry phone calls to the Pentagon to get an official launch go order and date, but the ABMA team was not the only team with a go date. In December, two months after Sputnik, Vanguard finally attempted to launch. It was a disaster that blew up on the launch pad. Medaris knew his team could do better and on January 31, 1958, Juno I, a multistage rocket derived from the Redstone missile, sent Explorer I, a scientific satellite weighing a mere eighteen pounds, into space.[33]

Before Explorer I ever got off the ground, the team at ABMA was already planning for manned spaceflight. They submitted a study to DoD titled "A Proposal for a National Integrated Missile and Space Vehicle Development

Program," laying out a twenty-three-year spaceflight program with launch vehicles ranging from small to huge. The missions included satellites ranging from small, unmanned scientific ones to a fifty-person permanent one. There were flights to the moon, interplanetary probes, and expeditions to Mars and Venus. DoD was not interested, but it did get people thinking that a long-term plan was needed. Another proposal, which had initial interest from both the Air Force and Navy, was Project Man Very High/Project Adam. This project was "to carry a manned, instrumented capsule to a range of approximately 150 statute miles: to perform psychophysiological experiments during the acceleration phase and the ensuing six minutes of weightlessness; and to effect a safe reentry and recovery of the manned capsule from the sea." Both proposals made their way around the halls of Washington. Project Man Very High/Project Adam even had significant interest from the Central Intelligence Agency, but ultimately neither project was to be funded and/or controlled by ABMA. The agency did get one piece of good news when Secretary McElroy announced that both Thor and Jupiter would be weaponized and deployed as part of the North Atlantic Treaty Organization's (NATO's) defensive shield.[34]

While the deployment of Jupiter sounded like a big win for the AMBA, the many details to be worked out forced Medaris to go toe-to-toe with the Air Force. He was so worried about relations with the air staff that he asked to deal directly with Strategic Air Command, the end user of the missiles. The first big hurdle was missile production. ABMA contracted with Chrysler to produce the missiles in Detroit, Michigan. When the timeline for deployment moved up, the retooling at the Chrysler plant did not get completed, forcing work to be done at ABMA. In order for the production timeline to be determined for both ABMA and Chrysler, the number of missiles to be produced and how fast they were to be deployed had to be determined. For two months Medaris went around and around on the numbers for production. When numbers did come down, the fight began over who was going to pay for the production. Since the Air Force was controlling the end product, the Army argued the money should come from them. The Air Force disagreed. The Army finally fronted the money to Medaris and continued the fight with the Air Force at higher levels.

Another problem arose regarding training for the crews manning the Jupiter missiles. They would have to be Air Force personnel trained by the Army at an Army location. Again, after many rounds of discussions, the squadrons were assigned by Strategic Air Command for training. To accommodate the new Army missile program and joint training with the Air Force, a new organization was stood up, the Army Ordnance Missile Command, with Medaris in charge. His new command swelled to include the Jet Propulsion Laboratory, White Sands Missile Range, the Army Rocket and Guided Missile Agency, test units at Point Mugu and China Lake in California, Fort Churchill in Nevada, Kwajalein Island in the Pacific Ocean, and an office in Paris, France, to deal with NATO concerns. Medaris was still fighting with the Air Force as his command expanded, spreading him thin while additional Explorers were being sent into space.[35]

When the Huntsville team finally got a launch date for Explorer I and Explorer II the following March, Medaris was elated. When the Vanguard launch ended in a blazing disaster, which the press mocked as "Kaputnik," the pressure on von Braun's team to deliver escalated. The Jet Propulsion Lab team joined the Huntsville team at Cape Canaveral to ready the Jupiter rocket for launch. The missile was on the launch pad readying for launch by January 24. Everything was a go until Mother Nature intervened. Wind shear on the launch date, January 29, forced launch boss Kurt Debus to delay it. Weather balloons released on January 30 again forced a launch delay. Medaris was not at all worried and spent the day golfing. He knew the team was ready; they just needed Mother Nature to play along. January 31 was the last date possible for launch before they would have to delay beyond another Vanguard attempt. Finally, the winds shifted, and all was a go for launch. On January 31 at 10:48 p.m. the Jupiter rocket ignited and lifted off. In a hangar five miles from the cape, two Huntsville scientists, Dr. Ernst Stuhlinger and Dr. Walter Haeussermann, waited to calculate the exact instant when a radio signal would ignite the second-stage rocket motors. They would have about three minutes to determine the exact apex of the rocket. If the motors fired too soon, the rocket would disappear into deep space. If they fired too late, the rocket would turn earthward and burn up. Thankfully, the second-stage motors fired as expected. A little over an hour later, radar indicated that Explorer I was in orbit circling the earth at intervals of 114.78 minutes. Joyous celebrations broke out all over the United States, especially in Huntsville, the Rocket City.[36]

Being the first with Explorer I and successful with five additional Explorer missions meant nothing when there were so many competing interests in space and space exploration. Both the Army and Air Force wanted to stay in the game. Senator Lyndon B. Johnson's inquiry into satellite and missile programs wanted civilian control. President Eisenhower pushed for rolling all space-related issues into the National Advisory Committee for Aeronautics, an organization that did not want to be left out. Nelson Rockefeller, special assistant to the president for foreign affairs, argued for separating military and civilian space missions. Medaris disagreed with any attempt to change the status quo with the Army and Air Force taking the lead. All the discussion ended when Congress passed the National Space Act on July 29, 1958, creating NASA. DoD would continue to control military space ventures, but all other exploration of space for peaceful purposes and research and development would be under civilian management. Two new committees, one in the House and one in the Senate, would work to sort out all the details of what this new division of space looked like in practice.[37]

After the National Space Act was passed, not much changed for Medaris and his team at Huntsville. Von Braun continued to come up with new proposals, including a more powerful rocket booster created by clustering existing rocket engines. Medaris persuaded DoD to build a tower to test the new booster design named Saturn, but no one openly talked about this being a step toward manned spaceflight. NASA came knocking in the fall of 1958. The director of NASA, T. Keith Glennan, commissioned a study to determine which Army assets should be transferred to NASA. The commission argued for the Jet Propulsion Laboratory and about half of von Braun's division. Medaris was incensed, feeling that separating people and resources would be a long, messy, and expensive process and a waste of the talented team built at Huntsville. Medaris rallied his supporters in the Army and Congress to attempt to derail the proposed move. He soon realized that his only option to stop the gutting of his staff was an anonymous leak to the press. On October 14 Medaris met with Mark Watson of the *Baltimore Sun* informing him of what NASA planned to do. The *Sun* broke the story the next morning, citing the threat to national defense if NASA got its way. Other papers took up the story, and the White House was forced to intervene. NASA got the Jet Propulsion Laboratory after it completed work for the Army, and the von Braun team would stay with Medaris and perform specific tasks for NASA on an as-needed basis. Medaris was happy, but some on the von Braun team were not. They were far more excited about the prospect of space exploration than of creating missiles for the Army. Some were also very worried about being transferred to another bureaucracy and having to fight the same fights all over again.[38]

Even with all the behind-the-scenes drama, the von Braun group continued to be successful. In March 1959 the Pioneer space probe traveled within 37,000 miles of the moon, and in May of that same year two monkeys, Able and Baker, traveled more than 16,000 kilometers an hour, withstood a G force of 38, and were recovered alive in a Jupiter nose cone. But relations with both DoD and NASA were strained. NASA cut programs and would only pay for work by the von Bruan team, not by contractors, and DoD cut the Saturn rocket booster because there was no need for it in the Army. In late August Medaris got word that the secretary of defense was giving the Air Force control of military space launches and most satellites. Once the team at Huntsville heard the word, an early morning meeting was held with Medaris and the senior engineers. They agreed that life with NASA was looking better and better. Medaris talked to von Braun, and he agreed NASA was better than being transferred to the Air Force. Medaris called Secretary of the Army Wilber Bruckner and recommended transferring the whole of the ABMA team to NASA. NASA director Glennan enthusiastically agreed, and Eisenhower announced the transfer on October 21, 1959, while the group at Huntsville was renamed the Marshall Space Flight Center.[39]

On that same day, Medaris announced his decision to retire, effective January 1, 1960. He was offered a third star and a Pentagon job but declined. He was asked by a reporter if he was resigning in protest of the transfer to NASA. He denied any such notion. He was retiring. He was tired of the stress and wanted to "make some money for his grandchildren." Huntsville celebrated "Big M Day" and gave Medaris a parade and a lavish dinner. The Army awarded him a second Legion of Merit medal for his command of ABMA. The Navy honored him with a commendation for service to the Fleet Ballistic Missile Weapons System. Even the Air Force, which Medaris bitterly fought against, awarded him the Air Force Legion of Merit for his leadership of the Jupiter Missile Program.

After all the accolades, Medaris wrote his controversial memoir, *Countdown for Decision*, and had numerous speaking engagements. He had some strong ideas about not only the military-industrial complex but also space exploration.

He lamented the move to a more bureaucratic military with civilians in all areas of DoD. He suggested eliminating 90 percent of all civilian positions, bringing the Army and Air Force back together again, and ending separate military and space programs, among other radical ideas. He, like Eisenhower, decried the rise of the military-industrial complex, arguing it would only bring about greed, corruption, and waste, with more civilian control of the new contracting system.[40]

After the publication of his book, Medaris spent months on a speaking tour. With all his experience, Medaris could have walked into a defense contracting job with any one of the many companies that did business in space. But Medaris did not think that was the right thing to do and actively avoided any such jobs. Instead, he took a position as president of Lionel Corporation of New York. After a few years, it was obvious that Lionel was in trouble, and Medaris and the company parted ways in 1962. Medaris and his wife relocated to Florida, where he started a management firm and built a new home in Maitland. Medaris became active in his local Episcopal church, the Church of the Good Shepherd, and by 1969 was ordained a deacon and in 1970 a priest. Once asked about a soldier being a priest, Medaris said, "A beautiful French author wrote years ago that an old priest and an older soldier get along well together. One spends his life defending his country on earth, the other defends his country in heaven." Medaris did both with gusto. In 1989 Medaris was finally inducted into the National Space Club, and the Smithsonian Institution honored him for his lifetime achievement in the promotion of and public awareness in the U.S. space program—a fitting, long overdue honor for a man who worked so hard to get the United States into space. Medaris died in Maitland, Florida, in 1990 and is interred in Arlington National Cemetery, Virginia.[41]

Conclusions

John Bruce Medaris was often a polarizing figure; people either loved or hated him. He was a "spit-and-polish" officer, requiring those serving under him not only to perform their best but also to look the part. He could be abrasive in his zeal to get something done but was always looking out for the combat troops he served. He was decisive but always took responsibility for his decision. He never threw his people under the proverbial bus. Medaris was honest, hardworking, and fair, requiring everyone to meet his high standards of conduct.

He never shied away from a challenge and was willing to do whatever it took to get the job done, even if that sometimes meant working outside regular Army channels. So when the Army needed someone to fix the ammunition problem during the Korean War or go toe-to-toe with the Air Force and Navy over missile production and space exploration, Medaris was the right man for the job. Medaris once said that he had "never been accused of trying to please everyone.... If you like what I have to say, fine. If you don't, I'm sorry about that," although many people doubted he was ever sorry about anything. He was absolutely sure of every decision he made, and he lamented officers who struggled to make a decision quickly. When at ABMA, every program was completed on or before schedule and at or below cost. He excelled at project management, and what greater project was there at that time than spaceflight? ABMA racked up many firsts under his leadership: first launch of a 1,750-mile missile, first object recovered from space, first U.S. satellite, first U.S. probe to the moon, and others. Medaris kept the Army in the missile and space race far longer than anyone anticipated. He fought to keep von Braun's team together, which only aided NASA in the long run. John Bruce Medaris earned the well-deserved moniker of "godfather of America's space program."[42]

Notes

1. Art Harris, "Touchdown for America's Pioneer Rocket Man," *Washington Post*, March 10, 1989, 1, https://www.washingtonpost.com/archive/lifestyle/1989/03/10/touchdown-for-americas-pioneer-rocket-man/a648f010-ade2-4ecb-9588-d6b1107142c0/; Gordon Harris, *A New Command: The Story of a General Who Became a Priest* (Plainfield, NJ: Logo International, 1976), 6.
2. Harris, *New Command*, 7–19; Maj. Gen. John B. Medaris, *Countdown for Decision* (New York: Van Rees Press, 1960), 11–15; Harris, "Touchdown," 2.
3. Harris, *New Command*, 19–20; Medaris, 15–16; Harris, "Touchdown," 2.
4. Harris, *New Command*, 24, 28; Medaris, 16.
5. Harris, *New Command*, 31, 34–36; Medaris, 16–17.
6. Harris, *New Command*, 38–39, 42–44; Medaris, 17.
7. Harris, *New Command*, 41, 45–53; Medaris, 17–78.
8. Harris, *New Command*, 54–55, 58–60; Medaris, 19.
9. Harris, *New Command*, 60, 65, 67; Medaris, 19.
10. Harris, *New Command*, 67–70, 76–77.
11. Harris, *New Command*, 80, 84–86; Medaris, 19; Harris, "Touchdown," 2.
12. Harris, *New Command*, 87–90; Medaris, 19.

13. Harris, *New Command*, 90–91.
14. Harris, *New Command*, 95–96.
15. Harris, *New Command*, 97.
16. Harris, *New Command*, 89–99.
17. Harris, *New Command*, 100.
18. Harris, *New Command*, 101–2.
19. Harris, *New Command*, 102–3.
20. Harris, *New Command*, 104–5; Harris, "Touchdown," 2; Medaris, 40–41, 48–49; Frederick I. Ordway III and Mitchell R. Sharpe, *The Rocket Team* (New York: Thomas Y. Corwell, 1979), 274.
21. Harris, *New Command*, 108; Medaris, 19.
22. Medaris, 19–20; Harris, *New Command*, 114.
23. Harris, *New Command*, 116.
24. Harris, *New Command*, 117–20; Medaris, 21.
25. Medaris, 21; Harris, *New Command*, 123–24.
26. Medaris, 21; Harris, *New Command*, 127–31; Maj. Peter J. Land, "Steel for Bodies: Ammunition Readiness during the Korean War" (Master's thesis, U.S. Army Command and General Staff College, 1990), 89, https://apps.dtic.mil/sti/pdfs/ADA416944.pdf.
27. Harris, *New Command*, 131, 133–34; Land, 39–44.
28. Harris, *New Command*, 136–37; Medaris, 21, 65.
29. Ordway and Sharpe, 372, 374; Medaris, 69; Harris, *New Command*, 137–39.
30. Michael J. Neufeld, *Von Braun: Dreamer of Space, Engineer of War* (New York: Alfred A. Knopf, 2007), 300; Harris, *New Command*, 145, 147; Neil A. Sheehan, *A Fiery Peace in a Cold War: Bernard Schriever and the Ultimate Weapon* (New York: Random House, 2009), 324; Medaris, 100, 105.
31. Walter A. McDougall, *... the Heavens and the Earth: A Political History of the Space Age* (New York: Basic Books, 1985), 130; Smithsonian National Air and Space Museum, "Jupiter-C Nose Cone," https://airandspace.si.edu/collection-objects/jupiter-c-nose-cone/nasm_A1959003100; Neufeld, 308.
32. Harris, *New Command*, 155, 157–58; Sheehan, 346, 358; Neufeld, 305.
33. Ordway and Sharpe, 382; Harris, *New Command*, 180, 191; Medaris, 154–55, 170; McDougall, 123, 131; Sheehan, 364; Neufeld, 313.
34. National Aeronautics and Space Administration, *Liquid Hydrogen as a Propulsion Fuel, 1945–1959*, part 3, "1958–1959, Large Engines and Vehicles, 1958," 11, https://history.nasa.gov/SP-4404/ch11-5.htm; Ordway and Sharpe, 383; Paul Drye, "Man Very High/Project Adam: Mercury before Mercury," *False Steps: The Space Race as It Might Have Been*, https://falsesteps.wordpress.com/; Harris, *New Command*, 183.
35. Medaris, 182–84; Harris, *New Command*, 191–93.
36. Harris, *New Command*, 185–90; McDougall, 168.
37. Harris, *New Command*, 196–97; Ordway and Sharpe, 387; Neufeld, 330; McDougall, 176.
38. Harris, *New Command*, 197–99; Medaris, 238; Neufeld, 335, 339.

39. Neufeld, 342; Harris, *New Command*, 201–2; Ordway and Sharpe, 388.
40. Harris, *New Command*, 205–6; McDougall, 213, 396.
41. Harris, *New Command*, 209, 221; Harris, "Touchdown," 2.
42. Harris, "Touchdown," 2.

4 PROBLEM SOLVER
Osmond J. Ritland
DAVID CHRISTOPHER ARNOLD

Osmond Jay Ritland was a pilot and engineer who witnessed and contributed to many of the most remarkable changes in aviation that took place during the course of his career. Ritland earned his wings in 1933 and flew early in his career under the direct command of Henry "Hap" Arnold and Carl "Tooey" Spaatz. Leaving the service in 1935, as many reserve officers including Bernard Schriever did, Ritland went to fly for United Air Lines until going back in the Army in 1939. He spent most of World War II as a developmental test pilot at Wright Field, Ohio. After the war, he returned to the test community, primarily working on the development of new jets and ejection seats. In 1950 the U.S. Air Force made him commanding officer of the 4925th Test Group (Atomic), the unit responsible for figuring out how to configure and use aircraft for the new nuclear mission, to include dropping live nuclear weapons at the Nevada Test Site.[1] Ritland later moved to the Pentagon, where he worked on nuclear issues and then as the Air Force project officer for the U-2 reconnaissance plane. In

1956 the Air Force assigned him to lead its space programs, first as Schriever's deputy commander and then as commander when Schriever left. Ritland finished his career as director of human spaceflight for the Air Force at a time when the air service was trying to figure out how to use pilots in space. All of this was done with what one newspaper writer said was the "confident nonchalance" with which Ritland went about his business, sure that he knew what he was doing but calm and relaxed even in the tensest of moments.[2]

Ritland's contributions to space were significant and based on his technical abilities and experience as a test pilot. Although he never completed his undergraduate education, Ritland was an engineer. As historian James R. Hansen put it in his book, *Engineer in Charge*, about the National Aeronautics and Space Administration (NASA)–Langley test site in the 1950s and 1960s, "The unwritten rule for the work of any engineer is to bring everything to bear on solving the problem of the moment. This means bending every effort, be it cut-and-try, experimental, theoretical, or any combination of the three."[3] This unwritten rule was the Ritland approach—he brought everything he had to "solving the problem of the moment." Or as he used to say when faced with a problem, "Let's get to work."[4]

Developmental Leadership

It was not enough to be a good engineer or scientist in the post–World War II world; one also had to be a good leader and manager. Ritland expressed this duality when he wrote in 1960, "One of the central factors of military affairs today is the interaction between strategy and technology.... At the same time, the reverse is true—the direction that military technology will take is being determined by the imperatives of military requirements and strategy."[5] Further, as historian Thomas Parke Hughes points out, government engineers and scientists, of whom Ritland was one, were not motivated by the pursuit of knowledge or money from their new systems. Instead, "Motivated by the conviction that they were responding to a national emergency, they single-mindedly and rationally dedicated the enormous funds at their disposal to providing national defense."[6] General Schriever tried to develop deterrence through strength and capability, not the threat of mutually assured destruction.[7] As Schriever put it in 1958, "Our job is to help our country and the whole free world to build that deterrent power which I have defined as military power used for the pursuits and purposes of peace. By this means we believe that we can help to gain the time and opportunity for the statesmen of the free world to work out conditions of peace by diplomatic, economic, political, psychological, and other measures."[8] For much of the last decade of his career, Schriever was Ritland's boss.

Ritland worked after World War II as a developmental test pilot flying hundreds of types and models of aircraft—from the smallest observation airplanes to the largest American bombers and the first jets. He described the aircraft laboratory at Wright Field as "a design group that theoretically designed new concepts [and] new airplanes."[9] Problems still existed to solve then because there were still a lot of airplanes on the books from the war, and budget pressures lengthened development schedules back to their slow, prewar pace. The late 1940s were a time when new jets like the B-45, B-46, B-47, and B-48 were all experimental and the B-36, B-49, B-51, and B-52 were all in development. But, he recalled, "It felt real good to get into the experimental aircraft business because I had flown all these things, and I was back in the old fold again."[10]

Significant, of course, is that many of these airplanes were bombers. Although the Atomic Energy Commission had responsibility for building nuclear weapons, the Air Force had responsibility for their delivery by airplane. By 1950 it had become "readily apparent," in Ritland's words, that the service needed a test organization to ensure its bombers could carry the weapons into combat. The B-36, for example, first delivered to Strategic Air Command in 1948, had not been designed during World War II as a nuclear bomber and needed to be retrofitted.[11] A test organization needed to work on developing ways to determine the ballistics of the various weapons and the procedures necessary to load and unload various aircraft with the bombs.

In February 1950 Ritland left Ohio to stand up and command the 4925th Test Group (Atomic) at Kirtland Air Force Base, near Albuquerque, New Mexico. His group was responsible for the development and testing of all equipment needed to create an Air Force nuclear weapons capability. "When they dropped the bomb on Hiroshima, only a few planes had been modified for such a mission," Ritland said. "They eventually decided that they wanted as many Air Force planes as possible modified to drop atomic weapons; so the job I had was to test the airplanes that were modified to carry nuclear devices and to get the ballistics on the bomb."[12] The problem of this moment revolved around how to defend the nation.

Responsible for sixteen nuclear tests, Ritland went along with the crews when they dropped a live bomb.[13] He was "actively engaged in developing the use of A-bombs for almost every type of plane the Air Force" flew, a reporter noted, including "jet fighter-bomber aircraft."[14] One officer who flew with Ritland remembered evidence of Ritland's direct leadership before a nuclear test. "At about 2 o'clock in the morning, a group of us were on a B-50 wing and the maintenance men were just finishing their work. The ladder started wiggling and it startled us all to see the boss peer up over a wing—he was simply checking to see what was going on in the hangar at that hour of the day. As I recall, we had some coffee, shot the breeze for a little while, and then went home for a few hours [of] sleep."[15] Said Ritland later, "I don't feel I'm doing my job unless I know the problems of my men."[16]

Ritland was largely on his own in developing test plans for dropping atomic bombs on Nevada: "No one ever reviewed anything that I was doing, ever.... The night before a drop, the people would meet at Kirtland and I would brief them on what they were going to see, but not for their approval."[17] Flight operations on the day of a test were extremely complicated, involving fifty to one hundred airplanes coming from all over the United States to a precise location at a planned time, requiring communications with all kinds of call signs and commands and air sampling without the automated control of today's systems and done each time without a single accident.[18] Ritland described the group's mission much more clearly: "Fly missions and drop these different weapon devices."[19] But Ritland did not always have the aircraft he needed, either, forcing him to borrow a crew and a B-36 from Strategic Air Command, for example. He got the crew pressure suits and told them to fly it up to 50,000 feet, an unheard-of altitude for a B-36. When Strategic Air Command told Ritland he could not take their airplanes above 45,000 feet, he replied, "Well, we're an R&D [research and development] command; we extend the capabilities of these aircraft."[20]

Following three years in command, Ritland attended a year at the Industrial College of the Armed Forces in Washington, DC. The Air Force then moved Ritland to the Pentagon in July 1954, where he worked on nuclear issues.[21] The job was not "firsthand, day-to-day operations" but rather staff work, and it did not make a big impression on Ritland.[22] However, from both his time in New Mexico and his work on the air staff, Ritland had already made a big impression on others.

U-2

On December 10, 1954, Lt. Gen. Donald Putt, deputy chief of staff of the Air Force for research and development, called Ritland into an E-ring Pentagon office where he met Lockheed's Kelly Johnson. They started discussing a new aircraft for which they wanted Ritland to be the Air Force's project manager, a plane eventually called the U-2. Recalled Ritland, "They outlined the purpose and preliminary understanding and definition of the total program, and it was going to be done in eight months, operational within a year, and over probably within a year and a half."[23] This was blinding-fast speed for a new aircraft program in the post–Korean War environment, but Ritland became special assistant to Putt while serving as the Air Force project manager for the U-2 and the service's liaison with the Central Intelligence Agency's (CIA's) Richard Bissell.

The CIA and the Air Force eventually worked out a management arrangement. The Air Force chose the pilots and trained them, providing weather, mission planning, and support. According to the CIA's own declassified history of the U-2 program, "The Agency was responsible for cameras, security, contracting, film processing, and arrangements for foreign bases, and it also had a voice in the selection of pilots."[24] Lockheed was to build and test the aircraft. The CIA remained in control of the program, but the Air Force was a key partner. Said Bissell later, "The Air Force wasn't just in on this as a supporting element, and to a major degree it wasn't in on it just supplying about half the government personnel; but the Air Force held, if you want to be precise, 49 percent of the common stock."[25] Recalled Ritland,

> We developed everything new, extending the state of the art, new fuel, new operating locations, total security, new concepts, new suits, new human factors, new training methods, you name it; it had never been done before. We just got the right kind of people. We didn't get a lot of them; we just got the ones that could do the job. I don't understand how we ever did it. We built a new field; got gas out there; got Shell Oil Company, through Jimmy Doolittle [Lt. Gen. James H. Doolittle, Army Air Forces], to build us a new nonvolatile fuel; delivered it out into the desert area; selected the field; flew for three years; and nobody ever knew the airplane existed.[26]

From December to the first test flight in August was only eight months.²⁷ The official CIA history of the U-2 program noted, "As soon as the U-2 began flying over the Soviet Union, its photographs became the most important source of intelligence available."²⁸ They showed the bomber gap and the missile gap did not exist. Author William Burrows quoted Defense Secretary Thomas S. Gates Jr., after the U-2 overflights ended: "From these flights we got information on airfields, aircraft, missiles, missile testing and training, special weapons storage, submarine production, atomic production and aircraft deployment . . . all types of vital information."²⁹ Clearly, the U-2 had helped solve the problem of what the Soviets were capable of.

Among the people involved in the U-2 program, Ritland met with the Air Force's civilian R&D chief Trevor Gardner "practically every day," Ritland recalled, while Gardner "was concurrently working with Schriever on the ballistic missile program in 1954." At one point, Gardner expressed that he wanted to run the ballistic missile program like the U-2 program: in a small organization with full and complete authority and responsibility to get the job done, which would hopefully accelerate missile development because that had not happened when Schriever had taken over the missile programs in 1954 and moved development to Los Angeles, away from Ohio and Washington, DC, but close to the aircraft manufacturing companies that would have to build the missiles.³⁰ The distance from Washington also meant Schriever was on a cross-country flight nearly once a week, which he recalled later was a little like being "a shuttlecock in a badminton game."³¹ He would work all day on the West Coast, fly overnight to Washington, meet with senior officials, and fly back to California, usually the same day.³² Understandably, Schriever wanted a deputy to run the programs while he traveled. In charge of his own organization with top-level interest, Schriever could and did get any officers he wanted for his staff. "I wanted them," Schriever said, "because they were smart and would tell me not what I wanted to hear, but what they really thought."³³ Said an officer years after working for Schriever, "Anyone not in good physical condition, who doesn't have a trigger-quick mind, had better not work for this general."³⁴

Missiles

In April 1956 Ritland got orders to become the deputy commander of the Western Development Division, the headquarters of the ballistic missile program in

California.³⁵ Many of Ritland's friends were skeptical of the position; he recalled them calling Schriever's outfit a "fly-by-night organization, said it would never fly, and that [Ritland would] be looking for a job one of these days." Putt observed later that "R&D personnel were sort of second-class citizens in the [Air Force]. ... Operational people looked down on them and what not. But I think as time went by and there was a greater and greater appreciation of the necessity for good research and development and what it could do for the operational forces, these attitudes changed."³⁶

It had become obvious to senior American leaders that the Soviets were making progress in ballistic missiles and the United States "had to do something," as Ritland put it.³⁷ In a 1958 speech to the National Security Industrial Association, he argued, "By 1960, Russia will be capable of blasting America with thermonuclear bombs equivalent to two billion tons of TNT. This capability will be composed of aircraft and some limited number of ballistic missiles. The death toll in the United States would be estimated at eighty-two million; an additional twenty-four million would be seriously injured [of the 174 million Americans in 1958]. These are the grim facts of life in the age of the hydrogen bomb, the ICBMs [intercontinental ballistic missiles], and thirty minutes of flight time.... In short, our capability must counter all Russian possibilities to strike."³⁸

Thus, when Ritland arrived in Los Angeles, the next "problem of the moment" to solve was the ICBM, except there was no ICBM hardware, only paper designs of the Atlas, Titan, and Thor missiles that would become vital for deterrence. Despite opposition from within the flying community of the Air Force and from Congress, which was loath to spend as much as was being requested, the missile programs eventually moved from paper to hardware using "concurrency" to speed up the design and construction of not just missiles but total weapons systems.³⁹ Space programs, an even less popular idea than ICBMs, generally remained on paper in the late 1950s, when there was only $4 million in the defense budget for space activities.⁴⁰

Concurrency is often criticized for being too expensive relative to the speed of system development. Recalled Ritland, "We were designing the operational environment along with the vehicles, the warhead, and the total weapons system all at once and together on a lead-time basis so that everything would fall into place." Doolittle recalled concurrency as developing systems in parallel and putting "the amount of money on each one of these unknown things, relatively unknown things, that would permit them all to be ready at a given time. This greatly shortened the time required and was probably the one thing that brought our missile program into actual operational use as rapidly as I think it was possible to do. Now, Benny Schriever was largely responsible for that concept, for working on that concept and the final implementation of it."[41]

But concurrency was nothing new. Ritland compared the concept to his test pilot days during World War II: "It's like we've always done in the past—we never built just one kind of fighter; we built two of them and made comparisons."[42] And he also described the U-2 "as the closest, in my experience" to a concurrent program: "Namely, from the day of the first meeting in December 1954, the estimate of completion and an operational date was identified; the locations of operations were not quite identified, but within a few months the locations of operations were identified; and actually a facility that was never used in England was ready, operationally ready, to receive the aircraft and operate before we were ready to send them there.... This is a much smaller example of concurrency [compared to] the missile program."[43] Later Ritland wrote that "as we see it, some of the fundamental principles that must be observed are the following: clearcut, vertical, and short management channels; a high degree of authority, delegated to the lowest possible operating management level; and centralized management responsibility for all essential elements of a program."[44]

Even so, the multiple missile program approach was expensive and complicated and became more so because three military services were working on missile programs at the same time. Furthest along initially were the Army's Jupiter and the Air Force's Thor intermediate-range ballistic missiles, both liquid-fueled, and the Navy's Polaris missile, for which they planned to use storable solid fuels. Ritland recalled the period as "a very difficult time." He later recalled that the big bureaucratic problem was

> who was going to get the money and who was going to win out from a role and mission point of view. It was a real deadly argument and I was kind of in the middle between Schriever and [Army Gen. John B.] Medaris.... They [the missile programs] both had all of these deficiencies so that the programs, from a strategic point of view, weren't quite as important. However, the personalities involved were vicious, and man, it was a real knock-down

drag-out battle of who was going to win. . . . I know that Schriever, one night at his house—there were several people there at his home in Santa Monica—he made an announcement to a few of us and he said, "[Ossie], I'm going to go after that Jupiter." Namely, he was going to attack the duplication of effort between the Army and the Air Force, and that he was going to win that battle. From that moment on, he worked on it with the press and with politicians in Washington. And of course, as you know, the program, the Thor, did in fact win out, but not because of any technical capability, because the Jupiter was performing equally as well.[45]

Space

Solving engineering problems could also be as tough a task as political issues because the cause of a technical problem was not always apparent. Sometimes it would be an obvious issue like a liquid oxygen valve stuck open, pouring fuel into the combustion chamber. But other times it would be more difficult as when five consecutive Atlas launches failed, not all from the same issue. "Of course," Ritland recalled, "we geared up and worked 24 hours a day on all of this business. Progressively, we thought we'd get something fixed, and we'd launch another missile and it would fail, but it wouldn't be the same failure." In another instance, a "tiger team" looked at a failure and made some recommendations: "[W]e made several corrections, changing the valves, some guidance changes, some propulsion changes, vent-valve changes. I can't recall all of those changes now, but I think we made about five or six changes and then started flying again. From then on, the missiles worked."[46]

Another problem to be solved had to do with the closed nature of the Soviet system and a lack of knowledge about just how capable the Union of Soviet Socialist Republics (USSR) really was. President Dwight Eisenhower was so concerned about the eventual detection of the U-2 that he personally approved flying over the most important targets in the USSR from the beginning of the program before the Soviets could counteract the U-2, which they eventually achieved in 1960 when they brought one down over the USSR, raising a new set of problems.[47] When it came time to approve space-based reconnaissance of the USSR, Eisenhower personally approved that program, too.

RAND had pitched its ideas about satellites at Wright Field to Air Force personnel in 1946, Ritland recalled, which was when Wright Field funded

RAND and its studies. Ritland said that RAND, and later the program office at Wright Field, "didn't necessarily describe [satellites] as a weapon or a capability, but they pointed out that it was possible to develop the so-called earth satellite, meaning you could have something that would be launched and go around the earth. Then, progressively, they proposed that if you did have a vehicle orbiting the earth, it could indeed be a reconnaissance vehicle."[48]

After he got to Los Angeles, Ritland became a proponent of the concept too, recognizing that space was another place to effect deterrence. The approach worked to keep space technologies in the discussions about how to spend Air Force money, but it still did not generate a lot of funding for the first few years of military space efforts until Sputnik in October 1957, which changed everything. Ritland then began to argue that space was in the future for the Air Force. For example, in a May 1959 speech in Denver, he maintained that the military's "space exploration" was simply "based on our recognized need for superiority in space power, as an extension of airpower. We consider space as only an extension of a medium in which we are already operating a deterrent power, and we know that superiority in space is a fundamental requirement for survival, and for maintaining the position, prestige, and welfare of the nation."[49]

In Los Angeles one day, Schriever, who knew about the U-2 program, called Ritland into his office. Said Schriever, "I want you to go to Washington and set up a program with [the CIA's] Dick Bissell on a new satellite system." So Ritland went to the offices in Washington where he had worked on the U-2 program before, where the same secretary still worked, and "drafted up policy statements with regard to what the country would do with regard to starting this kind of effort."[50] Sitting down with Bissell, they first created a cover story about studying the environment, for which they said they planned to use Thor intermediate-range ballistic missiles and Agena upper stages, which could not be hidden, to put satellites in orbit. The secret photoreconnaissance program, called Corona, was known to only a few people. The public face of the program Ritland and Bissell had created was that "as a preliminary goal [to getting humans in space], it should be demonstrated that you could launch small mammals, small vehicles in space, orbit the earth, and recover them," and that program was to be called Discoverer.[51] Organizationally, Ritland recalled, the program was an "exact repetition of the U-2 operation."[52] He and Bissell drew up a couple of memoranda authorizing the program to begin, which the Department of Defense's (DoD's) Advanced Research Projects Agency issued to the Air Force in California to authorize starting the reconnaissance satellite program.

Bissell recalled that they arranged the management of Corona using discussions on what military missions could be done in space:

> The program was started in a marvelously informal manner. Ritland and I worked out the division of labor between the two organizations as we went along. Decisions were made jointly. There were so few people involved and their relations were so close that decisions could be and were made quickly and cleanly. We did not have the problem of having to make compromises or of endless delays.... The program was handled in an extraordinarily cooperative manner between the Air Force and CIA. Almost all of the people involved on the Government side were more interested in getting the job done than in claiming credit or gaining control.[53]

The Corona program, recalled Ritland, "went for broke on the very first launch," as their "objective was to launch, orbit, photograph, deploy, and recover over the Pacific Ocean a capsule that contained film of the reconnaissance effort." The first launch attempt failed. "Starting from there, we had progressive failures on the launch pad, some horrible examples of early mistakes, miswiring, lack of technical surveillance, thorough inspection, and whatnot."[54] The launches came in rapid succession, too; flights three and four launched in June 1959, and flights five and six launched on August 13 and 19, 1959.[55] Ritland remarked on that feat in the press: "The successful orbiting of two Discoverer satellites in less than a week marks a tremendous forward step in the scientific study of space vehicles and their applications. The information and experience gained in the Discoverer series are preparing the way for launching of the first man in space."[56]

The first mission carrying a camera into space, Discoverer IV, failed to reach orbit because the Agena booster burned out too soon. Discoverer V's camera batteries failed on orbit, and other failures followed. The range safety officer blew up Discoverer X, and Discoverer XI's reentry vehicle ejected, but the spin-stabilization rockets exploded during reentry. Recalled Ritland:

Of course, the heat was beginning to get on when you had failure after failure after failure. This especially became true in the early Discoverer program where we were having too many failures. Schriever and I had to go into Washington into the Secretary of Defense's office and sit down with everybody and almost beg for one more chance before the program was canceled.... We went through the first batch of vehicles that I'd recommend[ed for purchase] with absolutely no success.... I think we got up to around 12 vehicles with absolutely no success. We had a big meeting in Secretary of Defense Gates' office to make a determination as to whether we would continue the program. We were given one more chance as a result of that meeting. Lo and behold it worked. The results of that first effort were so astounding and complete that beyond a shadow of a doubt this approach to satellite information gathering was a certainty.[57]

The first successful mission, Discoverer XIII, came in August 1960, a year and a half after the first launch. It was just a diagnostic mission and the pilots missed catching the return bucket, which had to be water-recovered by the Navy, but it was the first object launched into and recovered from space. Ritland was part of the team that presented the American flag the capsule carried in space to President Eisenhower in the Oval Office. The very next mission, Discoverer XIV, included the first air-recovered film capsule, returning more imagery of the USSR than the twenty-four previous U-2 flights combined.[58] CIA analysts reacted to the film recovery with "unbridled jubilation," according to National Reconnaissance Office historian Robert Perry. The CIA told Air Force Col. Paul Worthman the photos were "terrific, stupendous, and had confessed 'we are flabbergasted.'" Worthman's conservative report to General Ritland was that 'apparently design specifications on resolution have been met.'"[59]

In the declassified official history of the Corona program, Perry wrote,

In the context of its operational utility, exploitation of technology, and enhancement of the nation's fund of intelligence information, Corona had to be rated an outstanding success. Originally considered an interim system, and assumed to have, at best, three or four years of operational utility, Corona remained the sole source of overflight intelligence for the United States for nearly five years and was a primary source of basic information used to shape national defense policy for 12 years. Although designed as a search system, at the end Corona was providing better detail and resolution than several of the surveillance systems earlier touted to supplement it. Its eventual replacement, Hexagon, was six years in gestation and about five times as costly, [while] having an operational capability that Corona could never match.[60]

Corona achieved many memorable firsts in spaceflight history—from "first satellite in polar orbit" to "first dual-capsule reentry capability" to "first low-altitude satellite to utilize a solar array." Corona film buckets were the first objects launched into and recovered from space while its satellites were "the first to operate in stabilized flight, the first to be recovered from the water, the first to be caught in descent, the first to incorporate an engine restart capability, the first to carry a stereo camera (and, of course, the first to carry any camera at all), the first to perform orbit adjust maneuvers," and many others firsts.[61]

A key individual making that photography from air and space possible was Osmond Ritland. At Ritland's retirement dinner, the emcee read from a letter Bissell had written about Ritland's contributions to the U-2: "It was largely through your [Ritland's] efforts that this undertaking moved as rapidly and successfully as it did. Less than eight months after you went to work on this enterprise, there was a successful first flight. In this and in other activities, which I know intimately and firsthand, you made one of the largest individual contributions to the successful achievement of national objectives, the importance of which can hardly be exaggerated."[62] Lockheed's Kelly Johnson, who was not at the dinner, was quoted in another letter read at the gathering as saying, "This U-2 program would never have gotten off the ground without Ossie Ritland."[63]

Human Spaceflight

Satellites were an important part of the national space effort and helped in both deterring Soviet aggression and exploring the physical universe. But, argued Ritland in a March 1959 speech, "Let me emphasize the point that we are going to place man in space. We are not going to be content with merely sending instruments out there. Man will just have to go out there and see for himself. In such adventures, we expect that our Air Force Ballistic Missile Division

[AFBMD] will continue to have a constructive role to play."[64] AFBMD's role, with Ritland as its leader, was enormous, not just for missile programs but for national space programs as well.

Much is known about NASA's astronauts, but little is known about how the rockets they flew on were readied for flight. Many people worked on the Mercury and Gemini programs—numbers reached the hundreds of thousands by the time of Apollo—but few people were as important in getting astronauts in space after failures than Osmond Ritland. NASA's own official history of the Mercury program cited him as the "former test pilot in command of the Air Force Ballistic Missile Division" who, following a string of failed Atlas missile test launches, along with Bernhard Hohmann and Ernst Letsch, "assured the astronauts that their interests would not be sacrificed."[65] Ritland's technical background and his leadership abilities were critical in getting spacecraft ready for crewed missions in two important space programs needed to get to the moon. Again, it was all about solving problems.

Ritland recalled thinking it was a good idea to have test pilots as the first astronauts:

> Because of my heritage and experience as a test pilot, I always felt that, ideally, people with that experience in their so-called nerve ends, reaction time, operation under stress, unusual conditions, etc., would be ideal as astronauts. So I favored that kind of an approach, but I didn't have anything to do with it whatsoever. But I do know that when the astronauts were appointed and assigned to the task, one of the first things they did was to come out and visit us at [AF]BMD because they knew that we were going to be the provider of their booster system.[66]

At the first meeting with the astronauts in California, Ritland "emphasized that they were among friends because my experiences as a test pilot clearly showed that I understood their feelings and their approach to problems. I assured them that they could depend upon us, that we had their safety and keen interest at heart, as they would see for themselves."[67]

To make the Atlas missile safe for human spaceflight, a system to save the life of the astronaut in the event of a booster failure needed to be devised. Tension arose between AFBMD and Convair, the division of General Dynamics making the Atlas in California, that wanted an operational ICBM to deliver nuclear weapons on targets, and NASA's Space Task Group (STG) in Virginia and Florida that wanted a rocket to put humans in Earth's orbit. "STG engineers were far away, busy with other matters, and knew well how little they knew about the Atlas missile," NASA historians recorded.[68] Ritland appointed Hohmann in August 1959 to supervise the systems engineering of "a pilot safety and reliability program on the Mercury-Atlas series" at Space Technologies Laboratory in California. The company had broad "experience in systems engineering, missile development, and business management," but "STG had a deeper background in research and was directly responsible for the development of" the Mercury program.[69]

At a press conference in California in September 1959, Ritland sat with former test pilot and war hero Doolittle, who was often at AFBMD, and six of the Mercury astronauts to discuss the division's role in Project Mercury, part of a series of meetings he was having with the Mercury astronauts to update them on the division's progress. Ritland's team of government employees and contractors had responsibility for guiding the Atlas into orbit, developing the abort system, and helping with communications and tracking. The division was using its experience launching missiles that had started a few years before to complete these tasks. Ritland compared the work they were doing to get ready to put astronauts in orbit with the work of "the flight test business."[70]

One reporter asked Doolittle and Ritland to compare their test flight experience with preparing for a space launch. The response from the sixty-two-year-old Doolittle was typically self-deprecating: "I get the impression that my role here is one of contrasts, in order that you ladies and gentlemen may see the difference in appearance between an old aviator and six very attractive young astronauts." Ritland, though, pointed out that the relationship between pilots and engineers had changed in the years since he was in the test community: "I know that the engineers, test engineers, had me do things that I thought were pretty ridiculous and pretty crazy. I hope these people [the astronauts] have the sense that they can talk to the people who have an understanding of their problem." Doolittle simplified the comparison even further by saying, "I would say that that is the biggest difference between the early days of testing and today. There is complete rapport between the pilot today or the astronaut today and the individual who is making and testing his equipment."[71]

NASA's safety problem was that it wanted additional insurance built into the system that could warn of an impending booster failure and automatically abort the flight to save the life of the astronaut.[72] Hohmann's study, "General Aspects of the Pilot Safety Program for Project Mercury Atlas Boosters," "analyzed the differences between the ideas of reliability, quality control, and quality assurance before synthesizing them in a specific program adaptable to other areas of Mercury development."[73] Engineers eventually settled on two sensors to monitor for catastrophic indications in the liquid oxygen tank pressure, bulkhead pressure, booster attitude in all three axes, rocket engine injector manifold pressures, sustainer hydraulic pressure, and primary electric power. If any one of those systems failed or got out of tolerance, the abort sensing and implementation system (ASIS) "would by itself initiate the explosive escape sequence" to separate the capsule from the booster. The test conductor, flight director, range safety officer, or astronaut could also initiate an abort.[74] The result was ASIS, the only part of the Atlas missile system created solely for the purpose of putting a human being on top of it.[75] In NASA's words, "Together with Major General Osmond J. Ritland, former test pilot in command of the Air Force Ballistic Missile Division, Hohmann assured the astronauts that their interests would never be sacrificed."[76]

On July 29, 1960, the first launch of an Atlas rocket with a Mercury capsule on top occurred without an escape system because ASIS was still in development. After less than a minute of flight and reaching an altitude of just over eight miles, "the booster apparently suffered major structural failures," according to a NASA report from August 1960.[77] Ritland appointed Worthman to work with NASA's Richard Rhode on a team to figure out why the Atlas failed. They met in San Diego with Convair continuously during December 1960 and January 1961. The Atlas was a very thin-skinned missile, made up of steel only 0.01 inches thick. With pressurized internal fuel tanks, Atlas could survive the rigors of launch, but as it burned fuel, the rocket became less able to handle launch stresses. The solution was a "belly band" or "horse collar," basically an eight-inch steel band just below the adapter ring to reinforce the booster and the adapter between the Mercury spacecraft and the Atlas, which was where the rocket felt the highest stresses during launch.[78]

Following the successful launch of "astrochimp" Ham in January 1961, NASA felt it was time for another test of the Mercury-Atlas system intended to get an American in orbit. MA-2, on February 21, 1961, was also the first test of the ASIS system. Ritland was in Florida to witness the test from outside the control center that day, but with a press release "in his pocket making this shot a NASA 'overload' test in case of failure," he said. Fortunately, the Atlas roared off its pad at 9:12 a.m. for a successful flight of the new adapted booster and the ASIS system. According to Ritland's daughter, after the launch, Ritland and his fellow engineers cut their neckties in half to celebrate their success.[79] Although more tests followed, NASA was on its way to putting an astronaut in orbit. Ritland helped by pushing the contractors to keep the same crews at work on the Atlas missions so they would be extremely experienced when the time came to launch a human astronaut aboard an Atlas.[80]

By the time of John Glenn's flight in February 1962, ASIS was working as advertised, which no doubt gave him some confidence on his flight. Ritland recalled that as they got ready for launch, Glenn came down to the pad and spent the morning with the general and his team. Said Ritland, "We reviewed with Gilruth [Dr. Robert R. Gilruth, director of the NASA Manned Spacecraft Center, Houston] and the whole gang A to Z to see where we stood. We were just as critical as the devil. We had to have every pound of fuel in that thing or he would not make orbit. I can remember clearly that, when we got all through with the presentation, we asked John Glenn, 'How do you feel about it?' He said, 'If you guys say it's okay, it's okay with me.' And away we went."[81]

But NASA wasn't the only government agency interested in flying astronauts. The Air Force wanted to figure out what astronauts could do for the military, resulting initially in a program called "Blue Gemini" or simply "Gemini B." Ritland's Space Systems Division in Los Angeles, which was born from AFBMD's split into two space and ICBM divisions, developed plans in February 1962 to use NASA's "Gemini hardware as the first step" in a program to develop "a kind of military space station with Gemini spacecraft as ferry vehicles." Ritland wanted a more active role in Gemini because he felt the X-20 Dyna-Soar program was at least two years behind NASA's Gemini. Air Force Chief of Staff Curtis LeMay opposed Blue Gemini, which might threaten the X-20 program that he saw as the future of long-range aviation for the Air Force. And civilian Pentagon officials also were "skeptical" of military men in space because they saw it as destabilizing internationally. But as difficulties hit the X-20 and other programs, Gemini began to look even more attractive to Air

Force planners like Ritland who wanted to see if there really was a mission for military astronauts. NASA was generally supportive of Blue Gemini and even entered into a formal agreement to support DoD's use of Gemini hardware because, rather than fearing it would siphon off resources from their goal of reaching the moon by the end of the decade or complicate the "peaceful" nature of NASA's space program, they saw it as a useful addition. DoD was already providing the Titan ICBM as a booster, and now NASA saw an opportunity to infuse additional military funds into its own program. The two agencies also created the Gemini program planning board, cochaired by NASA and DoD, as an advisory body on military experiments for Gemini flights. This board became increasingly important as the Gemini program continued and the Air Force refined what it wanted for its human spaceflight program.[82]

Both NASA and Schriever's Air Force Systems Command had wanted a single person to act as a go-between on the bigger issues facing them both. Schriever simply could not attend all the meetings with NASA and run all of Air Force R&D at the same time. Air Force Secretary Eugene Zuckert requested DoD assign a liaison officer to NASA headquarters in December 1961, though it was not until May 1962 that Ritland assumed his new role at NASA. Ritland was simultaneously DoD's liaison to NASA and deputy commander of Air Force Systems Command for Manned Space Flight, starting May 15, 1962, responsible for all U.S. Air Force actions involved in American human spaceflight programs.[83] He had "broad responsibilities for the manned space flight effort, and for coordinating all Air Force manned space flight activities with NASA."[84] Schriever gave him an office at Andrews Air Force Base, where AFSC headquarters was located, and a staff of about thirty-five officers. This team of military officers worked directly with NASA spaceflight centers and worked with the NASA office of manned space flight. Ritland was the focal point for all space activity in AFSC and the Air Force's link into NASA.[85] The military was providing NASA with know-how on the Atlas and Titan boosters for the Mercury and Gemini programs, respectively, and NASA was providing the Air Force with its Gemini knowledge.

Ritland described his responsibilities as "staff management and direction of all tasks assigned to the Systems Command in connection with military space programs, booster developments, and space program support; formulation of Command policies for space activities and programs; and the programming and allocation of Systems Command resources necessary to accomplish these tasks." But Ritland also had an office at NASA headquarters in Washington, DC, where he worked with NASA's office of manned space flight director Brainerd Holmes to coordinate Air Force support for NASA's human spaceflight programs. He also managed the officers who were detailed to NASA under interdepartmental agreements.[86] But, recalled Ritland,

> In reality, my job was to be the staff monitor of all of the space activities that the Air Force was interested in and to support the now official organization, outside of the Air Force, of the reconnaissance programs. I essentially had the responsibility to work with NASA in the support of the manned space program, working with the Under Secretary of the Air Force in their special programs, and working with the Air Force and the Space Systems Division.... I also had the responsibility to further the command and control of the network up there at Sunnyvale, which was an Air Force responsibility, and to pursue the development of the Titan III which was the base space booster. The Titan IIIC was designed and developed for large space operations, and of course, was going to be identified later on as the MOL [Manned Orbiting Laboratory] booster launcher.[87]

Ritland's most public role at NASA was to serve as the Air Force's representative on the manned space flight experiments board, the entity that decided which experiments could fly and on which missions. Experiments were sorted as scientific, technological, or medical, and any NASA center or DoD could propose experiments. DoD experiments also had to go before the board, which decided if it would be "feasible to fly the experiment." But in the priorities of getting experiments and tasks completed on Gemini and Apollo missions, DoD requests were always considered at least second behind the requirements of the Apollo lunar landing program. Tasks needed for Gemini had to go first before the NASA–DoD Gemini program planning board, which was established to reduce duplication of effort and reported jointly to the NASA administrator and secretary of defense on Gemini flight operations, which of course were important for DoD aspirations for pilots in space. If the planning board recommended an experiment for flight, the manned space flight board still could reject it, although in practice it did not.[88]

Ritland was among those who espoused "the Need for a Man in Space," which he expressed, for example, in a February 1963 lecture to the Air Force School of Aerospace Medicine in Texas. He suggested two reasons for putting humans in space. The first was simply that "our culture is human centered; that machines, no matter how ingenious, are meant to extend human capabilities, not to replace them." The second reason for putting humans in space was that humans "have significant capabilities needed for complex space missions which can not be provided as well or at all by machines," using the X-15 and the Mercury programs as examples of the "superiority of the man-machine combination in terms of reliability and versatility." He acknowledged that there were still "numerous constraints and technical problems" to be overcome in "the greatest adventure men have ever attempted." Ritland also acknowledged what had been discussed since the beginning of the space race: "Somehow, it simply does not 'count' to send a machine to the moon; man himself must ultimately go there."[89] Ritland argued that "there are important potential military missions that are ideally suited" for human beings to do, emphasizing the need to track what the Soviet Union was doing in space. "We don't know too much about what they have up there—except what they tell us," he was quoted as saying.[90] In the end, Defense Secretary Robert McNamara did not like the concept because he felt it duplicated NASA's program, and NASA Administrator James Webb came to believe that the concept took away some of the "peaceful" character of the NASA program.[91] But by the time the nation made the decision to launch and deploy all its satellites on the space shuttle, there was a wealth of experienced military people who knew how to live and work in space.

Conclusions

In his interview with U.S. Air Force Lt. Col. Lyn R. Officer, an historian and scholar whose work in oral history is practically unmatched in the Air Force's archives, Ritland answered a question about whether technical competency or management ability is more important in actually running a project. Technical know-how and management ability are both required to lead at high levels, but it is really, he argued,

> a combination of both. I know of many, many technical people, scientists who are extremely competent and capable in their technical way, but who

have absolutely no interest in management. To me management is people; but they have no interest in people or managing. All they want to do is look at the end product. So that isn't the kind of guy you want running an organization. On the other hand, you don't want someone who is simply a manager that is managing people and looking at the ledger and the dollar value without knowing what he is building. So to me it has to be a combination of both.[92]

At Ritland's retirement dinner, Schriever said, "Ossie is one of the few people that has more friends than enemies in DC." The last few years on active duty had been rough, Schriever said, because Ritland was "always a man of action" and things move slower in Washington than in other places. But in a sign of just how many friends Ritland did have in government, NASA awarded him its Exceptional Service Medal for his contributions to the Mercury and Gemini programs.[93] At the same time, the Air Force awarded Ritland its first Distinguished Service Medal. And Nevada Sen. Howard Cannon entered Ritland's biography into the congressional record, praising Ritland as "one of this Nation's outstanding military leaders."[94] Osmond Ritland was that rare combination of technical expert and good manager who could solve the problem of the moment. The results were military capabilities that ultimately made the nation safer during the Cold War.

Notes

1. By 1951 he had accrued over eight thousand flight hours, including twenty-five hundred in test flying alone. Ritland biography, January 26, 1951, 128, box 2, misc. papers, Ritland papers, Edwards Air Force Base, CA.
2. Douglas Larsen, "Gen. Mills, Col. Ritland Key Men in Training A-Bomb Crews," *The Albuquerque Tribune*, May 2, 1952, 1, 10.
3. James R. Hansen, *Engineer in Charge: A History of the Langley Aeronautical Laboratory, 1917–1958* (Washington, DC: NASA SP-4305, 1987), xxxiii.
4. Kathleen Ritland Montoya, Zoom interview with the author, June 29, 2022.
5. O. J. Ritland, "Air Force Missiles," *Ordnance* 44, no. 238 (January–February 1960): 576.
6. Thomas Parke Hughes, *Rescuing Prometheus* (New York: Random House, 1998), 10.
7. Maj. Gen. Bernard A. Schriever, "ICBM—A Step toward Space Conquest," address to the Space Flight Symposium, San Diego, February 19, 1957, cited in David Christopher Arnold, *Spying from Space: Constructing America's Satellite Command and Control Systems* (College Station: Texas A&M University Press, 2005), 174n16. Or as Schriever put it in

2001, "I am looking for ways to avoid killing people.... We need to do something other than find ways to kill people better." Bernard A. Schriever, interview by author, Washington, DC, June 27, 2001, cited in Arnold, 174n17.

8. Bernard A. Schriever before the World Affairs Council, Los Angeles, November 7, 1958, in "Excerpts of speeches," Bernard A. Schriever papers, box 164, folder 6, Library of Congress (hereafter Schriever papers).
9. Maj. Gen. Osmond J. Ritland, oral history with Lyn R. Officer, March 19–21, 1974, Air Force Historical Research Agency (AFHRA), K239.0512–722, 65 (hereafter Ritland oral history).
10. Ritland oral history, 56, 58, 70, 93–95.
11. John M. Curatola, *Bigger Bombs for a Brighter Tomorrow: The Strategic Air Command and American War Plans at the Dawn of the Atomic Age, 1945–1950* (Jefferson, NC: McFarland, 2016), 72.
12. Lorine Flemons Wright, "Major General Blazes Path in Air Force History," *Rancho Santa Fe Review*, October 24, 1990, 8, provided by Kathleen Ritland Montoya.
13. Ritland oral history, 115–20.
14. Larsen, 10.
15. Letter, Col. W. J. Watkins to Maj. S. R. Kalmus, November 18, 1965, 2, Ritland papers.
16. Doyle Kline, "They Dropped Bomb in A-Test: 'Routine' Job, Local AF Men Say," *Albuquerque Tribune*, April 26, 1952, 4.
17. Ritland oral history, 128–30.
18. Ritland oral history, 125–26, 130.
19. Ritland oral history, 111–14.
20. Ritland oral history, 122–23.
21. Ritland oral history, 134.
22. Ritland oral history, 133–34.
23. Ritland oral history, 140.
24. Gregory W. Pedlow and Donald E. Welzenbach, *The CIA and the U-2 Program* (Langley, VA: Center for the Study of Intelligence, 1998), 60.
25. Bissell, quoted in Pedlow and Welzenbach, 61.
26. Ritland oral history, 142.
27. Pedlow and Welzenbach, 70.
28. Pedlow and Welzenbach, 322.
29. William E. Burrows, *By Any Means Necessary: America's Secret Air War in the Cold War* (New York: Farrar, Straus, and Giroux, 2001), 239.
30. Ritland oral history, 145–46. Ironic use by Ritland of the term "concurrently."
31. Gen. Bernard A. Schriever, "Military Space Activities: Recollections and Observations," in *The U.S. Air Force in Space: 1945 to the Twenty-first Century*, ed. R. Cargill Hall and Jacob Neufeld (Washington, DC: U.S. Air Force History and Museums Program, 1998), 15.
32. Jacob Neufeld, "Bernard A. Schriever," in John L. Frisbee, ed., *Makers of the United States Air Force* (Washington, DC: Office of Air Force History, 1987), 295.
33. Neufeld, 291.

34. Neufeld, 288.
35. Ritland oral history, 147.
36. Lt. Gen. D. L. Putt, USAF (ret.), oral history, interview by James C. Hasdorff, Washington, DC, April 1974, 24, 243, AFHRA, K239.0512–724.
37. Ritland oral history, 148.
38. Ritland, speech to National Security Industrial Association, April 28, 1958, 2, 7, Ritland papers. The text is in all capital letters that I have standardized for clarity.
39. Ritland oral history, 149–51.
40. Schriever, interview with the author.
41. Lt. Gen. James H. Doolittle, USAF (ret.), oral history, interview by E. M. Emme and W. D. Putnam, Washington, DC, April 21, 1969, 47, AFHRA, K239.0512–625.
42. Ritland oral history, 193.
43. Ritland oral history, 153–54.
44. Ritland, "Air Force Missiles," 578.
45. Ritland oral history, 159–61.
46. Ritland oral history, 190–91.
47. Philip Taubman, *Secret Empire: Eisenhower, the CIA, and the Hidden Story of America's Space Espionage* (New York: Simon and Schuster, 2003), 179–83; Curtis Peebles, *Shadow Flights: America's Secret Air War against the Soviet Union* (Novato, CA: Presidio Press, 2000), 122–23.
48. Ritland oral history, 230.
49. Ritland, May 26, 1959, quoted in "Excerpts of speeches," Schriever papers, box 164, folder 6.
50. Ritland oral history, 236.
51. Ritland oral history, 237.
52. Ritland oral history, 238.
53. Kenneth E. Greer, "Corona (The First Photographic Reconnaissance Satellite)," *Studies in Intelligence* 17 (Spring 1973): 6–7, https://catalog.archives.gov/id/7283860?objectPage=2.
54. Ritland oral history, 239.
55. Curtis Peebles, *The Corona Project: America's First Spy Satellites* (Annapolis, MD: Naval Institute Press, 1997), appendix 1, 272–73.
56. "U.S. Orbits 1,700-Pound Discoverer VI," *Stars and Stripes* (August 21, 1959), 1.
57. Ritland oral history, 240.
58. Peebles, *Shadow Flights*, 91; Dwayne A. Day, "The Development and Improvement of the Satellite," in *Eye in the Sky: The Story of the Corona Spy Satellites*, ed. Dwayne A. Day, John M. Logsdon, and Brian Latell (Washington, DC: Smithsonian Institution Press, 1998), 52–62. Discoverer XIII, a diagnostic mission, did not carry a camera, but the water-recovered capsule carried an American flag.
59. Robert Perry, *A History of Satellite Reconnaissance*, vol. 1 (Washington, DC: National Reconnaissance Office, October 1973), 99–100, https://www.nro.gov/Portals/65/documents/foia/docs/HOSR/SC-2017–00006a.pdf.

60. Perry, 219–20.
61. Perry, 221.
62. Schriever read the letter from Bissell, who was at the dinner. "The Ritland Fan Club," retirement dinner on or about December 1, 1965, digital audio recording, 27:00, provided by Kathleen Ritland Montoya.
63. Schriever read the letter from Johnson, who was not at the dinner. "The Ritland Fan Club," 28:00.
64. Ritland, March 20, 1959, in "Excerpts of speeches," Schriever papers.
65. Loyd S. Swenson, James C. Grimwood, and Charles C. Alexander, *This New Ocean: A History of Project Mercury* (Washington, DC: NASA, 1966), 255.
66. Ritland oral history, 267–68.
67. Ritland oral history, 267–68.
68. Swenson, Grimwood, and Alexander, 175.
69. Swenson, Grimwood, and Alexander, 174–75.
70. "Astronauts Press Conference," September 16, 1959, NASA release 59-230, NASA Archives, NASA Headquarters, Washington, DC.
71. "Astronauts Press Conference."
72. "Report of Ad Hoc Mercury Panel," April 12, 1961, in *Exploring the Unknown: Selected Documents in the History of the U.S. Civil Space Program*, vol. 7, *Human Spaceflight: Projects Mercury, Gemini, and Apollo*, ed. John M. Logsdon with Roger D. Launius (Washington, DC: NASA SP-2008-4407, 2008), 178.
73. Swenson, Grimwood, and Alexander, 255.
74. Swenson, Grimwood, and Alexander, 187–89.
75. Swenson, Grimwood, and Alexander, 175–76.
76. Swenson, Grimwood, and Alexander, 255.
77. Aleck C. Bond and S. A. Sjoberg, eds., "Post Launch Report for Mercury-Atlas No. 1 (MA-1)" (Washington, DC: NASA, August 2, 1960), 1, http://tothemoon.ser.asu.edu/files/mercury/mercury_atlas_1a_postlaunch_report.pdf. Eventually, NASA figured out the adapter connecting the spacecraft to the booster failed when the rocket reached MAX-Q, the region of maximum dynamic pressure, and the Atlas broke up, plummeting the uncrewed capsule into the ocean. NASA recovered more than 95 percent of the capsule but did not meet its test objectives. For the capsule parts that were donated to the Smithsonian, see "Fragments, Capsule, Mercury MA-1," https://airandspace.si.edu/collection-objects/fragments-capsule-mercury-ma-1/nasm_A19870191000.
78. Swenson, Grimwood, and Alexander, 307–20.
79. Swenson, Grimwood, and Alexander, 321–22; Kathleen Ritland Montoya, email to the author, Subj: "Re: O. J. Ritland," March 8, 2022.
80. NASA, "Transcript of MA-6 Press Conference," February 20, 1962, 7, NASA Archives.
81. Ritland oral history, 269–70. This story is consistent with a Ritland family story in which Ritland asked Glenn as he was boarding *Freedom 7*, "Are you sure, John?" Glenn asked back, "Would you [get in], general?" to which Ritland responded "YES!" Glenn

then said, "If it's good enough for you, it's good for me." (Susan Ritland Kosich, "Ritland Oral History questions" for the author, postmarked November 2, 2021.)
82. See Barton C. Hacker and James M. Grimwood, *On the Shoulders of Titans: A History of Project Gemini* (Washington, DC: NASA SP-4203, 1977), especially 122 and 140–43.
83. David N. Spires, *Beyond Horizons: A Half Century of Air Force Space Leadership* (Colorado Springs, CO: Air Force Space Command, 1997), 112.
84. "News Digest," *Aviation Week and Space Technology*, April 30, 1962, 35.
85. B. A. Schriever, memo for Gen. J. P. McConnell, Subj: "Continuity of Manned Space Flight Technical Exchange and Support," ca. November 1965, 2, Ritland papers.
86. "Biography of Major General Osmond J. Ritland, Air Force," Air Force Systems Command, March 1964, Ritland papers. See also General Ritland's unpublished submission for *Data* magazine publication (April 1965), Ritland papers (box 11). This draft article, written in a question-and-answer format, appears to be a follow-up to an article that appeared in the January 1964 edition of the journal. See also O. J. Ritland, "General Ritland Directs AF Space Effort: Coordinates with NASA," *Data* 9 (January 1964): 27–31.
87. Ritland oral history, 264–65.
88. "MSFEB charter," February 10, 1964, 1–2, Nation Archives and Records Administration, College Park, MD record group 255, records of NASA, "Manned Space Flight Experiments Board Minutes."
89. O. J. Ritland, "The Need for Man in Space," address to the fourth annual lecture series, Air Force School of Aerospace Medicine, Aerospace Medical Division, Brooks Air Force Base, Texas, February 5, 1963, Ritland papers. Also published in *Lectures in Aerospace Medicine*, February 4–8, 1963, 59–76. The cover of the book has a Titan III rocket with a DynaSoar spacecraft at the top.
90. "General Urges Space Stations," *New York Times*, February 6, 1963, 8.
91. Donald Pealer, "MOL Part I Manned Orbiting Laboratory," *Quest: The History of Spaceflight Quarterly* 4, no. 3 (Fall 1997): 4–5.
92. Ritland oral history, 302–4.
93. "The Ritland Fan Club," 6:50–13:00.
94. "Retirement of Gen. Osmond J. Ritland," *Congressional Record*, January 24, 1966, 1004–1005.

5 THE SPACE FORCE'S REVOLUTIONARY COMMANDER

Thomas S. Power

BRENT D. ZIARNICK

Most histories of the American space effort, especially the Air Force space program, make few mentions of Gen. Thomas Sarsfield Power. He will often be cited for a few scant pages in a book's index, but never as the primary subject of whatever section where he is mentioned. Conventional Air Force space history places Power squarely in a background role if he is noted at all. This situation is mostly due to the fact that he held no specific space posting in the Air Force. Yet Gen. Power made many positive contributions to the space program, and should be remembered by the Space Force as a visionary.

Early Years

Power was not a "space guy." Rather, he was a bomber pilot and nuclear commander. In the traditionally accepted Air Force history, Power toiled in an uninspired Air Corps career until he emerged as one of Maj. Gen. Curtis E. LeMay's B-29 wing commanders on Guam at the close of World War II. For an unknown reason, LeMay chose Power to lead the first low-altitude radar incendiary raid on Tokyo that presaged the wholesale firebombing of Japan and, ultimately, the nuclear strikes on Hiroshima and Nagasaki. After the war, the intellectually dim and sadistic but loyal Power rode LeMay's coattails to become LeMay's hand-picked successor as Strategic Air Command (SAC) commander. As SAC commander in chief, Power resented the rise of the intercontinental ballistic missile (ICBM) in his flying club and resisted this "ultimate weapon" until finally overcome by Gen. Bernard Schiever—the "father of the Air Force missile and space program"—and his brilliant successes in developing American spacepower.[1]

This conventional wisdom is a wildly inaccurate caricature of Power and a mockery of the true story of his career. Power was one of the only pilots with advanced instrument training and who flew all perfect missions on the East Coast during the Air Corps' ill-fated attempt to fly the air mail in 1934. While charged with training B-29 Superfortress aircrews in 1943 and 1944, Power pioneered the use of radar bombing techniques. As commander of the 314th Bombardment Wing, Power was the key innovator of the plan to strike Tokyo at low altitude with incendiary bombs using radar. LeMay chose Power to command the mission not only because Power devised the mission but also because Power ensured his wing was the unrivalled master of radar bombing before ever reaching Guam. After the war, Power spent much of his Air Force career in SAC as its deputy commander under LeMay (1948–54) and later as commander (1957–64) as they both overhauled the command from a sleepy backwater to one of the most professional and dominant military formations of the Cold War.[2]

Little of Power's childhood suggested that he would rise to become a great air commander. Born on June 18, 1904 (or 1905; the record is sketchy) as the third child to Irish immigrants, Tommy Power showed no particular interest in the military or flying as a child. He grew up in Mamaroneck, New York, in relative financial comfort. At school he was known as a gentle and studious boy who excelled in sports. However, when Tommy was halfway through high school, his father abandoned the family, and Tommy had to drop out of school and work in New York City construction to support his mother and two sisters.

Highly intelligent and hard-working, Tommy quickly moved up to become a construction foreman. It was at a company picnic that Tommy borrowed $10 from his construction crew to pay a barnstormer pilot with a World War I–surplus Curtiss JN-4 "Jenny" training plane to give him a ten-minute flight that included a few loops. Tommy was so enthusiastic about the flight the pilot took him up a second time, and Tommy "was hooked." Tommy found out that the Army Air Corps flying cadet program would train officers to fly for free, and he set out to pass the rigorous entrance exam. The flying cadet exam required applicants to demonstrate the equivalent of two years of college education, and Tommy, who probably never actually graduated from high school, spent every night after work and all day on weekends in the New York Public Library studying history, grammar, geography, mathematics through trigonometry, and physics for six months. In 1928 the high school dropout passed the test that was so rigorous that two of every three applicants failed. This raw intelligence, drive, and ambition served Power well to overcome great adversity to become a greater air commander. Even more impressive is that beyond his mighty contribution to airpower, Power took many early but critical steps to push the Air Force boldly into space, making him one of the great early space leaders as well.

Turning the Air Force from Missiles to Space
Power made his most important mark on Air Force space history in the time between his posting as deputy commander of SAC and taking full command. In this interim, Power served as commander of Air Research and Development Command (ARDC) from 1954 to 1957. ARDC was the Air Force's forward-looking scientific command responsible for the development of new technologies. The 1950s were also the beginning of the space age. Nuclear weapons and ballistic missiles were the talk of the Department of Defense and the nation, and many believed the first to unite these weapons into an ICBM would be the victor in the Cold War. The first satellites were being developed, and astronauts were moving inexorably from comic books and children's television shows to real life. The world watched and wondered if the communist or the free world would reach orbit first—with machine or man. As ARDC commander at this critical time, Power made a crucial impact on the American space effort that deserves to be remembered and earn him a place as one of the Air Force's most important space leaders.

When Power pinned on his third star and took command of ARDC in Baltimore, Maryland, in April 1954, he had been SAC deputy commander for six years, and he was ready for a command of his own. LeMay had previously served as deputy chief of staff for development and was keenly aware of the importance of research and development (R&D) to the future of the Air Force. LeMay may have helped place Power in ARDC specifically to ensure that SAC's interests would have first priority in R&D. Power's command of ARDC at this critical time was highly advantageous to SAC because it was then that the ballistic missile question—a technology that both threatened the manned strategic bomber and promised to open the space frontier—was becoming the paramount concern in the Air Force. With Power as the senior uniformed officer charged with the development of the ICBM, SAC was well positioned to develop the missile the way it wanted.

This did not mean, however, that Power had total authority over the ICBM project. The missile's development was a high priority in Washington, and many civilians made important decisions regarding it. One of the most important early civilian decisions was to establish an organization dedicated solely to ICBM development.

On February 26, 1954, Special Assistant for Air Force Research and Development Trevor Gardner—fresh from the so-called Teapot Committee that had reviewed the U.S. Air Force's strategic missile programs a few months earlier—argued the Air Force could not field the Atlas ICBM by 1960 under current management conditions. To do so, the Atlas program would have to be given top priority and be managed by a streamlined organization dedicated to the ICBM headed by a major general with the dual title of vice commander of ARDC and chief of missile development.[3]

Air Force Chief of Staff Gen. Nathan F. Twining agreed to Gardner's and the Teapot Committee's recommendations. On June 21, 1954, Lieutenant Lt. Gen. Donald Putt, deputy chief of staff for development, ordered Power to speed Atlas "to the maximum extent that technological development will permit" and to "establish a field office on the west coast with a general officer in command having authority and control over all aspects of the program, including all engineering matters." On July 1 Power ordered the establishment of the Western Development Division (WDD) in Inglewood, California, as an ARDC field office charged with developing and fielding the Atlas ICBM.[4]

Gardner originally wanted Maj. Gen. James McCormack, the current ARDC vice commander, to become chief of missile development, with Brig. Gen. Bernard Schriever as his deputy and industrial contractor coordinator.[5] McCormack, however, suffered a heart attack and retired from the Air Force a short time later. Schriever was instead elevated by the Air Force secretary to ARDC deputy commander and chief of missile development as commander of WDD.

From the beginning, Power was unhappy with this arrangement. Power knew Schriever primarily from earlier meetings at SAC headquarters when Schriever, then a colonel, argued with LeMay over support of the aircraft nuclear propulsion program. Schriever was against continuing the development of a supersonic nuclear bomber, LeMay's favorite R&D program at the time, and LeMay thought Schriever insubordinate. And in one rather tense meeting, Power—who held a black belt—asked Schriever if he would like to practice judo with him.[6]

A lingering distrust of Schriever aside, the practical problems were far more troubling to Power. The Teapot Committee had encouraged not only the development of the WDD but also the creation of a unique systems engineering management process that overturned the traditional Air Force approach of prime contractor acquisition. ARDC had begun the Atlas project in January 1951, and up to that time Convair had been the program's prime contractor. Gardner and Schriever were convinced that Convair lacked the engineering design skills to manage the complex ICBM project and instead chose the Ramo-Wooldridge Corporation (R-W) to manage the development of the entire system, leaving Convair to focus on manufacturing. This decision was met by furious objections from the aerospace industry in general and Convair in particular. Power did not agree that the ICBM provided such a significant challenge that existing processes would be ineffective. Worse than the R-W decision, however, was the fact that Putt's June 21 order gave Schriever command over all ICBM decisions but left Power with overall responsibility for the project's success. Power carried out the order but was not happy about it. The only decisions Power could make about the ICBM were organizational ones about the responsibilities of the WDD. Power did everything he could to use that limited authority to influence space and missile development, but the ICBM was only Schriever's game.

Power and Schriever met to discuss the WDD on July 17 at ARDC headquarters in Baltimore. This meeting was tense. Schriever had assumed Power would back him in his decision to abandon Convair in favor of Ramo-Wooldridge. Power instead disagreed with almost every decision that had been made on the Atlas program in the previous few months and with Schriever's actions in particular.[7] Worse for Schriever, Power let Bennie know it in "direct and brutal fashion."[8] After the meeting, Schriever wrote that Power thought that "we were attempting to tie [a] can to Convair and R&W [Ramo-Wooldridge] would grab off the prize." Power was further concerned he would not be able to supervise Schriever if the latter was in Los Angeles. Power felt that as a young brigadier general, Schriever would be "a country boy among the wolves" amid California's aircraft industry and that WDD should be in Baltimore with ARDC.[9] Schriever's explanation that the engineering talent to field the ICBM could most easily be found in California was persuasive, but just barely.

Schriever had told Gardner earlier that to deliver the ICBM on time, he had to be free to make decisions "without any interference from those nitpicking sons of bitches in the Pentagon." Power took Schriever's sentiment poorly. Schriever wrote that Power "made a point that he was senior to me and had much more at stake than I. . . . By his several allusions to my making big decisions on my own . . . he must feel that I am motivated by a personal desire for power. . . . He obviously does not trust me nor have confidence in me—very important factors when undertaking a job of this magnitude."[10]

Schriever left the July 17 meeting shaken but determined that he would "win over Tommy Power." As commander of WDD, Schriever wrote a report to Power every week on the division's progress, phoned or sent a teletype message to Power whenever a significant event occurred, invited Power to all significant meetings, and personally traveled to Baltimore to brief Power as often as his work permitted. By far the most important olive branch Schriever offered Power was arranging for frequent rounds of golf for the two men, for both were highly skilled aficionados of the game. Undoubtedly, the personal connection the two men developed on the links was vital to their effective relationship.[11]

Schriever's overtures to Power worked, aided immeasurably by Schriever's bureaucratic successes at WDD. Power listened to civilian experts such as John von Neumann regarding the ICBM and its importance. He also began to accept that the R-W systems management organization was working well and was impressed that Schriever had prevailed over Convair to continue the R-W management scheme. Power eventually realized "how badly he had misjudged [Schriever] in assessing him as a naive amateur."[12] In his April 1955 performance report on Schriever, Power wrote Schriever had "excellent staying qualities when the going gets rough. Professionally, he is characterized by his thoroughness. He has a brilliant mind and can be depended upon for outstanding work."[13] Less than a year after their horrible initial meeting as senior and subordinate, Power and Schriever were working with a mutual professional respect and personal trust. According to retired Gen. Bryce Poe II, who served as General Schriever's personal aide and chief pilot, Power routinely inquired of Schriever's well-being.[14]

This did not, however, stop Power's sternness. At one briefing, conducted by a colonel working for Schriever, Power grew angry and rejected the entire presentation.[15] Unfortunately, the briefing was very important to Schriever. When Poe told Schriever about the colonel's performance and Power's rejection of the plan, Schriever said, "I'll go in tomorrow and talk to him about it." Poe recalled that Schriever met privately with Power and got the proposal approved as originally put forth.[16]

It was important for Power and Schriever to develop a good working relationship because changing priorities in the Air Force and new opportunities were creating a need to confront new organizational decisions almost immediately. Moreover, the establishment of WDD and a new emphasis on developing an ICBM also meant that there might soon be available a rocket capable of placing a satellite in orbit. Many Air Force officers began to believe space-age weapons would shortly be operational, and the Air Force would have to develop an operational space capability. As Air Force historian Robert Perry noted, "To a great many Air Force planners it seemed obvious that only a military space capability could provide an effective counterweight to an intercontinental ballistic missile force."[17]

In May 1954 Headquarters U.S. Air Force directed ARDC to study the potential implications of a satellite program based on RAND's Project Feedback, which examined potential reconnaissance capabilities of spacecraft. On November 27, 1954, ADRC released system requirement no. 5, which requested industrial support to develop a reconnaissance satellite. RAND Project Feedback contributors presented many briefings to defense officials over the next few months. LeMay was an early enthusiastic supporter of the reconnaissance satellite, although his SAC staff was much more interested in manned bombers and refueling requirements.[18] Characteristically, Power was also a supporter, as he knew that pre-and poststrike intelligence of Soviet nuclear forces was of paramount importance to SAC war planning.

WS-117L

In October 1954 Trevor Gardner requested that the ICBM scientific advisory committee explore the ramifications of the satellite program, soon to be named weapon system (WS) 117L, and other rocket programs relating to the Atlas ICBM effort. The group concluded the review should be conducted by the Air Force, and a WDD staff recommendation on October 15 suggested WDD take responsibility for the management of the satellite, ICBM, and intermediate-range ballistic missile (IRBM) programs.[19] However, the von Neumann committee, a group that shared many members with Gardner's ICBM scientific advisory committee, argued in January 1955 that placing the WS-117L under WDD would put the rapid introduction of the Atlas missile into the Air Force inventory at unacceptable risk. Power evidently agreed with the von Neumann recommendations.[20] Schriever and Gardner both wanted WDD to stay away from WS-117L. In March 1955 Power placed WS-117L under the management of the Wright Air Development Center in Dayton, Ohio, the center in charge of managing Air Force air vehicle development.

However, pressure from ARDC, primarily from Power himself, began to build to place both the WS-117L satellite and the Thor IRBM in WDD. In June 1955 Gardner again called a meeting of the ICBM scientific advisory committee to discuss the issue. The committee unanimously agreed that "any Satellite program, Scientific or Reconnaissance, which is dependent on components being developed under the ICBM program, would interfere with the earliest attainment of an ICBM operational capability" and requested the committee chair write a letter to the secretary of the Air Force advising that such interference could inflict grave damage to the ICBM program.[21]

Historian Robert Perry criticized the findings of Gardner's group, writing that there "was no question of lack of foresight in such a decision. The group was overwhelmingly concerned with keeping the infant ballistic missile program alive and satisfying the critical need for an operational ballistic missile."[22] Perry admitted, however, that "there seemed slight prospect that the matériel and personnel resources then available to the Western Development Division could accommodate a major satellite program without diluting the effectiveness of its missile effort," nor were any additional resources likely to be forthcoming.[23]

On October 10, 1955, Power resolved the question of who was to manage WS-117L by playing his strongest card—his authority over WDD responsibilities—and placed the satellite program in WDD's jurisdiction.[24] Schriever was officially notified of this change on October 17 through the issuance of system requirement no. 5 from ARDC.[25] To understand why Power made this decision in the face of Schriever's and Gardner's contrary recommendations, it is perhaps best to explore exactly why Schriever did not want to manage the WS-117L or the Thor IRBM project, which Power gave to WDD with operations order 4-55, issued on December 9, 1955, though by then WDD had been unofficially working on the theater ballistic missile (TBM) for months.[26]

After the October meeting of the ICBM scientific advisory committee, Power requested Schriever and WDD study the potential relationships among the ICBM, TBM, and WS-117L satellite program. In an undated draft memorandum written by "R-W" and prepared as a staff study by Col. Charles Terhune, Schriever's deputy for ICBMs, in November, Schriever reported WDD's findings.[27] He opined that many of the technical problems shared between the ICBM and TBM "are virtually identical from 1,000 to 5,000 miles range. The sole and rather important exception is the aerodynamic heating problem." Schriever continued that the engineering "data required cover a broader range for the ICBM [system], but this range includes every condition which the TBM [system] payload meets on its re-entry into the atmosphere. Accordingly, work done for the ICBM [system] automatically provides the engineering basis for a sound design for the nose cone of the TBM [system], while the opposite is not necessarily true."[28] Schriever explained that the major difference between the ICBM and TBM programs was "the ICBM requires that all aspects of technology be pushed closer to the limit of the available art," while a "realistic program for the shorter-range missile would be based on a more conservative choice of all dimensions and performance requirements."[29]

Schriever made a forceful argument that the TBM program could be satisfied through the use of alternative approaches to the ICBM that WDD was then contemplating for Atlas. A single-stage TBM system could "look like a demagnified version of the one and a half stage ICBM [system]," or the TBM "could be looked at as a modification of the second stage of the ICBM [system]."[30] Schriever felt the single engine test vehicle and the reentry test vehicle equipment from his "ideally planned ICBM [system] development program" would "constitute minimum departures from the planned first or second stage of a two-stage final ICBM [system]," but "as part of the ICBM [system] program," they would "increase the chance of the TBM [system] vehicle's being automatically derived from the ICBM [system] program."[31] Instead of arguing against the TBM, Schriever attempted to use the TBM requirements to gain additional testing he needed to fund his ICBM program more robustly.

Schriever explained to Power a simple but significant fact concerning both the ICBM and TBM programs: "An ICBM missile can be attained by taking a short range missile and fitting it with a heavier booster that constitutes a first-stage to the shorter range missile's second stage."[32] Ultimately, Schriever argued that the ICBM should be explored in two configurations: a single tank one-stage system with detachable rocket engines (a 1.5-stage vehicle, which the Atlas would eventually have) and a two-stage configuration. Schriever recommended Convair proceed with the 1.5-stage approach, but that the "alternate [two-stage] approach should be carried out by some other airframe manufacturer . . . upon a full two-stage design. This approach is also ideal for incorporating the TBMS as a modification of a second stage."[33] The upshot of all this was that instead of seeing the TBM as a legitimate program in and of itself, Schriever saw it as a potential pathway to secure a much-desired second approach to fielding the ICBM.

When he examined the WS-117L satellite program, Schriever was just as protective of the ICBM. Although Schriever made an early distinction between the ICBM and what he called the "Satellite missile"—what we know today as a space launch vehicle—he nevertheless argued there were "enough elements in common between any project that contemplates bringing a noticeable mass up to sufficient velocity to orbit the earth and the ICBM to make it obvious that the closest of technical coordination will be necessary."[34] The problem, however, was larger than one of merely technical coordination. Schriever continued, "While it would be a grievous error if the two projects [the ICBM and satellite] were not properly associated with one another for mutual benefit, it would also be erroneous to conclude that the success of the Satellite missile is easily and directly assured by the success of the ICBM, for there are formidable technical problems associated with the Satellite vehicle that have no counterpart in the ICBM." Among these many problems were satellite power, terrain scanning, data storage, processing, and transmission, and launch vehicle trajectory control.[35]

Schriever noted that developing a space launch vehicle was a more difficult project than an ICBM, implying that his mission was to provide an ICBM and not a space capability given the time constraints he faced. Schriever was certainly aware there was considerable overlap between the two, but he argued that even a space program would benefit from the success of his ICBM program first, saying that the "major problems of propulsion, launching, structure, and guidance along the powered trajectory, by being solved in the ICBM program will save much time for the Satellite vehicle because of the great similarity of these problems."[36] In this and in most of his rationale, however, Schriever's concerns about the space mission seemed to extend only as far as it might interfere with his ICBM program: "By the time such satellite flights are practical, the ICBM program will either have attained or be close to attaining flights involving velocities near Satellite velocity with payloads probably comparable with the total weight to be carried by the satellite ... [but] it is not easy to see how the ICBM could mount its flight schedule during a period when the Satellite flights are being prepared for, without some substantial dislocation to the ICBM schedule."[37]

On December 20, 1954, Schriever sent a personal telex to Power describing why he felt the current Air Force TBM program would interfere with the timely, efficient, and successful completion of the Atlas ICBM. First, Schriever explained, "important elements of the industry [did] not make themselves available for the ICBM program" due to the TBM program. Schriever noted that Douglas Aircraft and Bell Labs had not participated in the Atlas study program because they were waiting for the Air Force to decide on the TBM. Schriever also claimed that if the TBM program went forward, his planned alternative approach to the ICBM (a two-stage tandem or in-parallel rocket) would probably not be approved due to significant overlap with the TBM. Second, Schriever worried that the shallow pool of ballistic missile engineering talent would be stretched too thin between two competing programs. Third, he was concerned the two programs might compete and cause friction in the Air Force, delaying decision-making for both programs significantly as well as adding "unnecessary duplication of technical programs and facilities." These problems could disrupt both programs so greatly that the resulting confusion could give detractors sufficient evidence to take all missile programs away from the Air Force and give them directly to the Department of Defense.[38]

Ultimately, Schriever concluded, "It is the opinion of R-W and the WDD technical staff that a ballistic missile having a range of 1,000–2,000 miles is one of a family of missiles which can evolve from the ICBM program" and that the Air Force TBM program could be best fulfilled by acting on the R-W recommendation to fund the "alternative configuration and staging approach" of a two-stage ICBM by a second airframe contractor.[39] Power eventually gave Schriever permission to develop this alternative configuration ICBM. It became Titan, and the program developed an ICBM as well as a fleet of space launch vehicles.

Schriever's hesitations about adding the TBM program to WDD's mission set are completely justifiable. His mission was to develop an operational ICBM as rapidly as feasible. The Thor IRBM, however, would be fielded before the Atlas ICBM (though by only a few months); and the Thor system became a mainstay of the American space effort, with its final descendent, the Delta II medium-lift vehicle, still in service as one of the world's most successful space launch vehicles. Although Schriever could not have known it at the time, his primary focus on the ICBM could have negatively influenced the American space program.

Power watched as Schriever, WDD commander, argued against an expansion of satellite programs. Schriever transmitted his original November 1955 "Interactions amongst Ballistic and Satellite Programs" memorandum to General Putt at Air Force headquarters to provide "in some detail both the technical and management reasons for the positions I have taken" (his opposition to the WS-117L satellite program) but warning Putt that "dissemination of this paper should be very limited." Schriever also made a point to tell Power he had sent the document to Putt.[40] On March 30, 1955, Schriever sent a memorandum to Power regarding intelligence on the Army's Redstone program and Army support of a "Scientific Satellite" and its "willingness to act in a contractor capacity to the Air Force." Schriever concluded, "I think that a joint effort of any nature would be a serious mistake.... First, it would be impossible for the Air Force to effectively manage a program carried out by another service. Secondly, it would be naive to think that the Army would develop a weapon and then turn it over to the Air Force to operate. Therefore, I strongly recommend that our relationship with Redstone remain on an exchange of information basis."[41]

Regarding the scientific satellite program itself, Schriever was even less enthusiastic, writing to Power that WDD's technical experts felt Air Force participation in the program "can contribute little if anything to the ICBM program." He felt that even "if successful, this program would contribute almost nothing in furthering a militarily useful satellite," and he recommended against any participation at all. "If other reasons are over-riding concerning Air Force participation in a short term satellite program," Schriever offered, "the Air Force should offer a separate program having greater payoffs."[42] Schriever then made clear he wanted no such separate program, either.

In mid-1955 it seemed clear that Schriever would lose and that both the TBM and the WS-117L would soon be given to WDD. In a memorandum to Terhune on April 15, 1955, Schriever wrote that the "Satellite Development Plan, if implemented beyond the study stage . . . is certain to interfere with the ICBM program. I feel quite certain that management of the satellite vehicle program, when it reaches the hardware development phase, must be under WDD in order to control the coordination which will be required among the several large rocket vehicle programs."[43] Schriever had seen the writing on the wall, and while he was still opposed to the satellite program for its danger of interference with the ICBM program, he began to believe his management of the program would be the best choice available in a bad situation. Even though WDD would not be officially tasked with the TBM program until October, on May 9, 1955, Power issued Schriever an order to manage some TBM business for ARDC.[44]

Overruling Schriever

If Schriever was against these transfers, why did Power overrule him and place the satellite and TBM in WDD? There are several possible explanations. From a purely bureaucratic standpoint, Power might have thought the merging of the three programs, however detrimental to the timely deployment of the ICBM, was simply inevitable. All three programs were dependent upon advanced rocket propulsion and guidance technology. Indeed, the RAND (then Douglas Aircraft Corporation) report *Preliminary Design of an Experimental World-Circling Spaceship*, which later became famous, envisioned a satellite vehicle as the rocket itself, not necessarily the payload of a launch vehicle as we know it today. The report explained, "There is little difference in design and performance between an intercontinental rocket missile and a satellite. Thus a rocket missile with a free space-trajectory of 6,000 miles requires a minimum energy of launching which corresponds to an initial velocity of 4.4 miles per second, while a satellite requires 5.1. Consequently, the development of a satellite will be directly applicable to the development of an intercontinental rocket missile."[45]

In this worldview, the spaceship *was* the launch vehicle, and most of the RAND report was on rocket engineering. As a result, the intellectual history of the ICBM, TBM, and satellites all sprang from the same source without distinction between a satellite and a missile. Perhaps intellectual inertia was simply too great to attempt to isolate artificially the ICBM from the desire to develop a space capability. It must also be stressed that Schriever himself was of two minds regarding the merger. He did not want the TBM and satellite to interfere with the ICBM, but he also felt that under WDD, both "inferior" projects would pose the least risk should the Air Force pursue them. Thus, Schriever's resistance to taking those two projects may have been rhetorically intense but practically very low. Schriever probably understood that, while he did not want the TBM or satellite, he should have responsibility for them.

Another reason that Power may have overruled Schriever was Schriever's successes at WDD. Although originally skeptical of Schriever's managerial skill, Power concluded in 1955 that Schriever was a highly capable officer. Even though Power knew Schriever wanted to focus on the ICBM to the exclusion of the satellite and TBM and that these projects had a high risk of undermining the success of the ICBM, Power might have nevertheless believed that Schriever could overcome those risks. Even with the danger, Schriever might have been the best man in the Air Force to take on these projects, and Power had confidence that Schriever could complete the mission successfully.

A final possibility should also be considered. Schriever was particularly enamored with the ballistic missile as a technology, and his association with

Trevor Gardner and John von Neumann in the beginning of the Air Force's ICBM effort attests to this deep, perhaps myopic, interest. Power, by contrast, was primarily an aviator and one of the leaders of the "bomber mafia," but he also had a keen interest in technology in general. As deputy SAC commander, Power defended the manned bomber from claims of obsolescence by the ballistic missile, and he was not convinced that the ICBM was the "ultimate weapon." Therefore, while Schriever might have seen the potential for space, he was primarily interested in the ICBM and regarded space as being little more than an interesting but nonessential side benefit.

Power, on the other hand, might have thought that the ICBM was an important project but that the real payoff of the technology was the possibility that it would open up space to the Air Force, a natural extension of the "higher, farther, faster" mantra of the airmen that later formed the basis of Thomas White's aerospace concept.[46] Power might have believed the Air Force's need for a space organization was greater than the delays imposed on deploying the ICBM by transferring the satellite and TBM projects to WDD. As an indication of Power's inclinations toward space, in 1954 he had approached industry to study problems regarding space, including manned craft and lunar probes, without Pentagon direction. There is little doubt that Power saw space as having the potential for being the next great Air Force frontier. There is also little doubt that he saw the ICBM as the initial gateway to that future rather than an end in itself. This could well have been a primary motivator of aligning the three major space development programs under WDD.

Most likely, Power's motivation was a combination of all three rationales. Thinking the Air Force needed a dedicated space organization, that such an organization was necessary due to existing bureaucratic inertia, and that Schriever could accomplish all these tasks in a reasonable time were not contradictory beliefs. A combination of all three reasons was possibly why Power made the decision to turn WDD into a space organization. By doing so, on October 10, 1955, Power put the United States and the Air Force on the path to spacepower.

Just as Power accepted Putt's order to establish the WDD with Schriever in command despite his own misgivings, so did Schriever accept Power's order to incorporate both the WS-117L satellite and Thor IRBM with the Atlas program under WDD against his better judgment. And just as Power soon realized his worries were unjustified, so did Schriever soon realize the wisdom of

Power's decision to make WDD into a space organization rather than simply a ballistic missile organization.

Schriever quickly embraced the satellite as well as the rocket in a unified Air Force space effort through his "concurrency approach," by which he developed both the satellite and the missile in parallel, including launch site construction, installation, and checkout, flight testing, and crew training following overlapping and accelerated schedules.[47] This approach dramatically increased risk and cost but was "revolutionary for the R&D community" and saved an enormous amount of time, ultimately propelling the Air Force to obtain a great many operational space capabilities in the 1960s.[48] Schriever did have some space vision. Perhaps with Power's tutelage, as early as January 1955 Schriever was boasting that the goal of the ICBM was not fighting war but conquering outer space.[49]

Unfortunately, Schriever was not totally converted to Power's vision of aerospace—that the air and space were operationally indivisible and that Air Force crews needed to fly in both. Schriever accepted the WS-117L and IRBM into WDD but rejected adding the Wright Air Development Center's BOMI (bomber-missile) spaceflight project to the WDD's portfolio in November 1955.[50] BOMI was an early prototype of a "boost glide" spacecraft designed by the renowned German aerospace engineer Walter Dornberger. Meant to travel into space on a rocket (boost) and use aerodynamics (glide) to maneuver to a landing site, BOMI was a precursor to the space shuttle and the direct antecedent to the Dyna-Soar (later X-20) Air Force manned spaceplane program. Schriever's flat rejection of BOMI in 1955 presaged his later lukewarm attitude toward human spaceflight when he was commander of the ballistic missile division and Air Force Systems Command. With the BOMI decision, Schriever hinted that under his leadership, the Air Force space program would focus on "space and missiles," not the heavy manned space program that Power would eventually strongly support.[51]

Power did not push BOMI on Schriever, so Schriever did not take it. Although Power made WDD into a space organization, he did not force Schriever to make it a truly aerospace one, perhaps to the ultimate detriment of Power's space vision. However, as always, history is not quite as clear-cut as simple narratives suggest. While Power advocated that WDD should manage both missiles and space vehicles (including the satellite and BOMI), he did not always push for all space activities to be transferred to WDD. In July 1956, with responsibility for WS-117L, the ICBM, and the TBM firmly under his control, Schriever requested that primary responsibility for managing nuclear rocket studies be transferred to WDD. Power replied that WDD should stay focused on developing and operationalizing the vehicles at hand and that advanced studies should remain at ARDC under the deputy commander for weapons systems.[52]

The debate over adding WS-117L and the IRBM to WDD has long been neglected in Air Force history. David Spires, in his otherwise excellent history *Beyond Horizons: A Half Century of Air Force Space Leadership*, succumbed to the notion that Schriever was the father of the Air Force space program and claimed Schriever gained WS-117L for WDD over Power's implied objections, based on Power's initial support of keeping WDD focused on the ICBM following the von Neumann committee recommendations as stated above. This is an inversion of reality.[53] With his decision to turn WDD into an inclusive space organization rather than simply an ICBM bureau, Power established the Air Force's first organization dedicated to collect, investigate, and manage the development of American spacepower. WDD became the Air Force's center of space expertise. As Spires himself wrote, "The late fall of 1955 arguably [marked] the beginning of what would evolve into a space subculture within the Air Force."[54] But contrary to popular belief, this milestone was not due to the "father of the Air Force space program" Bernard Schriever but rather to Thomas Power.

Creating an Air Force Space Organization

Forming WDD into a true space organization was perhaps the greatest single contribution Power made to the Air Force space program, but it was not the only one. Power, throughout his tour as ARDC commander, stressed that the command's main responsibility was to retain and expand America's qualitative superiority in weapons relative to its adversaries, especially the Soviet Union. Speaking about ARDC's role in the Cold War, Power believed that in "their determined quest for world domination, the Soviets have unscrupulously resorted to a seemingly inexhaustible variety of hot and cold war techniques. Since the end of World War II, they have placed increasing emphasis on a third type of warfare—the 'slide-rule' war. As a result, the United States has been forced into an all-out struggle with the Soviet Union for technological supremacy."[55] To win this slide-rule war, ARDC stood ready to play its part.

Power argued, "As I have explained in several recent addresses, we can remain ahead of the Soviets in the development and production of new weapons. I am confident that continually advancing the state-of-the-art; by an aggressive development program, utilizing the latest findings of basic research; and by applying principles of management which are possible only in a free economy such as ours and which are far superior to any advantages the Soviets might derive from their system of dictatorship, we can maintain our qualitative supremacy for as long as is needed and can do so within the limits of our economic capability."[56] Nowhere did Power apply this method with more enthusiasm than in determining the role of space in the Air Force of the future.

On October 7, 1955, Power requested that his newly established board of officers on guided missile development "be bold and imaginative in its concept of the scope and importance of future space vehicle development programs."[57] A few months earlier, ARDC proposed a feasibility study of a "manned ballistic rocket research system." Major aircraft companies and other interested organizations were briefed on the study. Because ARDC had no money to support a study on its own, they were also urged to conduct independent investigations of the problem. The firm AVCO studied a manned satellite, and RAND, a strong proponent of reconnaissance satellite systems since 1947, reported on space vehicles for other than reconnaissance purposes. In May 1956 RAND also proposed a "lunar instrument carrier" that circulated through ARDC and the Air Force.[58]

The Air Force needed many studies to assist in planning during the technological revolutions that took place in the 1950s, including exploratory, feasibility, analytical, and design investigations. But money for such inquiries was lacking. An ARDC review for fiscal year 1956 indicated that the fifty-five studies ARDC contemplated required $13,678,000, but only $4,357,000 existed in the current budget. To bridge the gap, Power established a weapons system requirements release program in late 1955 to communicate "future weapon system requirements to industry sooner than heretofore" and to encourage contractors "to conduct voluntary, unfunded studies which will be used for planning purposes."[59] Rather than keeping industry at arm's length until a contract was awarded, ARDC would instead "let industry in on what used to be ARDC secrets."[60] Thus emerged the study system requirement (SR) program that sought from interested private corporations "a statement of an anticipated requirement for a weapon or supporting system, including a definition of the problem area or need, and all considerations having a bearing on the problem and its solution, such as background, intelligence information, present state-of-the-art, related development, etc."[61] Claude Witze believed the SR program offered both the government and industry a distinct advantage: "At no time in history has there been closer co-operation between industry and the government.... The secret is that the System Requirements study program should improve industry's capability before the final weapon system requirement becomes urgent. Technical knowledge, placed on the shelf as it sometimes will be, will shorten the engineering learning curve when the project gets hot. The same holds true for the USAF: with better material upon which to base decisions, the decisions should come more quickly and have more merit."[62]

In December 1956 Power established the guided missile and space vehicle working group, which sought to use the SR system to study many advanced space concepts. The group lasted far beyond Power's tenure as ARDC commander but retained his mission to study advanced space issues. In December 1957 the group issued a "Special Report concerning Space Technology" that laid out an "ARDC five year projected astronautics program." These included a "manned lunar-based intelligence system," with a projected first flight in 1967. In January 1958 the Air Force initiated Program 499, a "lunar base system," and by March the Air Force was formalizing plans for a "manned lunar base study."[63] The SR studies in 1959 were numerous and wide-ranging and included SR 126 (Boost Glide), SR 178 (Global Surveillance System), SR 182 (Strategic Interplanetary System), SR 183 (Lunar Observatory), SR 184 (24-Hour Reconnaissance Satellite), SR 187 (Satellite Interceptor System), SR 199 (Advanced Ballistic Missile Weapon System), SR 79500 (Intercontinental Glide Missile), and SR 89774 (Recoverable Booster Support System).[64]

However, the year's two most remarkable studies, SR 192 and SR 181, shed light on one Air Force long-range vision for space activity. SR 192, Strategic Lunar System, was "implemented to explore the strategy of potential military application in the lunar area." The very comprehensive study considered the "potential offensive, defensive, reconnaissance, and support (communications, weather, logistics, etc.) aspects of the lunar area, all integrated into one system concept."[65] SR 183, Lunar Observatory, was an allied study to SR 192 intended to determine an optimal approach for establishing a manned intelligence station on the moon. SR 183 was considered especially critical to national security because it was believed a moon base would provide unparalleled surveillance of hostile space vehicles and the earth's surface. It was also felt that a military base on the moon might provide the ultimate high ground in space supremacy and provide a highly effective deterrent if armed with ballistic missiles. The SR argued the "military and political effect of earth circling satellites might be nullified by the control of the moon with the accompanying control of cislunar space."[66]

SR 181, Strategic Orbital System, explored an integrated, mature earth orbital military space force that might have existed in the 1965 to 1980 timeframe. The study considered relationships of potential offensive, defensive, reconnaissance, deterrence, and support systems for the orbital military force. The study specifically addressed both manned and unmanned systems, as well as conventional and exotic systems and their potential impact to military operations, including "potential methods of offense and defense using other than nuclear bombs." This study was meant to be holistic, considering political, military, and economic dimensions of military space activity. It also aimed to identify new areas that deserved further in-depth study in the SR system.[67] The study spawned numerous others in its wake, including SRs exploring advanced expendable boosters, military space stations, space logistics, maintenance, and rescue facilities, and various space weapons systems.[68]

The internal dynamics of the Air Force/contractor relationship under the SR system were displayed by a letter from Convair Astronautics to the Air Force Special Weapons Center. In this letter, Convair asked the Air Force for a number of papers written by General Atomics under contract to the Air Force that ranged from "Hemholtz Instability over a Shallow Layer of Fluid" to "Trips to Satellites of the Outer Planets" by noted scientist Freeman Dyson, so that it might use them in preparation for its own study for SR 181.[69] Although prepared under contract from a competitor organization, there is little doubt that Convair received these papers. This conclusion is evident in the fact that two years later, Convair delivered an SR 181 report of breathtaking scope and breadth with the General Atomics program an important centerpiece.

SR 181 was, surprisingly, not the furthest-looking study of the SR series. SR 182, Strategic Interplanetary System, intended to "determine probable military applications in interplanetary space; recognize and outline state-of-the-art advances which are prerequisites to these applications; and to indicate the type and phasing of research vehicle and test programs required to attain and support an interplanetary weapon system concept."[70] The SR studies could be the most advanced and forward-looking space thinking accomplished under the U.S. Air Force banner. Unfortunately, most of them are still classified, not for their risk to national security but rather because they are still corporate proprietary information, even though some corporations such as Convair have been defunct for over a half-century or absorbed into other companies. Regardless, it was Thomas Power, not Bernard Schriever, who supported these advanced and visionary space studies. In time, Power attempted to act on them.

Power left ARDC as a full general and became the third SAC commander in chief on July 1, 1957. Trading his responsibility for the future of the Air Force for the immediate need to deter Soviet aggression in the Cold War with the most powerful striking force ever created on constant alert, Power nonetheless kept his interest in expanding the service into space. Most importantly while at SAC, Power operationalized the ICBM by seamlessly incorporating the new weapon into the command's inventory. Through Power's and SAC's tireless efforts, spacepower in the form of the Atlas and Titan—and a few of the new solid-propellant Minuteman—ICBMs were at President John F. Kennedy's call during the Cuban Missile Crisis in 1962.

However, Power also pursued much more visionary goals that were consistent with many of the SR studies' advanced space concepts. Power codified his space vision in a letter to Air Force Chief of Staff Thomas White detailing SAC's space policy, which advised moving SAC's nuclear deterrence mission into space, stressed the supreme importance of space activity to economic growth in the twentieth and twenty-first centuries, and insisted upon the critical need for mastering human spaceflight for the success of both.[71] Power used SAC's influence to push for both the Air Force's Dyna-Soar spaceplane bomber and Project Orion nuclear pulse propulsion spacecraft (a quantum leap in space propulsion that might have allowed for spacecraft the size of World War II light cruisers as early as the 1970s) projects as far as he could. He spoke to the public about the importance of space to the United States, fighting Department of Defense censors to speak of these advanced programs to fulfill his dream of a Strategic Aerospace Command to deter aggression with a mix of manned bombers, ICBMs, and crewed spacecraft on orbital alert. Although

Power was supported by LeMay, soon to be Air Force chief of staff, and Secretary of the Air Force Eugene Zuckert, Power's vision was strangled by opposition from Secretary of Defense Robert McNamara and his deputy director of research and engineering, Harold Brown, and the relative indifference of General Schriever and his myopic fixation on the ICBM.[72]

Conclusions

Ultimately, beyond successfully operationalizing ICBMs into SAC and the Air Force, Power's space efforts at SAC were generally unsuccessful. Military historians generally remember Thomas White's "aerospace concept" as an Air Force sales gimmick, but they also have completely forgotten or ignored SAC's 1958 space policy that suggests the logic of a serious proposal.[73] Schriever's ICBMs and WS-117L–derived satellites formed the backbone of the Air Force's—and the new Space Force's—fielded space systems. Power's preferred Dyna-Soar spaceplane was cancelled, and the Orion space platform was almost entirely erased from official military history. It is little wonder that today Power is known more as being LeMay's somewhat pale shadow than as a forceful and determined space visionary. Forgetting Power is also convenient when elevating Bernard Schriever to the secular saint of the Air Force space program and the Space Force, when in actuality Schriever's specific actions and efforts were geared almost entirely to ICBMs.

Today, the Space Force culturally defines itself more by its systems—satellites—than by its medium of outer space. It "supports the joint warfighter" with information by operating constellations of satellites hundreds or thousands of miles away while working in climate-controlled work centers almost equally distant from the front lines. Power must be credited as an important father of satellite spacepower for his role in elevating WDD from a mere ICBM shop into the world's first military space organization, ignoring Schriever's opposition. In doing so, Power opened the way for the Air Force to develop the satellite systems that have changed the world as well as the battlefield. However, with innovative companies such as SpaceX and Blue Origin teasing at the prospect of a revolution in human spaceflight and an explosion of new economic space activity, some Space Force personnel (especially its energetic and ambitious junior officer corps) may begin to see Schriever's satellite Space Force as a half-finished revolution.

When Space Force officers start seeking to leverage the commercial space industry to launch crews into space to perform critical military missions physically present in the medium they defend, as every other armed service does, they will find little support for their efforts in the legacy of Schriever. Instead, if they search history for allies, they will find Gen. Thomas S. Power—the man who really created the Air Force space effort and charged the defense industrial base to study space forces, moon bases, and the American conquest of space—warmly welcoming them to the fight to complete the spacepower revolution.

Notes

1. For a brief but illuminating example of this uncharitable characterization of Power, see Stephen Budianski, *Air Power: The Men, Machines, and Ideas that Revolutionized War, From Kitty Hawk to Iraq* (New York: Penguin Books, 2004), 366.
2. See Brent D. Ziarnick, *To Rule the Skies: General Thomas S. Power and the Rise of Strategic Air Command in the Cold War* (Annapolis, MD: Naval Institute Press, 2021).
3. Jacob Neufeld, *Ballistic Missiles in the United States Air Force 1945–1960* (Washington, DC: Office of Air Force History, 1989), 104.
4. Neufeld, 107.
5. Neufeld, 104.
6. Neil Sheehan, *A Fiery Peace in a Cold War: Bernard Schriever and the Ultimate Weapon* (New York: Random House, 2010), 157–58.
7. Sheehan, 251.
8. Sheehan, 250.
9. Sheehan, 252.
10. Sheehan, 252.
11. Sheehan, 253.
12. Sheehan, 260.
13. Sheehan, 260.
14. Gen. Bryce Poe II, U.S. Air Force oral history interview, November 7, 1987, Office of Air Force History, 143, Air Force Historical Research Agency (AFHRA) K239.0512-1729. General Poe recalled that after a T-33 (the trainer version of the F-80 fighter) flying a brigadier general to Air Research and Development Command (ARDC) headquarters crashed, killing both aboard, Power told Poe (then an aide to Schriever meeting with Power), "Schriever is flying coast to coast all the time in that T-33, and he has all that [Atlas] program in his head. Tell me who he is flying with!" Thinking that Power didn't trust Schriever's pilot, Poe responded, "His aide is flying with him." "How much time has he got?" "About 300 hours of jet time, but we have a guy out there with several thousand hours of jet time [Poe] that doesn't know how to do anything else." At that point, Poe found himself as both the aide to General Schriever and his personal pilot, too!

15. Poe interview, 157.
16. Poe interview, 157.
17. Robert L. Perry, *Origins of the USAF Space Program 1945–1956* (Los Angeles Air Force Base [AFB], CA: Air Force Space Systems Division, 1961), Air Force Systems Command Historical Publications Series 62-24-10, 41.
18. Perry, 42.
19. Perry, 42.
20. Perry, 43.
21. Perry, 44.
22. Perry, 44.
23. Perry, 44.
24. Perry, 44.
25. ARDC system requirement no. 5, October 17, 1955, in *Document History of WS-117L 1946 to Redefinition* (Los Angeles AFB, CA: Air Force Systems Command, n.d.), no. 68, AFHRA K243.012-34v1.
26. Operations Order 4-55, HQ Air Research and Development Command, December 9, 1955, reprinted in David N. Spires, ed., *Orbital Futures: Selected Documents in Air Force Space History* (Peterson AFB, CO: Air Force Space Command, 2004), 1:518–19.
27. Memorandum, Col. Terhune to Col. Sheppard, Subj: Visit of Majors Green and Rieppe WADC, to WDD, November 3, 1954, in *Document History of WS-117L 1946 to Redefinition*, no. 35.
28. Memorandum (draft), Schriever (WDD) to Power (ARDC), "Interactions amongst Ballistic Missile and Satellite Programs," undated, 1–2, AFHRA 168.7171-82.
29. Memorandum, "Interactions," 3–4.
30. Memorandum, "Interactions," 5.
31. Memorandum, "Interactions," 6–7.
32. Memorandum, "Interactions," 13–14.
33. Memorandum, "Interactions," 15.
34. Memorandum, "Interactions," 16.
35. Memorandum, "Interactions," 16.
36. Memorandum, "Interactions," 18.
37. Memorandum, "Interactions," 19.
38. Schriever, Memorandum for Record, "Subj: Interaction of TBMS with ICBM," December 30, 1954, AFHRA 168.7171-75.
39. Schriever, Memorandum for Record.
40. Letter, Schriever (WDD) to Putt (DCS, Development HQ USAF), February 4, 1955, in *Document History of WS-117L 1946 to Redefinition*, no. 41.
41. Memorandum, Schriever (WDD) to Power (ARDC), "Subj: Redstone–Scientific Satellite," March 30, 1955, AFHRA 7171-82.
42. Schriever, "Redstone–Scientific Satellite," emphasis added.
43. Memorandum, Schriever to Terhune, April 15, 1955, in *Document History of WS-117L 1946 to Redefinition*, no. 47.

Chapter 5

44. Memorandum, Power (HQ ARDC) to Schriever (WDD), "Subj: Tactical Ballistic Missile," May 7, 1955, AFHRA 168.7171-82.
45. Douglas Aircraft Company, "Preliminary Design of an Experimental World-Circling Spaceship," report no. SM-11827, May 2, 1946, 10.
46. Thomas D. White, "The Inevitable Climb to Space," *Air University Quarterly Review* 10, no. 4 (Winter 1958/59): 3–4.
47. Neufeld, 201.
48. David M. Rothstein, *Dead on Arrival? The Development of the Aerospace Concept 1944–58* (Maxwell AFB, AL: Air University Press, November 2000), 54.
49. Sheehan, 266.
50. Rothstein, 54.
51. See Roy F. Houchin II, *U.S. Hypersonic Research and Development: The Rise and Fall of "Dyna-Soar," 1944–1963* (New York: Routledge, 2006), for more information on the Dyna-Soar project.
52. Alfred Rockefeller, *History of Evolution of the AFBMD Advanced Ballistic Missile and Space Program 1955–1958* (Baltimore, MD: Air Research and Development Command, February 11, 1960), 3.
53. David N. Spires, *Beyond Horizons: A Half Century of Air Force Space Leadership* (Peterson AFB, CO: Air Force Space Command, 1996), 37–38.
54. Spires, *Beyond Horizons*, 37–38.
55. Lt. Gen. Thomas S. Power, "The Air Atomic Age," in *The Impact of Air Power on National Defense*, ed. Eugene Emme (Princeton, NJ: D. Van Nostrand Company, 1959), 686–67.
56. Power, 690.
57. Letter from General Power to General Yates, "Board of Officers on Guided Missile Development," October 7, 1955, quoted in Rockefeller, 5, 14; *History of ARDC*, vol. 1, *July 1–December 31, 1955* (Baltimore: Air Research and Development Command), V-178, AFHRA K243.01v1.
58. Air Force Systems Command, *Chronology of Early Air Force Man in Space Activity 1955–1960*, AFSC Historical Publications Series 65-21-1 (Los Angeles AFB, CA: Air Force Systems Command, 1965), 1.
59. *History of ARDC*, V-178.
60. Claude Witze, "Industry Role in New Weapons Increases," *Aviation Week* 65, no. 6 (August 6, 1956): 86.
61. *History of ARDC*, V-186.
62. Witze, 86, 89–80.
63. Dwayne Day, "Take Off and Nuke the Site from Orbit (It's the Only Way to Be Sure...)," *The Space Review* (June 4, 2007), https://www.thespacereview.com/article/882/1.
64. Tab E–Air Force Study Program, n.d., AFHRA K168.8636-4, 46/00/00-60/02/15.
65. ARDC LRP 60-75, *Air Research and Development Command Long Range Research and Development Plan 1960–1975* (U), n.d., C9-100548, AFHRA K243.8636-1, IRIS 486895, B2-12.
66. ARDC LRP 60-75, B5-9.

67. ARDC LRP 60-75, B2-13.
68. ARDC LRP 61-76, *Air Research and Development Command Long Range Research and Development Plan 1961–1976* (U), n.d., CO-85017, AFHRA K243.8636-1 1961–1976, C2-5.
69. Letter from Louis Canter, chief librarian, Convair Astronautics, to commander, Air Force Special Weapons Center, attn. Capt. Donald M. Mixson, May 20, 1959, Air Force Research Laboratory History Office, Kirtland AFB, NM, Orion Archives.
70. ARDC, LRP 60-75, B7-1.
71. Letter to Chief of Staff Gen. Thomas White from Gen. Thomas Power, "Strategic Air Command Space Policy," August 13, 1958, in Spires, *Orbital Futures*, 27.
72. See Ziarnick, *To Rule the Skies*.
73. See Rothstein, *Dead on Arrival?*

6 FATHER OF THE SPACE COMMAND

James V. Hartinger

MARGARET C. MARTIN

Gen. James V. Hartinger, nicknamed "Father of the Space Command," was the first commander of Air Force Space Command upon its activation on September 1, 1982. Hartinger's selection as the inaugural commander marked his final assignment in a career that began with his service in the Army during World War II. A cursory read of his career might call into question how he ended up in command of the Air Force Space Command, the first separate command among the services dedicated to operational control of space. He had an accomplished record as a pilot, with multiple overseas assignments to include combat in Korea and Vietnam, several different airframes, and command at the wing  and numbered air force levels. Without investigating further, it appears that he spent nearly all his career in arenas conducting the traditional Air Force

[144]

business, only arriving on the space scene for four critical years at the end of his four-decade career.

A closer read of his record reveals an accomplished athlete and thoughtful student of leadership who not only achieved personal success during his career but also leveraged experiences from across his career to emerge as the right leader at the right time to see the Air Force Space Command fully realized as a major command. With an extensive flying background that ranged from air defense missions to combat to a test environment, Hartinger emerged as a process-oriented operator who appreciated tactical and technical expertise. He valued competence, thoroughness, accountability, and adherence to rules and regulations, as well as energy, action, and personal involvement. When placed in positions of leadership, he instilled those values in the organizations around him. Hartinger's time in staff positions reinforced his understanding of the value of thoroughness and expertise. He learned to trust his instincts as an advocate for programs he determined were in the best interest of the Air Force. He developed a reputation for being willing to challenge those who had not done sufficient analysis to make an informed decision. Instead of being viewed as insubordinate, Hartinger was viewed as effective and built trust with his commanders, who in turn opened doors to new career opportunities.

Hartinger attacked any leadership challenge with energy. He earned a reputation as an officer who could turn a failing organization around; he was a fixer. He carried the importance of reform and improvement with him into different roles, and even as a staff officer, he was willing to engage with tough issues of organizational reform—not just of processes or programs but in areas of staff alignment and training as well. By the time Hartinger was named the first commander of Air Force Space Command, he had commanded multiple units and had served in roles that exposed him to the increasingly overlapping realms of air and space that shaped the eventual emergence of a separate space command. His assignments flying air defense missions, experience as a young general officer serving on the dually tasked North American Air Defense Command (NORAD)/Aerospace Defense Command (ADCOM) staff, and finally as the commander in chief of NORAD leaders, positioned him to lead the first major command for the space domain as the Air Force, Army, and Navy grappled with the best way to organize for warfare in and through space.

Early Years

Born in 1925, Hartinger was the third of five children raised by his parents in Middleport, Ohio.[1] He was a farm kid whose family moved into town for his high school years. He had a successful high school career, which included athletic and academic accomplishments. Hartinger was the high school quarterback and, during his senior year, led his team to an undefeated regular season and a league championship. His football success earned him college scholarship offers even as he played on the school basketball team and town baseball team. Academically, he thrived in topics from Latin to algebra and even taught advanced algebra during his senior year after the assigned teacher quit. Although academically prepared for college and a recruited athlete in 1943, Hartinger was pragmatic about the ongoing war and expected to be drafted following high school graduation.[2]

Hartinger reported to Fort Hayes, Ohio, in July 1943, for duty in the Army. Although he departed for Fort Benning, Georgia, in September, he was stationed at Fort Hayes long enough to find success on the company softball team and earned recognition as an all-city player in municipal softball and baseball leagues. He credited his athleticism for easing his transition to Army life, especially for his success enduring the rigors of basic training.[3] Basic training also quickly revealed Hartinger's academic aptitude. His high performance on an Army intelligence quotient test earned him a spot in a writing and speaking course at Stanford University as part of the Army Specialized Training Program.[4] From Stanford, it was on to infantry training.

Hartinger's enlisted time in the Army was limited to training. He spent just over a year with the Seventy-First Infantry Division stationed in Fort Hunter Liggett, California, and trained back at Fort Benning. Hartinger eventually earned a second stripe, and then he was selected for an appointment to the United States Military Academy at West Point.[5] Hartinger recalled the appointment to West Point as a surprise—the application had been initiated by his father. He eventually decided to accept the position, which kept him out of combat but laid the groundwork for the rest of his military career as he established a network of professional colleagues, including two future general officers who would be important to the inception of the Air Force's space command.[6]

Four years at West Point proved to be a navigable challenge for Hartinger. Preparatory work at Amherst College, assigned by the Army before his arrival at West Point, and his stint at Stanford readied him for the engineering curriculum. Although his academic record was mixed—high marks in math and the lowest marks in language—he finished his cadet years academically ranked forty-fifth in a class of 576. His prior infantry training dulled the intensity of "beast barracks" and helped him navigate the cadet environment. He continued to excel athletically, boxing and playing both football and lacrosse, earning all-American honors in the latter for three years.[7] Despite an affinity for the infantry, in part because of his natural athleticism, Hartinger elected to serve in the Air Force and to attend pilot training.[8]

Along with 25 percent of his West Point class, Hartinger entered the Air Force in 1949. He started training in the T-6, where he again felt that coordination from athletics helped him overcome a lack of knowledge about airplanes and aviation.[9] Follow-on training took Hartinger to Williams Air Force Base in Arizona, where he completed his first solo ride in a jet aircraft. After graduating in the top part of his class, the Air Force assigned him to the 23rd Fighter-Bomber Squadron at Fürstenfeldbruck Air Base in Germany.[10] Although he missed a direct assignment to Korea, during his time at in Germany he was assigned to a variety of ancillary duties that allowed him a moment in his career to develop his leadership penchant for, in his words, "cleaning up the place."[11]

In 1952 the Air Force sent Hartinger to the 428th Fighter-Bomber Wing at Kunsan Air Base, Republic of Korea, as part of a temporary duty program to provide combat experience to pilots. After accumulating about thirty-five combat missions, he was reassigned to Williams as part of the 3526th Pilot Training Squadron.[12] Hartinger's reflections on his time as an instructor pilot provide insight into how he approached future leadership opportunities, especially the value of understanding a system. He shared that after his tours in Germany and Korea, he had thought himself an experienced pilot, but that it was only after serving as a training command pilot that he had really learned about airplanes. "Instructing in flying at a training base," he recalled, "is like an insurance policy. You really learn how to fly an airplane there."[13] Hartinger's observations on the value of mastering a skill point to his expectations of others; he expected airmen to be thoroughly competent in their given area. As he moved on to leadership roles, that expectation—and what it meant for problem-solving—appeared often as a key element of his approach to leadership and problem-solving.

Hartinger's instructor pilot tour was followed by a semester at the Air Force Institute of Technology as part of a West Point–sponsored faculty pipeline program. He quickly determined that the degree program (nuclear physics) was not for him and successfully lobbied for return to a flying assignment. Although stationed at Stewart Air Force Base in New York to fly the F-86 once again, he was sent to squadron officer school (SOS) before settling into the assignment. Hartinger leveraged his academic and athletic talents, leading his section to an undefeated flag football season and earning recognition for writing the best airpower report. He credited his success at SOS in giving him an advantage in his newest squadron, which allowed him to earn additional opportunities to lead and innovate.

Once back in his operational home, Hartinger settled into a midlevel leadership role in the 331st Fighter-Interceptor Squadron as an assistant operations officer. The squadron was part of the Air Force's air defense mission, maintaining two aircraft on five-minute alert, twenty-four hours a day, to guard against the Soviet bomber threat. Hartinger's SOS success earned him influence in the squadron and allowed him to take on projects such as implementing a training aid designed like a tic-tac-toe board to boost operational knowledge on topics ranging from gunnery to weather. Success at the personal level, such as being named the first "expert" in the eastern air defense sector, coupled with squadron improvements put Hartinger in line for an assignment to the Pentagon.[14]

Hartinger's Pentagon assignment allowed him to develop a programmatic perspective on the air defense mission. He arrived as a captain, a relatively junior rank for such an assignment, and spent four years in the directorate of requirements working on projects to compare aircraft and to approve aircraft modifications. In his own estimation, he found success cultivating support for projects in two ways. First, he found it necessary "to tell it like it is." Although this tactic often caused him moments of discomfort, it also earned him respect for knowing all sides of an issue and being willing to defend what was best for the mission. This approach garnered him the support of director of operations requirements Gen. James Ferguson, who selected Hartinger as his assigned instructor pilot. This role gave him regular access to General Ferguson and then to Gen. Bruce K. Holloway and Gen. William W. Momyer, who also employed Hartinger as an instructor when they occupied the director role. These relationships gave him exposure to senior-level thinking and grew a professional network that would open doors throughout his career.[15]

Similarly, Hartinger's second great lesson of staff work centered on learning about those around him and using his understanding of personalities to finish projects. His athletics background gave him an avenue to connect with staff members, and he used that athleticism to cultivate off-work relationships that translated into connections in the Pentagon. He also proved adept at understanding who in the building had equities in certain projects, encouraging him to route staff packages to those favorable to an initiative first, so as to imply consensus when engaging with actors who were initially less supportive. Hartinger's willingness to expand his networks and to think critically about building consensus allowed him to enjoy success despite his relatively junior rank. The Pentagon assignment earned him the endorsement "future general officer" from General Momyer and launched him back to an operational assignment with an upward career trajectory.[16]

Leadership Challenges

Following his tour at the Pentagon, Hartinger was assigned to Pacific Air Forces from 1962 to 1965 in the air defense section of the general planning division. His time there overlapped with the beginning of sustained American involvement in the Vietnam War. In his first year, Hartinger traveled as part of the mutual assistance program to countries such as the Republic of Korea, Japan, and the Republic of Vietnam. He was responsible for briefing out the findings of each visit, which made him revisit what he was learning about the region. Following the Gulf of Tonkin incident, he was tasked to create the air order of battle and turned in a preliminary draft of his plan under twenty-four hours. He credited his work on the product for earning him widespread recognition, and although it is not explicit in his autobiography, the work likely contributed to an early promotion to lieutenant colonel and to selection for senior service school at the Industrial College of the Armed Forces in 1965.[17]

After completing his school program, Hartinger retrained in the F-4 on the way to what he expected to be a squadron command in Vietnam. His hopes were dashed, however, by another early promotion, this time to colonel. Instead of command, Hartinger ended up in a command post where he scheduled and executed missions over North Vietnam. In this role, he again reported to

General Momyer, who trusted Hartinger's advice on whether a specific strike mission should execute or not. Momyer also allowed Hartinger to fly missions out of multiple bases and in different types of aircraft. Hartinger used his own mission experiences to help him with his command post responsibilities, specifically controlling rescue missions.[18]

After his year in Vietnam, Hartinger went to the tactical fighter weapons center at Nellis Air Force Base, Nevada, as an aircraft test director, where he tested and evaluated the new F-111, which had been one of his programs during his time in air defense requirements. Although the most versatile aircraft in the inventory in 1968, the F-111 experienced a slew of problems that caused pilots to doubt its effectiveness. True to form, Hartinger flew the plane and put it through its paces. He used his own experience to demonstrate the aircraft's versatility in an attempt to validate its proposed employment. His oversight of the program also led his commander, Gen. Ralph Taylor, who had also led him at the Pentagon, to appoint him the investigating officer of an F-111 accident at Nellis. Conducted in the wake of multiple accidents in Thailand, Hartinger's investigation revealed a failed seal in the hydraulic actuator of the aircraft control system. As he explained, when the seal malfunctioned, the system failed and caused the pilot to lose control of the aircraft. An inspection of the full inventory of aircraft showed that the seal was failing on roughly a third of the fleet; changing the seal eliminated the problem and prevented additional crashes.[19]

Hartinger's next opportunity was orchestrated by his mentor, General Momyer, who in 1969 was the commander of Tactical Air Command. After just one year at Nellis, Momyer pulled Hartinger over to McConnell Air Force Base, Kansas, to command the 23rd Tactical Fighter Wing, which flew F-105s, a plane Hartinger had never flown. Momyer had fired the previous wing commander in the wake of eight training accidents and the loss of three pilots and an "accidentally" supersonic flyover of the Air Force Academy that caused $300,000 worth of damage to windows. He instructed Hartinger to "clean up that place."[20] Momyer picked Hartinger despite his never having served as a squadron operations officer or squadron commander or on a wing staff.[21] Hartinger's success in his wing command reflected the habits he had developed in his career up to that point.

As a wing commander, Hartinger could best be described as a hands-on leader. In keeping with his past performance, it was mastery of the mission requirements, an active physical presence, direct feedback at the individual level, and a high level of discipline that turned the wing around. Hartinger began with a deep dive into the existing issues; he spent his first days as a commander reading the accident reports and flying as often as possible to learn about the plane and the training program. He qualified as an instructor pilot in the minimum time required and used his own experiences in the aircraft, along with observations of related operations ranging from flight line taxiing to performance at the air-to-ground gunnery range, to restructure standard operating procedures. To make his expectations clear, he personally corrected lapses in basic customs and courtesies to cultivate a climate of compliance and professionalism. He shifted disciplinary practices to hold supervisors accountable for sloppy work and went so far as to assign pilot error to instructors who allowed preventable student mistakes. His reforms had the desired effect. The wing saw a reduction in accidents, which allowed the commander to pivot to improvement in other areas, from maintenance operations to base services.[22] Attention to detail was the key to earning General Momyer's praise as the "best wing in [Tactical Air Command]."[23] Hartinger's first command lasted only a year but was likely the basis for his selection to brigadier general.

NORAD

In June 1970 Colonel Hartinger began a three-year posting to NORAD as the deputy chief of plans. His mentor, General Momyer, wanted him to continue in Tactical Air Command, but Hartinger's former group commander from Germany, Gen. Seth J. McKee, who was commander in chief of NORAD, laid claim to the brigadier general–select. Hartinger's arrival at NORAD coincided with an overall decline in the Air Force's support for air defense. The Soviet Union's deployment of intercontinental ballistic missiles meant changes for the primary strategies for security from nuclear attack. Interceptor aircraft were effective counters to bombers but were not useful for defense against ballistic missiles.[24] There was some hope that arms control agreements could reduce the risk of nuclear war, and the Strategic Arms Limitation Treaty I interim agreement on strategic offensive weapons was one attempt to cap offensive weapons. The initiative fell short of imposing real arms limits, and both the United States and the Soviet Union realized that it would be nearly impossible to protect their countries from attack. An alternative idea to discourage large-scale attack was to unilaterally limit antiballistic missile systems; in theory, limited defense disincentivized offensive actions because it all but ensured mutually assured destruction in the event of war. Against this backdrop, NORAD's primary mission shifted to surveillance and warning versus planning an active defense against a nuclear strike. Correspondingly, ADCOM reduced in size to accommodate the lower requirement for an aircraft-based air defense system.[25]

During his tenure as chief of plans, General Hartinger chaired an ad hoc committee to reorganize NORAD by combining it with Army Air Defense Command (ARADCOM) and ADCOM. Hartinger later recalled that the plan successfully reduced the size of the staffs by approximately one thousand personnel.[26] This reorganization reflected the continued consolidation of the traditional air defense mission and hinted at the increasing importance that space-based assets would have in the future. The final reorganization approved by the joint chiefs of staff consolidated the headquarters of the Continental Air Defense Command with the headquarters of ADCOM, its Air Force component.[27]

Maj. Gen. Hartinger departed NORAD in 1973 to serve as commandant of the Air War College at Maxwell Air Force Base, Alabama. While at the Air War College, Hartinger taught a leadership course and developed a leadership continuum model to help his students identify where they fell on a range of issues. The idea was to encourage leaders to embrace their natural tendencies across several categories such as style (autocratic to democratic), decision-making (centralized to decentralized), and supervision (intense to assumed). Hartinger placed himself solidly in the far-left column of each category—autocratic, centralized, and intense—as he characterized it, in the "black hat" column.[28] Then, between 1975 and 1980, he commanded two numbered Air Forces: Tactical Air Command's Ninth Air Force (Shaw Air Force Base, South Carolina, 1975–78) and Twelfth Air Force (Bergstrom Air Force Base, Texas, 1978–80) as a lieutenant general.[29] Over the years, he identified and embraced his leadership style, which continued to yield positive outcomes in each of his commands and upon his return to NORAD in late 1979.

Although the roles and missions of NORAD and air defense more generally had been under scrutiny during Hartinger's earlier tour in Colorado Springs, many of the organizational changes that were on the horizon in the early 1970s had been fully implemented upon his return as commander in chief of NORAD and the Air Defense Command. Primary among those changes was the disestablishment of Air Defense Command as a major command and the reduction of the commander in chief position from four-star to three-star general. In fact, the change in rank requirement became effective on the same date as Hartinger's assumption of command: January 1, 1980. As it was Hartinger's third straight three-star command, he admitted he had some trepidation about assuming the role.[30] However, the organizational changes reflected the current understanding of nuclear threats to the United States and, to a certain extent, the increasing importance of space-based assets to the continental defense mission. Over the next four years, Hartinger drove important reforms to operations within NORAD, and his leadership ensured that when the Air Force activated a space command, it would be situated alongside NORAD.

The prestige of the NORAD position was affected by the Air Force's reorganization of its air defense mission as it sought to assert itself as the lead service for space operations. The Air Force found encouragement in President Jimmy Carter's Presidential Directive 37, which highlighted defense priorities as a basic principle of the nation's space program. Paired with the subsequent Presidential Directive 42, "U.S. Civil Space Policy," the president provided a clear vision for a long-term space program.[31] Additional factors influenced the air staff's thinking about the aerospace defense mission. Throughout 1976 and 1977, Congress criticized the high costs that appeared to stem from unnecessary redundancies in the management of the mission. Moreover, Air Force Chief of Staff Gen. David Jones was interested in cost savings along with organizational changes to strengthen Strategic Air Command's (SAC's) claim to the missile and space defense mission.[32]

General Jones' influence on the Air Force position on space was best articulated in a policy memo that introduced three tenets for an Air Force space policy. Those points included the assertion of the Air Force's responsibility to conduct military operations in space; the need for close coordination with the National Aeronautics and Space Administration on projects of national security; and the requirement to maintain the freedom of space through space defense.[33] Into the 1970s, NORAD and ADCOM had exercised control over both orbital and ground-based systems; however, as reliance on space systems increased throughout the decade, and the importance of centralized management and control of capable space systems shaped national thinking about air defense, the viability of ADCOM as an independent command came into question.[34] General Jones directed an air staff study in 1977 that yielded a "Proposal for a Reorganization of USAF Air Defense and Surveillance/Warning Resources." Gen. Daniel "Chappie" James Jr., ADCOM's commander, was unsuccessful in preserving existing organizational structures, despite offering alternatives. Instead, the findings of the study, colloquially known as the "green book," became "the blueprint for disestablishing ADCOM."[35]

By the time Hartinger took command of NORAD, the transition of ADCOM was nearly complete. Despite the efforts of his predecessor, Gen. James E. Hill, to preserve ADCOM's role as the major command with responsibility for space operations, the substantive parts of its operational units and major installations related to air defense were transferred to Tactical Air Command in 1979, with units and installations responsible for missile and space defense scheduled for transfer to SAC in 1980. The official deactivation of ADCOM as a major command took place on March 31, 1980, with the Aerospace Defense Center designated as a new direct reporting unit in advance of ADCOM's transition to a specified command.[36] Thus, Hartinger assumed a role in transition on January 1, 1980, as the "Commander in Chief of NORAD, ADCOM (specified command) and Commander of the Aerospace Defense Center (direct reporting unit)."[37]

In addition to assuming command on the tail end of organizational upheaval, one of Hartinger's first challenges was to deal with the aftermath of a computer fault that had impacted operations on November 9, 1979. For Hartinger, the event was evidence of low morale, low-performing personnel, and a loss of confidence in NORAD, all of which needed to be fixed if the four-star command were ever to be restored. He pledged to reform NORAD to make it "elite."[38] Hartinger had successfully "fixed" the 23rd Tactical Fighter Wing; in some respects, his challenge at NORAD was similar but on a grander scale. The organizational changes that he effected while in command, along with the policy developments related to the space mission, positioned Colorado Springs and NORAD to be the home of a future independent space command.

In a March 1980 letter to Chief of Staff Gen. Lew Allen Jr., Hartinger provided the results of the operations review board investigation of the November 1979 fault and relayed the command's intent to establish an offline test facility in order to prevent any future interference to operational systems.[39] Hartinger leveraged support from the office of the secretary of defense and the expertise of the NORAD J-6 to see that an offsite test facility was up and running in roughly six months.[40] Despite his attentiveness to fixing the computer fault, a second fault plagued the Cheyenne Mountain system on June 3, 1980, and sent out false alerts of missile attacks.[41] Hartinger's response to this event was to work through the chain of command to set the conditions to duplicate the fault. He also guaranteed coordination across different units to ensure no forces reacted to the expected false alerts. The duplication effort was successful in identifying the cause of the fault—computer chip failure—but prior coordination failed to ensure the lack of response Hartinger envisioned; some number of SAC bombers began their alert sequence, which caught the attention of the media.[42] Hartinger's previous experience analyzing the system failure in the F-111 undoubtedly prepared him for this investigation as well.

In the wake of the two system faults, several agencies conducted investigations and issued reports to address the problems at NORAD. Hartinger recalled that the report from the inspector general recommended that the NORAD commander be designated as the entity responsible for oversight of technical integrity of the system. Such a designation alleviated any confusion over who was responsible and placed the NORAD commander in charge of the system.[43] The Air Force scientific advisory board also delivered a report called the "Summer Study of Space" in 1980, which concluded that the Air Force had been successfully conducting space operations over a fifteen-year period but that it was actually not properly organized to further exploit space or to successfully integrate space systems. Those conclusions propelled the Air Force Systems Command (AFSC) to conduct its own study and to recommend the formation of a consolidated space operations center (CSOC).[44]

Hartinger's willingness to make extensive change to fix processes, programs, and organizations was trained on more than just the technological lapses revealed in two system malfunctions. He also turned his attention to energizing the staff and upgrading other facets of the infrastructure. He had been given some latitude to make by-name selections to the staff, and he used that leeway to bring in capable officers. Hartinger also recognized that others possessed functional expertise he himself did not have and trusted his capable staff officers to run command, control, and communications and other computer issues with relative autonomy. He empowered his system integration office to develop a building-block approach to system replacement so that the command replaced key components of its system in a modular fashion, without ever losing system capability. He also oversaw the replacement of the power plant at Cheyenne Mountain to ensure an uninterruptible power supply after having experienced disruptions from commercial suppliers. Hartinger also emphasized morale and professionalism through what he called his "Better Mountain" program. As he had done during his wing and numbered Air Force commands, he challenged people to identify problems and propose solutions to improve the work environment. He tackled everything from facility security to the dining facility to bus schedules and basketball courts to instill pride and focus among those who worked in NORAD and ADCOM.[45] As early as April 1981, Hartinger had confirmation that he had indeed restored NORAD's reputation when General Jones, chairman of the joint chiefs of staff, declared, "NORAD has its act together."[46]

Air Force Space Command

As the Air Force edged toward a separate space command, the air staff formed a space operations steering committee to oversee studies related to space operations. A 1978 study, "Space Missions Operations and Planning Study," recommended that the Department of Defense (DoD) designate the Air Force as the executive agent for space. This conclusion spurred a group of advocates to coalesce around the idea of an operational organization for space, perhaps even a major command (MAJCOM).[47] For the first time, in 1979 the Air Force also identified space operations as one of its operational missions in Air Force Manual 1-1, *Functions and Basic Doctrine of the Air Force*. Although some still asserted that space was a "place" and not a "mission," the addition remained.[48]

The decision to follow through with AFSC's suggestion to build the CSOC as "a centralized facility for operating all Air Force satellites on orbit" was another of the important steps to the emergence of the MAJCOM.[49] The Air Force settled on Colorado Springs for the location of the CSOC, and construction started in fiscal year 1982 at a location that became known as Falcon Air Force Station, east of Colorado Springs.[50] As the CSOC concept was moving forward, Lt. Gen. Jerome O'Malley assumed the director of operations position on the air staff. An avid supporter of a space command, he was receptive to input that a reorganization to better support space operations was one of two critical initiatives to move the Air Force forward. General O'Malley's interest in reorganizing space operations coincided with initiatives coming from the MAJCOM commanders who had their own ideas on space operations.[51]

In 1981 the commanders of SAC (Gen. Bennie L. Davis), AFSC (Gen. R. Thomas Marsh), and Hartinger at NORAD entertained the idea of raising the concept of a separate space command at a gathering of Air Force senior leaders.[52] Hartinger recalled shaping the plan with his old West Point classmate, Marsh, and using the four-star executive sessions at the 1982 conference to socialize the issue. By the last session, he had guidance from Chief of Staff Allen to present a proposal by April.[53] The two months of work that ensued highlighted some key differences of opinion between key stakeholders—Aerospace Defense Center, AFSC, Space Division, and SAC. The plan that AFSC presented in April was a more complete proposal than what had been presented but stopped short of an independent MAJCOM. While debating the plan, General Hartinger reintroduced the option of a MAJCOM, which would be coequal to AFSC under the chief of staff and would exercise operational control over the CSOC. His suggestion was well received and permanently shifted planning efforts to a separate space command.[54]

The Air Force's prioritization of a separate space command matched the drumbeat of activity from external entities. Late in 1981, Undersecretary of the Air Force Edward C. Aldridge spoke at the American Astronautical Society and suggested it was time for a space command. Representative Ken Kramer from Colorado Springs advocated publicly for renaming the Air Force as the Aerospace Force. The Air Force did not like the renaming suggestion, but it served as a reminder of the Air Force's responsibility to the space mission. In 1982 the General Accounting Office published a report criticizing DoD for "poor management" of space systems. In the same report, it identified the CSOC as a "potential nucleus" for an emergent space force or space command.[55] Against this backdrop, and preempting a DoD announcement of the conclusion of its own space study, on June 21, 1982, General Allen announced the formation of the Air Force Space Command, which at the time was abbreviated SPACECOM to prevent confusion with the Air Force Systems Command.[56]

The Air Force declared SPACECOM activated on September 1, 1982, and General Hartinger became its first commander. It was the first time in over thirty years the Air Force had formed a new major command.[57] The Air Force placed operational control of the service's on-orbit space assets with the command, and AFSC retained oversight of research and development of space systems. The new commander was expected to exercise "operationally related cognizance over the separate space medium and mission area."[58] General Hartinger's previous command experience with two different numbered air forces gave him credibility as commander of the new MAJCOM and its operational focus.[59] Hartinger understood the focus on operational control of space and at his assumption of command noted that the new SPACECOM was a "crucial milestone in the evolution of military space operations."[60]

In Hartinger's two years in command, he grew the nascent SPACECOM into a mature organization, reestablished oversight of space systems that had previously belonged to ADCOM, and acquired oversight of systems new to the command. Leadership traits that had served him earlier in his career, such as attention to detail, personal involvement, and emphasis on training and process, helped set SPACECOM on solid footing. As an operator, he valued repeatable processes and procedures, which influenced his leadership approach.[61] Also, during his time in command he advanced discussions and organizational relationships that moved forward the opportunities for a unified command for space. As the first Space Command commander, Hartinger's immediate impact on the Air Force's operational space mission was profound.

The official history of SPACECOM's first full year of operations reads like a laundry list of major reorganizational steps; Hartinger's team accomplished a significant amount of work transforming Space Command into an effective operational MAJCOM. The team clearly understood its task and articulated that in its historical record when it asserted, "The new major command fulfilled its missions of managing and operating assigned space assets, consolidating planning, defining requirements, providing operational advocacy, and ensuring close interface between research and development activities and operational users."[62]

During its first year, SPACECOM received the assets ADCOM had ceded to SAC, including over fifty space and missile warning systems, bases, units, and projects. Moreover, the command began to exercise its responsibilities by acquiring the appropriate oversight role for key technologies and programs of record. Of note, in 1983 SPACECOM became the operational manager for the Milstar satellite communications system and assumed the resource management for the global positioning system (GPS). Importantly, highlighting the expected civilian and military program coordination, DoD appointed Hartinger its manager for space shuttle contingency operations.[63]

To properly administer these and other programs, Hartinger paced the staff through a series of personnel reforms, instituted training requirements, and developed plans for new facilities. The substantive staff changes took place internal to operations. Realignment of roles, the elimination of obsolete offices, and the creation of new positions all focused either on integration (creating a focal point for interface and integration, enforcing standards, and developing test and training procedures), understanding and aligning new and existing operational requirements, or establishing directorates to oversee operational missions. Similar efforts at inside communications also focused on mission support and integration.[64]

SPACECOM's educational agenda focused on the development of its own staff by assigning mandatory readings. Hartinger even gave pop quizzes at his staff meetings and posted the scores by name. The command's manpower and personnel team was also empowered to coordinate with external entities, including the U.S. Air Force Academy, the University of Colorado, Colorado Springs, and the Air Force Institute of Technology to develop space-related curricula. The growth of the staff also necessitated the creation of a directorate of training, standardization/evaluation, and instructional systems development to ensure alignment between systems training, standards, and evaluation. Finally, understanding the command's unique and substantial training and education requirements, the command requested, and was granted, authority to establish a directorate of training and education, which liaised with institutions of higher education, developed space curricula, and designed formal training for personnel assigned to specific centers under SPACECOM.[65]

The command's growth also required new facilities. In just under a year, SPACECOM's authorized personnel grew threefold, and the anticipated influx of personnel and mission required new facilities. The two major initiatives begun in 1983 included a new test, development, and training center and a new headquarters building.[66] The new center was envisioned as a three-building complex on Peterson Air Force Base that would eventually replace the off-site test facility built in 1980. Designed to house off-site the models of various command and communication systems, two of the three proposed buildings, with estimated building costs of $5.2 and $4.46 million respectively, were formally planned and funded by the end of 1983.[67] General Hartinger was less successful in having the proposed $19.5 million headquarters project included in the fiscal year 1984 military construction program, but he was successful with pushing ahead on the design process. Ultimately, $25 million was set aside for the headquarters in the 1985 military construction program.[68]

As the structural pieces of Space Command fell into place, Hartinger also personally advocated for operational oversight of missions, systems, and programs—providing what he called "the operational pull" that complemented the traditional "technology push" of the space mission.[69] In June 1983 Hartinger negotiated the terms of managing GPS with General Davis at SAC. They concluded that operational control would remain with SAC and resource management would transfer to SPACECOM. During the year, both he and the SPACECOM assistant vice commander spent time in public and before Congress advocating for an orbital antisatellite system. Hartinger also established the transition timeline and process for the transfer of DoD's contingency support of space shuttle operations to SPACECOM. Finally, alongside local officials and under a light dusting of snow, General Hartinger also presided over the groundbreaking for the CSOC.[70]

In his last eight months on active duty, General Hartinger continued to preside over a maturing Space Command. As was intended, the command continued to act as a focal point for space operations, managing assets and operational programs and coordinating with research and development initiatives. As the command continued to refine its organizational structure, it also added offices, including that of a political advisor. Hartinger had requested a foreign service officer be assigned as a political advisor, citing two reasons. First, Space Command's growing involvement in "national and international issues relating to space and space systems must be considered as we plan for future operations." And second, with units dispersed across eight countries, a political advisor was essential to help navigate a "diversity of political and diplomatic challenges for which sound counsel is a must."[71] The requested position was filled in the summer of 1984.

As the command evolved its structure and improved training, it also continued to oversee programs related to missile warning, space defense, and air defense. As it was chartered to do, SPACECOM continued to manage and operate assets for strategic defense in support of NORAD and ADCOM.[72]

Hartinger also used his position to advocate for a multiservice unified space command, an organization that could consolidate direction of all military space operations. The initial announcement of the formation of the Air Force's Space Command included language indicating that the Air Force, as a service, was supportive of a unified command, although Hartinger was one of the most vocal officers promoting the idea.[73] He believed that with an existing space command, the Air Force was positioned to take the lead coordinating space operations as the core component of a unified command. His plan called for the commander of SPACECOM to serve as the commander of the unified command, the Air Force MAJCOM, and as NORAD commander.[74]

As a supporter of the unified command construct, Hartinger necessarily supported the Navy's authorization of a Navy Space Command in 1983; he understood that a unified command required at least two service components.[75] Hartinger's advocacy for a unified command was in line with changing policy and program initiatives, especially President Ronald Reagan's Strategic Defense Initiative, colloquially referred to as "Star Wars." As the joint chiefs of staff grappled with how best to support the initiative, Hartinger again offered the solution of a unified command. In early 1984 Air Force Chief of Staff Gen. Charles Gabriel and Air Force Secretary Verne Orr issued a joint statement supporting a unified command. By the end of the year, and following General Hartinger's retirement on August 1, 1984, the secretary of defense and the joint chiefs of staff recommended to President Reagan that he establish a unified space command, and on September 23, 1985, after much study, DoD activated United States Space Command.[76] (The Air Force redesignated its space command as Air Force Space Command in November 1985 to distinguish it from U.S. Space Command.[77])

Conclusions

Upon retirement, General Hartinger was recognized with many accolades. He received the Gen. Thomas D. White space trophy, and the National Security Industrial Association named its outstanding achievement in the military space mission award after him. Lengthy career accomplishments as a pilot and commander gave way to the defining moments of his career: four years of transformative leadership at NORAD and the Air Force's space command that left a lasting impact on the oversight and execution of the service's space operations.

At the end of his career, he was recognized as a pioneer, indeed, as the father of the Space Command. His leadership style, self-characterized as autocratic, centralized, and intense, paved the way for the formation of the Air Force Space Command and fostered momentum for the United States Space Command.[78]

Notes

1. James V. Hartinger, *From One Stripe to Four Stars* (Colorado Springs, CO: Phantom Press, 1997), 9.
2. See Hartinger, 21, 25–28, for a summary of his high school years.
3. Hartinger, 29–32.
4. Hartinger, 32–33.
5. Hartinger, 35–37.
6. James V. Hartinger, interview by Barry J. Anderson, September 5–6, 1985, no. 416, transcript, United States Air Force Oral History Program, USAF Academy, CO, 10–11 (hereafter Hartinger interview). Hartinger relates that he graduated with Gen. Robert T. Marsh, USAF, and Lt. Gen. Donald R. Keith, USA.
7. Hartinger, 39–48, for a summary of his West Point years.
8. Hartinger humorously cited the lure of more pay as the key element to the "comprehensive decision-making process" that solidified his choice to choose the Air Force, in Hartinger, 49. See also Phillip S. Meilinger, *Airmen and Airpower Theory: A Review of the Sources* (Maxwell Air Force Base, AL: Air University Press, 2001), 80.
9. Hartinger interview, 11–13.
10. Hartinger interview, 18–20.
11. Hartinger interview, 22.
12. Hartinger interview, 29–32. Hartinger states he flew "about thirty-five" combat missions.
13. Hartinger interview, 34.
14. Hartinger interview, 35–40.
15. Hartinger interview, 41–45.
16. Hartinger interview, 43–45.
17. Hartinger interview, 48–50, 52.
18. Hartinger interview, 56–58, 62. Hartinger flew F-4s and RF-4s out of Ubon, Thailand, and Da Nang and Cam Ranh Bay, Republic of Vietnam, and A-1s with the Vietnamese Air Force out of Tan Son Nhut, Republic of Vietnam.
19. Hartinger interview, 66–69.
20. Hartinger interview, 69–70, 72. The accidents took place in the last half of 1968.
21. Hartinger, 135.
22. Hartinger, 135–47.
23. Hartinger interview, 77.
24. Hartinger interview, 88–89. See also David N. Spires, *Beyond Horizons: A Half Century of Air Force Space Leadership*, rev. ed. (Maxwell Air Force Base, AL: Air University Press,

2004), 178–79. At the time of his first assignment, NORAD stood for North American Air Defense Command. In 1981 it changed to North American Aerospace Defense Command ("A Brief History of NORAD as of 13 May 2016," Office of the Command Historian, NORAD, https://www.norad.mil/Portals/29/Documents/History/A%20 Brief%20History%20of%20NORAD_May2016.pdf?ver=2016-07-07-114925-133).
25. Spires, 178–79. In various sources, Aerospace Defense Command is abbreviated ADC or ADCOM. Air Defense Command was reestablished as a major command on January 1, 1951 (first established March 21, 1946), and redesignated Aerospace Defense Command on January 15, 1968. (See AFHRA Fact Sheet, "Air Defense Command," n.d., https://www.afhra.af.mil/About-Us/Fact-Sheets/Display/Article/433912/air-defense-command/.
26. Hartinger, 161.
27. Edward J. Drea et al., *History of the Unified Command Plan 1946–2012* (Washington, DC: Joint History Office, 2013), 17, 18, 20, 25, 30. The Continental Air Defense Command (CONAD), activated September 1, 1954, was headquartered at Headquarters USAF Air Defense Command, Ent Air Force Base, CO. The commander in chief (CINC) of CONAD was dual-hatted as the commander, USAF Air Defense Command, a component command. The dual command was reversed in 1956 and CONAD was directed to establish a separate headquarters. With the establishment of NORAD in 1957, CINC-CONAD was redesignated CINCNORAD. Also in 1958, CONAD was designated a unified command. In 1963 CONAD had operational control over a U.S. Army Nike Zeus unit when it was employed to shoot down Soviet satellites. The 1973 reorganization restored the dual-hatted command of CONAD/ADC and left intact the relationship to NORAD, while CINCCONAD/commander ADC remained CINCNORAD.
28. Hartinger interview, 100–2.
29. Hartinger, 11.
30. Hartinger interview, 140–41.
31. Spires, 190–91. Presidential Directive 37 was dated June 20, 1978, and Presidential Directive 42 was issued October 11, 1978.
32. "History of ADCOM/ADC, 1 January–31 December 1979," NORAD History Office, 4.
33. Spires, 191–92.
34. Spires, 187.
35. "History of ADCOM/ADC (1979)," 4; Spires, 193.
36. "History of ADCOM/ADC (1979)," 27; Maurice C. Eldredge, "A Brief History of 'ADTAC': The First Five Years" (Maxwell Air Force Base, AL: Air Command and Staff College, 1985), 4, https://apps.dtic.mil/sti/pdfs/ADA256294.pdf.
37. Eldredge, 4.
38. "Missile Alert a False Alarm," *Washington Post*, November 10, 1979, 4; Hartinger interview, 141–42. In this interview, he implies that the restoration of confidence in NORAD contributed to the dual-hatted command structure of Space Command and NORAD.
39. Gen. James V. Hartinger to Gen. Lew Allen Jr., March 14, 1980, https://nsarchive2.gwu.edu/nukevault/ebb371/docs/doc%2011.pdf.

40. Hartinger interview, 144.
41. Earl S. Van Inwegen III, "The Air Force Develops an Operational Organization for Space," in *The U.S. Air Force in Space, 1945 to the Twenty-first Century*, ed. R. Cargill Hall and Jacob Neufeld (Washington, DC: U.S. Air Force History and Museums Program, 1998), 139.
42. Van Inwegen, 139; Hartinger interview, 146–47.
43. Hartinger interview, 147.
44. Van Inwegen, 139–40.
45. Hartinger interview, 150–57.
46. Hartinger interview, 161.
47. Van Inwegen, 138.
48. Van Inwegen, 138.
49. Benjamin S. Lambeth, *Mastering the Ultimate High Ground: Next Steps in the Military Uses of Space* (Santa Monica, CA: RAND, Project Air Force, 2003), 29, https://www.rand.org/content/dam/rand/pubs/monograph_reports/2005/MR1649.pdf. The Scientific Advisory Board had recommended a consolidated space operations center in 1980.
50. Lambeth, 29; "History of Space Command/ADCOM/ADC, 1 January–31 December 1982," NORAD History Office, 72. General Hartinger, along with his vice commander, Lt. Gen. Richard C. Henry, chose the name Falcon due to the proximity of town of Falcon, CO, as well as a nod to the bird of prey, which was also the mascot of the United States Air Force Academy.
51. Van Inwegen, 140.
52. Van Inwegen, 140.
53. Hartinger interview, 166–67. Historian David Spires offers a slightly different account and credits General Marsh with the key ideas of reorganization but stopping short of recommending a major command. He proposed to have AFSC's Space Division commander also serve as ADCOM's deputy commander for space, essentially serving under NORAD/ADCOM commander General Hartinger. Regardless of details, the senior leader gathering was a turning point to solidify an actionable plan for the reorganization of space operations in the Air Force. Spires credits Hartinger's own assessment of "good personal relations" among West Point classmates (Hartinger, Marsh, and Henry) as influential to the outcome (Spires, 203).
54. "History of Space Command/ADCOM/ADC (1982)," 7–8.
55. Van Inwegen, 141.
56. Spires, 205.
57. "History of Space Command/ADCOM/ADC (1982)," 2.
58. Lambeth, 29.
59. Lambeth, 29.
60. Hartinger's speech quoted in "History of Space Command/ADCOM/ADC (1982)," 10.
61. William D. Sanders, "'Space Force Culture': A Dialogue of Competing Traditions," *Air and Space Operations Review* 1, no. 2 (Summer 2022): 31–32. Sanders describes the operator mentality as "procedurally focused perfectionists" and critiques the approach for

substitution process for deep technical training, which allowed the Air Force to sideline engineers in place of less-trained operators. It was not clear in 1982 that this was the trade space between operators and engineers, nor is the point that Hartinger preferred operators to engineers. Rather, the phrase does capture his proclivity to instill reliable and repeatable processes to facilitate mission success.

62. "History of Space Command/ADCOM," January 1–December 31, 1983, NORAD History Office, xiii.
63. "History of Space Command/ADCOM (1983)," xiii, 2–5.
64. "History of Space Command/ADCOM (1983)," xiv.
65. "History of Space Command/ADCOM, (1983)," xv; for standardization/evaluation, 9; for education and training, 14–15; Hartinger, 251–52.
66. "History of Space Command/ADCOM (1983)," xv–xvi. The exact increase of authorized positions was from 1,963 to 6,083 personnel.
67. "History of Space Command/ADCOM (1983)," 22.
68. "History of Space Command/ADCOM (1983)," 22–23.
69. "History of Space Command/ADCOM (1983)," 2.
70. "History of Space Command/ADCOM (1983)," 67, for transfer of SAC assets; for antisatellite, 69; shuttle operations, 75; consolidated space operations center, 81.
71. "History of Space Command/ADCOM (1983)," xiii; for political advisors, xv; for excerpts of Hartinger's letters, 26.
72. "History of Space Command/ADCOM, January 1–December 31, 1984," NORAD History Office, iv–v, xiii.
73. "History of Space Command/ADCOM (1984)," 7.
74. Spires, 218.
75. "History of Space Command/ADCOM (1983)," 13.
76. Spires, 218–19.
77. Margaret Ream, "Space Operations Command (USSF)," n.d., https://www.afhra.af.mil/About-Us/Fact-Sheets/Display/Article/2886917/space-operations-command-ussf/.
78. Hartinger, 247, 270–71. Hartinger was also the first recipient of the National Security Industrial Association award in 1984.

7 FATHER OF THE SPACE FORCE
Thomas S. Moorman Jr.
WILLIAM D. SANDERS

Thomas Samuel Moorman Jr. did not fit the classic image of a staid and aloof general officer. His affable nature was not an affectation; he was kind to all, making lifelong friends with congressmen, service secretaries, and junior officers alike. He liked people, and the feeling was mutual. His smile radiated atop his tall frame, a lighthouse of warmth. At social events, crowds gathered around him not because he was a general officer, for there were many of those, but because he could tell a story. And he had substance to match his style.

Under the veneer of an archetype gentleman was a scholar with a warrior's sensibilities. He was passionate about history and ensured the Air Force documented the development of space operations. Even though he possessed a formidable intellect, he had no formal technical education. He thrived in space operations through a ready grasp and urgent curiosity. Moorman was also a generous mentor and never missed an opportunity to teach others, often late into the night. That did not mean Moorman was easy to work for. He demanded perfection. On his first review of a staff package he might write, "YGTBSM," shorthand for one of his favorite colorful sayings, "You've got to be shitting me." In preparing briefings or papers, Moorman insisted his staff refine them until the last minute, long past the point of "good enough." But to Moorman, good was the enemy of perfect. His tendency toward perfectionism was a symptom of his intense drive. Moorman was a fierce competitor, as anyone who played pickup basketball with him knew. Like a battlefield commander, he was self-assured and seldom wavered in his convictions. And nowhere was that conviction stronger than his feeling that the Air Force should have a greater role leading space operations.

Thomas S. Moorman Jr. had a hand in Air Force space operations for nearly a quarter of a century. He played a key role in establishing a separate Air Force Space Command, then leading that organization through the crucible of the Gulf War. Moorman became the first, and only, career space officer to rise to the second highest post in the Air Force, vice chief of staff.

Early Years

On November 16, 1940, Thomas "Tom" S. Moorman Jr. was born in Bethesda, Maryland. Moorman Jr. would be a third-generation military officer. His eponymous grandfather was an Army colonel. His maternal grandfather Maj. Gen. Allen W. Guillion was the highest-ranking lawyer in the Army during World War II. Thomas S. Moorman Sr. reached the rank of lieutenant general in the Air Force.[1]

In 1954 Moorman Sr. took command of the air weather service, taking the family to Maryland. At Suitland High School, Moorman Jr. excelled in academics and had a reputation as a young man of high character. He and Steny Hoyer, future U.S. representative from Maryland, formed their own student body party, called the "Rampublicans," to contest the incumbents, the "Ramocrats," a play on the school's mascot, the rams. Moorman was well-liked and had a knack for politics. In 1956 Hoyer ran Moorman's successful campaign for student body president, about which Hoyer said, "Tom's political career at that point looked much clearer than mine."[2] Moorman did not contemplate a political career, however. He "enjoyed the military culture" and decided to follow in his father's footsteps in the Air Force.[3] Moorman Jr. desperately wanted to attend the new Air Force Academy and become a pilot, but he did not have the perfect eyesight that the service academy required. Years later, his father served as its three-star superintendent.

Moorman Jr. attended the only other school to which he applied, Dartmouth College, graduating in 1962 with a double major in history and political science and a commission from the Air Force Reserve Officers Training Corps as a second lieutenant.[4] He was off to an auspicious start as a distinguished graduate of ROTC from the Ivy League school. With characteristic humility, Moorman suggested that grades must not have been a part of the determination.[5] Despite coming from a lineage of successful officers, Moorman questioned if he was cut out to be a career officer. In 1962 a nonflier with a political science degree was destined to become an intelligence officer, with few paths to the higher ranks.[6]

In summer 1962 Lt. Moorman made his way to Schilling Air Force Base (AFB) in Salina, Kansas. Schilling might have been in a sleepy flyover town, but the Strategic Air Command (SAC) base was a beehive of activity at the height of the Cold War. Just a month after Moorman's arrival as a new imagery analyst, the Soviet Union placed nuclear missiles in Cuba. One day during the crisis, he heard the Klaxon sound and watched in terror as aircrews scrambled to their planes, indicating possible nuclear war. Although the nations avoided a nuclear exchange, Moorman recalled the "razor's edge alert posture."[7] During the thirteen-day crisis, SAC bombers flew 2,088 missions, amassing nearly 50,000 hours of flying time.[8] The Air Force's intelligence apparatus was similarly strained, attempting to locate the Soviet surface-to-air missiles that posed a threat to SAC bombers.

Contributing to the backlog of intelligence, the National Reconnaissance Office's (NRO's) Corona satellite, the nation's first space-based photoreconnaissance system, provided a windfall of imagery and overwhelmed the Air Force's analytic capacity. To expand capacity, the Air Force cleared one intelligence officer from each SAC base to analyze Corona imagery. Moorman, chosen as the officer from Schilling, quickly became "fascinated" with the world of overhead reconnaissance, despite the demands of the additional duty. He would be a general officer, he claimed, before he worked as much as he did in that first assignment.[9]

Moorman's boss, Col. John Neal, convinced him to go to Offutt AFB for an interview with a secretive organization. With nothing but a phone number and a name, Moorman headed to the heavily guarded bowels of SAC headquarters and learned he was to process imagery for the SR-71 high-altitude strategic reconnaissance program.[10] Believing his upward mobility was limited, he welcomed a move to Beale AFB, California, where he could leave the Air Force behind and pursue plans to attend Stanford law school.[11] But Moorman was drawn to the quality of people assigned to sensitive programs. Indeed, of the officers assigned to that initial wing, over a dozen became general officers, with multiple future four-star generals—one of whom, Jerome O'Malley, would later play a crucial role in Moorman's career and the Air Force's future in space.[12]

In the fall of 1966 Moorman deployed to the 432nd Reconnaissance Technical Squadron in Thailand. There, he analyzed tactical reconnaissance images from the RF-4s to track adversary movements in Vietnam. As the Rolling Thunder bombing campaign continued, Moorman recoiled at losing friends as they flew "the same times and same basic ingress and egress routes."[13] Although Moorman was frustrated with how leaders prosecuted the war, he witnessed the importance of sound intelligence. For the first time, the nation used space assets to support a regional conflict. The defense meteorological satellite program provided an important weather picture to air commanders, and the Defense Satellite Communications System (DSCS) allowed Washington to transmit and receive critical real-time messages.

In 1967 Moorman was assigned to the 497th Reconnaissance Technical Group in Schierstein, West Germany. When Soviet-led Warsaw Pact forces invaded Czechoslovakia in 1968, seeking to end reforms in Prague, satellite reconnaissance imagery captured the entire invasion on film. Moorman was the resident expert on Corona imagery and briefed the results to dozens of senior officials. He believed those images unequivocally demonstrated the need for near–real time electro-optical satellite reconnaissance systems, which the country later pursued.[14] In 1970 Moorman transitioned to the NRO processing facility at Westover AFB in Massachusetts, where he enjoyed learning the technical aspects of film processing for the Hexagon and Gambit satellite reconnaissance systems.[15] The work captured Moorman's imagination, and he committed to a career in the Air Force.[16] The decision made, Moorman knew the Air Force expected him to have a graduate degree, so he took night classes and completed a master's in business administration degree at Western New

England College. The assignment at Westover was the last time Moorman would have hands-on experience.

In 1973 Moorman became a major and was selected to attend the Air Command and Staff College at Maxwell AFB in Montgomery, Alabama. Instead of taking a respite, Moorman opted to attend graduate school at night at Auburn University. The decision effectively doubled his coursework so he could earn a master of political science degree. Despite the extra work, he still managed to achieve distinguished graduate honors from the command and staff college. Moorman recalled, "Some people relax well; I don't relax well.... at Maxwell, I only went home to sleep. I went to blue suit school during the day, Auburn for political science at night."[17]

Space Policy

After graduating, Moorman was assigned to the office of space systems, the secret NRO headquarters in the Pentagon. After a year as an executive officer for the director, Moorman became the NRO's deputy director of plans and programs. It was a dynamic time for policy. As systems matured, U.S. officials recognized the growing importance of space to the American way of war. The Soviet Union also realized the shifting dynamic and looked to challenge the United States on another front in the Cold War. The conditions left U.S. policymakers concerned about space system resilience. It was a formative time for space policy, and it was a formative time for Moorman. Few Pentagon officials were interested in space policy, so the NRO staff was in the unusual position of representing national security space equities for the entire Department of Defense (DoD).[18] Thus, Moorman found himself as the Pentagon's working-level representative to the interagency group that drafted national space policies in the Gerald Ford and Jimmy Carter administrations. The policies were thoughtful but reflected tension in space cooperation with the Soviet Union.

That tension manifested most clearly in the nation's policies regarding antisatellite (ASAT) weapons. President Ford attempted to demonstrate restraint to the Soviets while continuing U.S. ASAT research.[19] In February 1976 the Soviet Union resumed testing its co-orbital ASAT program, executing four tests in short succession, belying the administration's attempt to hedge. Thus, in the final days of his administration, Ford ordered the Pentagon to develop an operational ASAT capability.[20] President Carter was similarly conflicted but upon taking office was hopeful he could negotiate an ASAT arms control agreement.[21] In 1978 Carter issued Presidential Directive (PD) 37, which Moorman called "one of the best-written documents on space policy ever written."[22] Much of PD-37 represented Moorman's perspective on key space issues, but it demonstrated a muddled approach toward ASATs. PD-37 directed DoD to "vigorously pursue development of an antisatellite capability" while also seeking a verifiable ban on them.[23] The attempted ASAT ban stalled under pressures of the larger bilateral relationship and the inherent contradiction of such an approach.[24] In most respects, however, the policy coherently stated U.S. interests in space. For one, it declassified that the United States would use space to support conventional military forces, signaling a larger role for space systems than previously imagined. The policy also acknowledged that space systems were vulnerable and directed the U.S. government to make them more survivable, commensurate with their need in time of conflict. Moorman could not believe his good fortune. He shaped policy as a relatively junior officer, and the experience shaped him in return.

Despite how much he enjoyed the policy work, Moorman felt separated from the Air Force. He had spent most of his career behind the "green door," often not even wearing a uniform.[25] Moorman's fears were confirmed when, during a chance meeting in the halls of the Pentagon, a friend of his father, Maj. Gen. Jerome O'Malley, lectured the younger Moorman on the need to get back to the "mainstream" Air Force.[26] Perhaps in an effort to bring Moorman back into the fold, O'Malley saw that Moorman was appointed to the air staff's Space Mission Organization Planning Study.[27] The study was chartered largely in response to PD-37 to examine space mission management, organization, and resourcing.[28]

Ostensibly, Moorman was assigned to the study to represent the NRO.[29] His work on PD-37, however, meant that he had a broader perspective than the other staff officers despite having little personal in-depth exposure to the Air Force's "white" space systems. According to one of the other action officers, then-Maj. Robert S. Dickman, "He was probably the quickest study I'd ever known, within a couple weeks, he was not only conversant with all the Air Force and other service nuances ... he was the principal drafter of the SMOPS [Space Mission Organization Planning Study] report!"[30] The study concluded, among other things, that the Air Force should centralize space operations within the service and seek designation as the DoD executive agent for space. The bureaucracy did not embrace the findings as Moorman would have liked, but it would later prove to have been, as he called it, an "extraordinarily important socialization process."[31]

Moorman had learned that sound policymaking required not only an understanding of the issues and positions but also a willingness to compromise. For the first time in his career, Moorman began to crystallize his own thinking on space policy issues. That thinking received attention from Dr. Hans Mark, director of the NRO and undersecretary of the Air Force. Moorman recalled that he would often argue with Dr. Mark over policy issues, "as much as a Lieutenant Colonel can argue with an Under Secretary."[32] But he was Dr. Mark's "policy guy," and Moorman said, "For some reason, he took a liking to me."[33]

After a year at the National War College, in summer 1980 Moorman became the deputy military assistant for Mark, now secretary of the Air Force, putting Moorman's career, once at risk of jumping the tracks, decidedly on the fast track.[34] His duties required long hours and frequent "redeye coach flights between Washington and Los Angeles," reminiscent of Bernard Schriever's hardships.[35] The cost was worth it. In Moorman's view, Mark propelled Air Force developments in space through his "strong convictions" guided by a "sense of history."[36] Historian David Spires credits Mark with leading the "charge for an operational space commitment ... in the face of reluctant or overly cautious senior leaders."[37] Moorman was grateful to continue his tutelage under Mark, and though he did not adopt Mark's positions wholesale, he was certainly aligned with the secretary's campaign to break through the Air Force's ambivalence about space.[38]

Mark's most pressing issue was space launch, which unfortunately was an area where his strong convictions led him astray. Mark gambled that the space shuttle would provide cheap, reliable space access. According to Spires, Mark sought to make the shuttle "cancellation-proof."[39] Mark had convinced the Carter administration to declare in PD-37 that the shuttle would "service all authorized space users."[40] And in an effort to convince Congress the shuttle would bring cost savings, Mark sought to close other launch vehicle production lines.[41] Mark's single-minded focus on the shuttle eventually proved disastrous for the Air Force, and later in Moorman's career, he saw it undone.

In February 1981 Verne Orr replaced Mark. Moorman believed that Mark convinced Orr of his "vision for space."[42] One thing Mark imparted was his conviction that Colorado Springs should be the future of Air Force space operations.[43] Mark was not alone in his assessment. Colorado Rep. Ken Kramer and Sen. Gary Hart were also staunch advocates.[44] Among Air Force officers, too, there was a burgeoning movement to consolidate space operations. If there were two hubs of activity for reorganizing and consolidating space operations, one was in Washington, DC, and the other in Colorado Springs, in part because that is where Colonel Moorman was headed.

Space Operations

After nearly twenty years in uniform, Moorman had his first Air Force space operations assignment, becoming the director of space operations for North American Aerospace Defense Command (NORAD) inside Cheyenne Mountain in 1981. As events unfolded, his operational time was short-lived. After years of Air Force studies, space organizational initiatives began to congeal. Moorman's knowledge of the plans and politics of the effort meant he was far more valuable as a staff officer than as an operations officer.

In his first meeting with the NORAD commander, Lt. Gen. James Hartinger, Moorman raised the possibility of a "space organizational initiative," to which Hartinger said, "Why the hell would you say that?"[45] Moorman said it was like "hitting General Hartinger with a hammer."[46] Hartinger was apparently unaware of the efforts or was being coy, but Moorman sparked his interest, and a conversation planned for thirty minutes lasted two hours.[47] Hartinger had good reason to be interested. NORAD had been downgraded to a three-star command, and Hartinger had been a three-star general for five years. He was on a mission to get the command's fourth star back, and bringing more space missions under his authority would help his cause.[48]

Hartinger knew that he did not have many space experts on his staff. Most were like Hartinger himself: pilots in their first space assignments. As a result, he activated a new plans organization, with Moorman as the deputy to Col. Wes Clark, who had led the Space Mission Organization Planning Study in the Pentagon. It was a stroke of good fortune to have people so intimately involved in the air staff's space efforts on Hartinger's staff during a decisive period.

In February 1982 the Air Force's top leaders gathered and discussed the future of space in the service. Chief of Staff Gen. Lew Allen directed Gen. Robert Marsh of Systems Command to work with Hartinger to formulate space reorganization proposals.[49] The task fell to the space division of NORAD, whose members understood the politics and bureaucratic positions as well as anyone in the Air Force.[50] Within days of receiving the task, the space division had different courses of action defined.[51] Moorman evaluated and diagrammed all of the personalities and their respective views on the issue and worked every weekend for months to prepare Hartinger.[52] With Moorman's help, Hartinger crafted an approach to engage two key leaders—O'Malley, whom Moorman called "the most powerful three-star in the Air Force," and Air Force Undersecretary Edward "Pete" Aldridge.[53] O'Malley, deputy chief of staff for operations, plans, and readiness, favored moving space systems out of research and development (R&D) and into the hands of operators.[54] The R&D community, which stood to lose in that proposal, objected.

In mid-April 1982 at General Allen's request, ADCOM and Systems Command presented the joint pitch for a single space organization to Allen and his staff. With support from O'Malley, Hartinger shared the space division's proposal for a space command. It resonated with Allen, who concluded, "That's the way I think we ought to do it." On June 21, 1982, just days before Allen's retirement, he and Aldridge announced that an Air Force Space Command (AFSPC) would be formed effective September 1, 1982.[55] Hartinger got his fourth star, and Moorman's star continued to rise. Like the dog that caught the car, Hartinger had to decide what his new command would do. Hartinger asked Moorman to head his new commander's action group, saying flatly, "This space stuff isn't my strong suit."[56] Determining the command's vision and priorities was Moorman's first order of business, which he developed into what Hartinger would call the "perfect pitch." According to Moorman, all of Hartinger's subsequent initiatives stemmed from that work.[57] For the next two years, Moorman accompanied Hartinger everywhere he went, building and evangelizing the institution.

Moorman's reward was to become the vice commander of the 1st Space Wing at Peterson AFB, Colorado. Moorman believed that he would soon become the one-star wing commander but learned the position was to be downgraded to a colonel. So when NRO asked Moorman to come back as the director of its staff, he agreed. Thus, Moorman left Colorado Springs, still having never held a command position but on his way to brigadier general nonetheless.

In April 1985 Moorman pinned on his first star and was about to face a host of challenges in the launch business. Just months after Moorman's arrival, in January 1986, the space shuttle *Challenger* broke apart shortly after takeoff.

Already over budget, the disaster further contradicted Mark's earlier promises that the shuttle would provide cheap, reliable access to space. To make matters worse, in April a Titan 34D carrying an NRO satellite exploded shortly after liftoff.[58] The following month, the National Aeronautics and Space Administration (NASA) also lost a rocket.[59] The space program was grounded, and as Spires wrote, "The nation confronted an ailing space industry and a space program in disarray." After nearly three decades of budgetary independence, NRO had to adjudicate competing priorities. Moorman's boss Aldridge described a situation in which the programs had more technology than money, leading to "a period of unhealthy competition."[60] By Moorman's own accounting, he did not play a "great leadership role" during this time but rather orchestrated the staff and presented options to Aldridge.[61]

In truth, Moorman navigated the NRO staff through turbulent waters. One staff member recalled that the environment that Moorman created "was among the very best in my career. . . . it was purposeful; it was mission focused; it was an environment where our leadership played to each individual's strengths." Moorman, she said, appreciated "candor, openness, integrity, and loyalty. . . . the culture thrived and I've never seen an organization that was better led or more fun to be a part of."[62]

Space Acquisitions

Meanwhile, the Air Force's ambivalence toward space continued. By the mid-1980s the space shuttle and Titan launch failures led to higher costs at a time when DoD believed it should have been fielding more space technology to keep pace with Soviet developments. The Air Force questioned whether it was living up to its aspiration since 1983 to exercise "full and exclusive responsibility for all space launch, on-orbit control, space defense . . . and space acquisition activities."[63] The activation of the strategic defense initiative office in 1984 and the subunified U.S. Space Command (USSPACECOM) in 1985 created space equities and expertise outside the Air Force and challenged Air Force primacy in the space mission.[64] Moreover, the Army and Navy did not want to fund space programs under the Air Force's control and so engaged in what Air Force officers called "space activism."[65]

This was the situation Moorman faced in fall 1987 when he took his first space acquisitions assignment as director of space and space defense initiative programs. Air Force Chief of Staff Gen. Larry D. Welch, who recognized the disconnect between the service's words and actions, asked Moorman to bring him up to speed on space issues. Moorman described what he believed to be "systematically wrong in space—everything from career management to requirements, to education and training, to integration."[66] Welch gave Moorman a platform to explain his concerns to the top Air Force brass when Moorman led a daylong space-focused symposium. At the end of Moorman's remarks, Welch declared, "The Air Force's future is inextricably tied to space."[67] In recognition that the service was not living up to that vision, leaders directed a blue ribbon space panel to review the issues systematically. Though he might not have known it at the time, when Moorman shared his vision, he was building the foundation for his most important work as a general officer.

In 1988 the blue ribbon panel concluded that the Air Force should remain the principal agent for military space activities, but not to the exclusion of other services.[68] It was a practical move that acknowledged the Air Force did not have the money to oversee all space systems, and it was a savvy bureaucratic move that gained advocacy from the other military departments, which would have contested a land grab from the Air Force. The panel also advocated for AFSPC to have a stronger position relative to USSPACECOM. Perhaps most importantly, the panel recognized a lack of space expertise, commitment, and appreciation throughout the Air Force.

To reflect the findings of the study, Secretary Aldridge and General Welch issued an "Air Force Space Policy," declaring, "Spacepower will be as decisive in future combat as airpower is today."[69] The policy charged the Air Force to rewrite doctrine to include space operations, to change personnel management to seed space experts throughout the Air Force and bring "other operational expertise" to Air Force Space Command, to establish air components as centers of space expertise in respective unified commands, and to consolidate space "requirements, advocacy, and operations" in AFSPC.[70] Moorman's advocacy yielded important intellectual groundwork for the maturing space organization within the Air Force in general and for Air Force Space Command in particular. His executive officer at the time, future three-star general Michael A. Hamel, recalled that Moorman had established himself as "the most knowledgeable and trusted senior space expert and advisor in the Pentagon."[71] So it was little surprise when Moorman became the heir apparent to AFSPC.

In early 1990, armed with the new policy, Moorman began building consensus for his agenda at AFSPC. His most urgent goal was to transfer launch from the R&D community to AFSPC.[72] For years, the incumbent AFSPC commander, Gen. Donald J. Kutyna, had been lobbying General Welch to do just that.[73] Where Kutyna proposed a wholesale transfer, Air Force Systems Command (AFSC) favored an incremental approach to the transfer, and Welch was reluctant to pick a side.[74] Moorman was deeply involved in Kutyna's plans and understood AFSC's reservations. He returned to skills that had served him so well as a staff officer. He understood the issue deeply and knew the centers of gravity were not only in U.S. Space Command and AFSC but also in Headquarters Air Force. He found the crucial decision-maker, built a relationship with him, and ultimately persuaded him. In this case, that person was Lt. Gen. Ronald W. Yates, who was on his way to becoming commander of AFSC.[75] Moorman's engagement with Yates paid off when Welch agreed to transfer launch to the operational command.[76] With the AFSC opposition tamped down, the hard part was done by the time Moorman took over AFSPC.

In March 1990, one month before his promotion to lieutenant general, Moorman began his command of Air Force Space Command. In the eight years since Moorman had helped create the organization, it had gained missions and was gradually becoming the center of excellence for space expertise Moorman had imagined. Looking outward, Moorman set out to "increase the status and influence" of AFSPC, better integrating it into the Air Force and DoD.[77] Looking inward, Moorman wanted to inspire space operators toward a "cultural transformation."[78] He believed centralizing space operations gave the command the opportunity to build a cadre of people who thought about space. At the end of the Cold War, Moorman believed space's most important contribution would be "more effective employment of military forces."[79] To that end, Moorman sought to operationalize space launch because "it was so integral to the process of getting space to the warfighters."[80] When AFSPC finally received responsibility for the space launch mission in October 1990, Moorman called it a "landmark event. . . . I believe this transfer is part of the natural evolution of the Air Force space program. It is a testimony to how our thinking about space operations has matured. . . . The decision to transfer the launch mission was based on the beliefs that placing satellites into orbit has matured to a point where it should be considered an operational task, and that Air Force Space

Command had sufficiently matured where it could assume the responsibility."[81] The launch transfer was overshadowed, however, by another landmark event, this time on the world stage.

First Space War

On August 2, 1990, Iraq invaded its neighbor Kuwait. During the five months of Operation Desert Shield, the U.S.-led coalition established a foothold in the Middle East with an impressive logistics tail critical to any effort to eject Iraq from Kuwait. The timing meant that Moorman was not able to achieve his vision for the command before the conflict.

Nevertheless, Moorman did everything in his power to ensure the space infrastructure in the theater was prepared. As Moorman recalled, the Air Force and unified command deployed "a robust mix of user sets, mobile terminals, and portable receivers for receiving and disseminating space-based surveillance, weather, communications, and navigational data."[82] It was no small feat. In one example, during the early days of the buildup, there were only a few hundred global positioning system (GPS) receivers in theater. Joint and combined forces quickly realized such receivers would be invaluable in navigating the featureless Arabian Desert. With Moorman's help, by war's end, there were more than forty-five hundred GPS receivers in theater.[83] In another example, as demand for secure, long-haul communications grew, Moorman directed the movement of a Defense Satellite Communications System satellite over theater. It was the first time a DoD satellite had been moved to support combat operations, demonstrating what Moorman saw as the "inherent flexibility of our sophisticated space systems."[84]

The Gulf War also revealed challenges for the young command. Moorman believed that, ideally, operations should drive headquarters and acquisitions activity through what he called "ops pull."[85] This simply was not possible in the buildup for Desert Storm because of the paucity of space experience in theater. Thanks to Kutyna and Moorman, by the end of Desert Storm, the unified command had a space cell at U.S. Central Command, and Air Force Space Command had a cell at Central Command Air Forces.[86] In one example of operational pull, space weather proved its utility in new ways in the Gulf War, but the Air Force had no plans to get weather in the hands of fliers on tactical timelines. It was a crucial oversight because the precision-guided munitions of the Gulf War were laser-guided (GPS would not be integrated into precision-guided munitions until years later), and laser designators worked best in clear weather. But necessity is the mother of invention, and late in the war, AFSPC introduced prototype backpack receivers to make weather data more accessible to forward-deployed air and maneuver units.[87]

Another example of operational pull was a reimagining of the Defense Support Program, a Cold War system designed to detect Soviet intercontinental ballistic missiles. From the beginning of the Gulf War, AFSPC planners were aware of the threat posed by Iraq's Scud missiles and optimized the Defense Support Program satellite system to help protect coalition forces.[88] The preparations during Desert Storm to make the program more tactically responsive resulted in alerts allowing for forces to don their chemical protection suits and for Patriot batteries to engage missiles.[89]

In the decisive victory of the Gulf War, space played a crucial role. As important as Moorman's efforts were to employing space in a theater conflict, he was dissatisfied in the end. He graded space's contributions to the conflict as a "C+," believing that space still had a long way to go to integrate into the joint fight.[90] Given that it had not really studied for the test, AFSPC performed admirably. For his efforts, Moorman was awarded the 1991 National Geographic Society's Gen. Thomas D. White U.S. Air Force Space Trophy as the "individual who has made the most outstanding contribution to the nation's progress in space."[91] Moorman was happy with what the command had learned during the conflict but was dismayed that they had to learn it all. Space operations had neither doctrine nor theory, the 1988 blue ribbon panel had identified. Moorman put a fine point on the situation when he said the Air Force had never "exercised space in any comprehensive way. There was no operations plan or annex. . . . we literally were catching up in all areas."[92] Space proved it would be an essential element to future warfare, and by war's end, there was more operational pull than AFSPC and the larger Air Force could handle.

Blue Ribbon Panel

Moorman had a clear vision for the role space would play in future warfighting and believed AFSPC should lead those efforts, so it came as a great disappointment when he learned he was to move from commander of AFSPC down to vice commander. It was an unusual move that might signal to some that Moorman was found wanting. It was not personal; Moorman was a casualty in the chief of staff's campaign against organizational inefficiency in the Air Force.

When Chief of Staff Gen. Merrill McPeak raised with Moorman the possibility of consolidating U.S. Space Command, AFSC, and NORAD under a single four-star general, Moorman was not surprised.[93] The Air Force had already made the decision to merge tactical and strategic forces into Air Combat Command and dual-hat Air Mobility Command with U.S. Transportation Command. Moorman's immediate response to McPeak, later supported by exhaustive data, was there would not be significant cost savings in consolidating space commands.[94] Moorman believed such a move would "emasculate" the Air Force's role in space, since the unified commander, even if an Air Force officer, would have to be "above the fray" and represent joint interests.[95] Moreover, Moorman himself was not eligible to lead the unified command because the position was rated only for flying officers, a requirement he believed was "bogus."[96] Moorman pleaded his case, but in March 1992, McPeak consolidated the commands under fighter pilot Kutyna, and Moorman became AFSPC vice commander. Moorman was disappointed but believed he could still contribute to the development of spacepower.

That opportunity came sooner than he might have imagined when, later that year, McPeak led the effort to include space in the Air Force's vision and mission statements. For the first time in its history, the Air Force recognized space as a coequal mission to air. It was an important step, but McPeak recognized that it would not be enough without a commensurate effort on practical matters. He therefore directed Moorman to run a post–Gulf War space study, giving him a chance to prophesy about spacepower. In fall 1992 Moorman convened the "Blue Ribbon Panel of the Air Force in Space in the 21st Century" at Maxwell AFB, Alabama. He was keen to capture the Gulf War's rich practical lessons on space integration and approached the panel with thoughtfulness and urgency. The way Moorman ran the panel provides a miniature case study on his leadership style.

For starters, to build credibility for the panel's findings, Moorman insisted on having more than just "space gonks" participate. He gathered thirty representatives from all the major commands, including civilians and officers from various career fields.[97] To account for the group's diverse perspectives, Moorman taught seven hours of space academics every day on top of his duties leading the study.[98] Through academics, Moorman not only educated those less familiar with space but also shaped their thinking. One of the panelists was future AFSPC commander William Shelton, who recalled that Moorman was "clearly the space expert" and concluded, "We all wanted to be like General Moorman."[99]

Moorman held the group to a high standard and worked them hard over the two-month study. About Moorman's approach to accountability, John Hyten recalled that Moorman could "say something with humor and grace that would make you know you screwed up and know that you needed to do better."[100] Moorman demanded panelists work long hours, six days a week. Even Sundays were not truly days of rest; they were for team building. Moorman hosted sporting events where the "pencil-neck geeks," as McPeak called them, could take on the fighter pilots.[101] Moorman was tenacious on the basketball court and expected the same from his fellow pencil-neck geeks. Hyten recalled just how seriously Moorman took those games, saying that he was "the most competitive person I'd ever met."[102] In any event, Moorman's investments paid dividends; he could honestly say the final report, which represented his vision, also represented the consensus of the panel.

The blue ribbon panel was most effective in recommending things over which AFSPC, and by extension Moorman, already had control: space integration, education, and training. The report noted a lack of "in-depth understanding and detailed knowledge of how to employ space capabilities in the prosecution of military operations."[103] As one example of a remedy, the panel advised the creation of a space warfare center, which Moorman ensured was underway before the final report was even issued.[104] If the command wanted to create tactically focused and cognizant space professionals, the warfare center was an important development in its own right, but it also paved the way for a space tactics school and, eventually, a space division at the Air Force weapons school.[105]

When the panel's recommendations exceeded the control of the Air Force, the recommendations faced intense criticism. For instance, the panel recommended the Air Force be the "single manager" of space launch and operations and that "other Service space operations commands should be eliminated."[106] The Army and Navy saw this as the latest iteration of an Air Force power grab. Moreover, the House Appropriations Committee questioned why the Air Force already controlled 90 percent of the military space budget when it generated a far smaller proportion of the requirements and further questioned whether the Air Force could meet the needs of the other services.[107] Despite efforts from General McPeak and Air Force Secretary Sheila Widnall to assuage the concerns, the Air Force could not overcome the bureaucratic inertia, and the recommendation was not enacted.

In its assessment of the future security environment, the final report concluded, "Today's operations are significantly enhanced by U.S. Space superiority—tomorrow's will be nearly impossible without it. Success in future battles will require us to establish information dominance: providing essential information to friendly forces, denying it to the enemy and exploiting it to nullify or destroy the enemy's ability to control his forces."[108] After Desert Storm, few could deny that space would be important to any future conflict. The idea, however, that the United States needed to establish space superiority had a complicated legacy. The thinking had a clear lineage to air superiority and certainly matched the view of McPeak, who had complained publicly that "our ability to prevent hostile use of space is virtually nonexistent."[109] Since 1978, national policy had supported development of ASATs, but largely due to congressional opposition, the United States did not have an operational capability.[110] This time, lawmakers and policymakers alike believed pursuing ASATs in support of space superiority to be premature absent any adversary threats, and the effort stalled.

Why did Moorman, who demonstrated a keen eye for bureaucratic dynamics and politics, sponsor a document whose core conclusions were unlikely to be adopted? Moorman knew that his boss, General Kutyna, had to represent joint interests and could not (or would not in any case) be so strident to recommend cutting the other services out of the space business. Yet Moorman was aligned with McPeak's desire to establish the Air Force as the lead service in space and to normalize the idea of space superiority. Regardless of the prospects for implementation, Moorman seized the opportunity to describe his vision for military space operations, and time would prove him a herald of spacepower.

Sensing the tectonic shifts in the nation's approach to space, Moorman had an eye to the past, too, so he asked Air Force historian Dr. Richard Hallion to "add a history of the Air Force in space to his program's book writing plans."[111] The effort resulted in David Spires' book, *Beyond Horizons: A Half Century of*

Air Force Space Leadership, essential reading for any space historian, military or otherwise.

Interagency Leadership

By 1994 Moorman was the foremost space expert in the general officer corps, and he had a reputation for consensus-building and solving tough problems. Recognizing this, Defense Secretary Les Aspin directed him to resolve one of the most vexing issues still facing the national space program: launch.[112] Moorman's space launch modernization study was a daunting task. Past attempts at rocket replacement programs had been fraught with difficulty because the Air Force, NRO, and NASA had different requirements and strongly held views. Moorman was at his best in such circumstances: "I find ... understanding the needs of your customers and their perspectives is absolutely essential. If you have understanding and consensus, lots of things are possible."[113] Moorman's personal relationships helped, too. He was friends with NASA administrator Dan Golden and considered NRO deputy Jimmy Hill one of his "closest friends."[114] Those relationships, combined with Moorman's long tenure in NRO and senior positions within the Air Force, meant he was widely trusted by both national security and civil space organizations. His leadership of the small cadre of officers supporting the study was equally important. One officer, Victor Hillard, recalled Moorman's leadership style, saying he had an "immense capacity to stay intensely and tirelessly focused," to complete a study of "vital implications for the future of national security space."[115] Moreover, Hillard recalled Moorman's "unique ability to clearly see the path to successful coordination through myriad organizational sensitivities."[116]

What Moorman clearly saw in this instance was NASA's need for heavy launch vehicles and DoD's interest in a family of vehicles to account for a variety of missions. He managed to satisfy both organizations. His final report proposed that "NASA oversee reusable launch and the military oversee expendable launch."[117] Where the blue ribbon panel reflected an Air Force position, the modernization study reflected interagency compromise; as a result, the latter was codified in President Bill Clinton's national space transportation policy. Spires wrote that with Moorman's proposal accepted, "the service now had a clear path to what promised to be a responsive, reliable, and affordable family of [evolved expendable launch vehicles] in the twenty-first century."[118]

Chapter 7

In July 1994 Moorman became the twenty-sixth vice chief of staff of the Air Force and pinned on the rank of general a few months later. At his promotion ceremony, General McPeak, never one to mince words, commented on how unusual it was to have a nonrated officer become vice chief. Martin Faga, former assistant secretary of the Air Force, recalled the chief's words on the occasion: "We have not done this since 1967, so we don't do this often, nor should we."[119] Yet McPeak believed it was "very easy" to make Moorman the vice chief because of how important space was to the nation and how capable Moorman was.[120] Moorman's time as vice chief of staff was marked by futile efforts to consolidate management of space under the Air Force. The time was also marked by a fissure between the military and civilian leadership of the department, ultimately resulting in Gen. Ronald Fogleman's early resignation from his post as chief of staff of the Air Force.

When Moorman arrived as vice chief, McPeak's time as chief of staff was numbered in months. Moorman's previous argument to centralize the Air Force's control over space appealed to McPeak's sensibilities about efficiency. In September 1994 McPeak had an opening to raise the issue in the congressionally mandated "Commission on Roles and Missions of the Armed Forces." He channeled Moorman's blue ribbon panel when he recommended the Air Force be appointed the DoD space acquisition executive because there was "no need for multi-Service involvement in space."[121] McPeak later explained that each service building expensive, unique space systems in the post–Gulf War era of austerity was "no way to run a business."[122] The commission ultimately agreed, and its final report recommended that DoD "assign the Air Force primary responsibility for acquiring and operating multiuser space systems."[123] Despite the commission's finding, and despite the fact that McPeak, Moorman, and Widnall were aligned on what ailed national security space, DoD did not make the change due to ever-present opposition from Navy and Army leaders.[124]

In fall 1994 Moorman had a new boss when General Fogleman became chief of staff. Fogleman was expected to maintain the status quo and let the Air Force settle from McPeak's whirlwind of changes.[125] But he had different ideas, believing he had a charge to "restore the soul of the Air Force."[126] As a result, Fogleman felt there was an imperative to make principled decisions, and he stood by those decisions at the cost of his career. High-profile disciplinary issues with Lt. Kelly Flinn and Brig. Gen. Terry Schwallier dominated the chief's headlines, but he also had substantive disagreements with superiors on matters of force structure.[127] Fogleman's struggles were a challenge for Moorman, too. For one, he shared the burden with his boss, but on a personal level, Moorman took pride in his ability to get along with others. And yet Moorman never wavered on matters of principle, so he likely sympathized with Fogleman for doing what he felt was right.

In August 1997 Moorman retired from the Air Force after thirty-five years in uniform, having left an indelible mark on the nation's space capabilities and organization. Days before Moorman's retirement, his friend Steny Hoyer recounted some of Moorman's numerous awards and honors on the floor of the U.S. House of Representatives, concluding that "General Moorman's greatest contribution has been his leadership related to the space programs.... He has played a pivotal role in establishing national and Defense Department space policy and developing improved space capabilities."[128]

Whether or not it was intended as such, Moorman's appointment as vice chief was a consolation for him after he was denied the chance to be the four-star commander in charge of space. Nevertheless, in making the rank of general, Moorman surpassed any expectations he had as a junior officer, or even two years earlier when he was made the AFSPC vice commander. He also surpassed his father, whom he had always referred to as "the real General Moorman."[129] It certainly would have seemed impossible to a younger Moorman who thought so little of his prospects in the Air Force that he planned to leave the service and go to law school. Moorman was the first and only career space officer to ever become vice chief, which might have signaled a growing importance of space in the Air Force, but it was eight years before another career space officer reached the rank of four-star general.

In retirement, Moorman turned his attention back to space. In one sense, he followed the normal trajectory of a retired general officer. He was a senior executive advisor and partner with the consulting firm Booz Allen Hamilton, in charge of the firm's Air Force and NASA accounts. He also served as board chairman of the Space Foundation and as a member of the National Security Space Association board of advisors. In another sense, Moorman plotted his own trajectory and stayed involved in military space issues, notably as one of the most influential members of the Commission on National Security Space Management and Organization. Chaired by Donald H. Rumsfeld, the commission included former members of Congress, former officials from the earlier George H. W. Bush and Bill Clinton administrations, a former NASA deputy administrator, and Army, Air Force, and Navy flag officers. Spires argued that despite that the commission's august membership, it was Moorman's vision that came through most clearly in its findings.[130] Moorman had been raising the core issues over the previous two decades. The findings zeroed in on the Air Force because it had the preponderance of space personnel, budget, assets, and infrastructure.[131] It was disconcerting, but not surprising, that the commission found that the Air Force was not a good steward of space missions.[132] The commissioners believed that "despite official doctrine that calls for the integration of space and air capabilities, the Air Force does not treat the two equally. As with air operations, the Air Force must take steps to create a culture within the service dedicated to developing new space system concepts, doctrine, and operational capabilities."[133]

The commission's report followed Moorman's thinking in important ways. First and most importantly, it recommended building a more professional space cadre and a "stronger military space culture, through focused career development, education and training, within which the space leaders for the future can be developed."[134] It was Secretary Rumsfeld's view that the ideal solution to resolve many of the cultural challenges might have been creation of a "space service" and that there was a "general consensus within the Commission" to do just that. But given the bureaucratic and political difficulties of such a change, the final report stopped short of recommending a separate service but left open the possibility for a "Space Corps" and, in the longer term, a military department for space.[135] Second, it endorsed Moorman's argument for the mission of space superiority when it recognized a need to "develop and deploy the means to deter and defend against hostile acts directed at U.S. space assets and against the uses of space hostile to U.S. interests."[136] Third, Moorman's influence was clear in the recommendation to separate the command of USSPACECOM and NORAD from command of AFSPC. Though it was over a decade too late for Moorman, he must have been pleased when nonflier Gen. Lance Lord became AFSPC commander in April 2002 and the Air Force finally had a senior space professional in charge of the mission.

Ten years after the space commission, Gen. William Shelton, commander of AFSPC, recounted its impact on national security space: the president had established space as a national security priority, the Air Force was the executive agent for space, the AFSPC commander was a distinct four-star command with oversight of space acquisitions, and AFSPC had created the National Security Space Institute.[137] The following year, Shelton dedicated the Moorman Space Education and Training Center. On the occasion, he said, "We commit this center to the development of deep space expertise and capability within our Air Force and we name it after someone who has dedicated his adult life to that same objective. May all who pass through the doors of the Moorman Space Education and Training Center be inspired by the strong character and renowned expertise of the center's namesake, General Thomas S. Moorman Jr."[138]

Shortly before Moorman's passing, he witnessed two events that represented so much of what he had advocated for. In August 2019 the Pentagon reestablished USSPACECOM as the joint warfighting command for space. The event signaled broad political recognition of a contested space domain. And in December 2019, the U.S. Space Force became the nation's first new military branch in seventy-two years.

On June 18, 2020, Thomas Moorman Jr. passed away in Bethesda, Maryland, one mile from where he was born, at Walter Reed Army Medical Center.

Conclusions

In April 2021 during a memorial for Moorman, Vice Chief of Space Operations Gen. D. T. Thompson said, "Just as [Brig. Gen. Billy] Mitchell and [Gen. Henry] Arnold are written into the history of the United States Air Force . . . [Gen. Bernard] Schriever and Tom Moorman will be written into the history of the United States Space Force."[139] Thompson was not alone in his praise. Vice Chairman of the Joint Chiefs of Staff Gen. John E. Hyten declared that Moorman deserved the title "father of the Space Force."[140]

Moorman himself might have rejected the claim, as he once rejected the need for an independent space service.[141] Moorman almost certainly would have acknowledged that his efforts to establish a bigger place for space in the Air Force, and to raise awareness for national security space more broadly, set the conditions for an independent service. Moorman played a crucial role in the founding of Air Force Space Command, and he led that command through the Gulf War. Despite having very little Air Force space operations experience, he was unquestionably the most important uniformed thought leader on space for more than two decades. In retirement, Moorman was one of the most important commissioners on the influential space commission. Thus, if literary license permits, Tom Moorman was a father of the U.S. Space Force.

Given his enormous influence on the development of military space issues, it is worth considering what made Moorman successful. As a leader, Moorman was genial and warm. He did not shy away from holding others accountable, but he did so with charm and good humor. Moorman was intellectual without pretense, and his lack of formal technical education was more than offset by his curiosity and aptitude. Moorman had almost no practical space operations experience by today's standards; he never operated a satellite, for instance. And yet, whether as an imagery analyst or staff officer, Moorman's mastery of new topics and sheer industry garnered notice from peers, subordinates, and superiors.

Moorman achieved policy victories not by virtue of his position but through his power of persuasion. Whenever he could, he built consensus toward his ideas. But he never backed down from his convictions, fighting for the independence of Air Force space forces and, eventually, an independent space service. He was an enthusiastic historian, writing contemporary history documents and empowering the Air Force to write its own space history.[142] Above all of that, General Moorman was a leader of leaders and a generous mentor, influencing many of the other stars in space. Hyten argued that neither he nor General Raymond, General Thompson, or General Shelton would be where they were without Moorman's influence. Hyten emphasized the point, saying, "I would not be the Vice Chairman of the Joint Chiefs of Staff without General Moorman."[143]

Moorman's most enduring legacy might be his role as a thought leader and advocate for military space operations. He led or participated in the most significant studies concerning the Air Force's role in space operations, planting seeds for trees that continued to bear fruit decades after his retirement. In Moorman's twilight years, after quietly fighting cancer, and a career of not so quietly fighting for a greater place for space in the Air Force, he became an important advocate for an independent Space Force.[144] He could rest easy knowing he lived well, and national security space was better for his efforts.

Notes

1. Moorman's father did not adopt the convention of referring to himself as "Sr.," but I have done so for clarity.
2. Steny Hoyer, Gen. Thomas S. Moorman memorial service, April 14, 2021, National Security Space Association, https://www.youtube.com/watch?v=00kzF0fqDZY.
3. Thomas S. Moorman Jr., oral history interview by R. Cargill Hall, June 4, 1997, transcript of interview at Center for the Study of National Reconnaissance, Chantilly, VA, 6.
4. Moorman interview with Hall, 6.
5. Moorman interview with Hall, 6–7.
6. Moorman interview with Hall, 7.
7. Moorman interview with Hall, 9.
8. Eric Schlosser, *Command and Control: Nuclear Weapons, the Damascus Accident, and the Illusion of Safety* (New York: Penguin Books, 2014), 297.
9. Moorman interview with Hall, 7.
10. Moorman interview with Hall, 14.
11. Moorman interview with Hall, 14.
12. Moorman interview with Hall, 15.
13. Moorman interview with Hall, 17.
14. Moorman interview with Hall, 28.
15. Moorman interview with Hall, 33; J. Kevin McLaughlin, Moorman memorial service.
16. Moorman interview with Hall, 33.
17. Moorman interview with Hall, 34.
18. Moorman interview with Hall, 36.
19. Paul B. Stares, *The Militarization of Space: U.S. Policy, 1945–1984* (Ithaca: Cornell University Press, 1985), 172–73.
20. James Clay Moltz, *The Politics of Space Security: Strategic Restraint and the Pursuit of National Interests* (Stanford: Stanford Security Studies, 2008), 179–80.
21. Moltz, chap. 9.
22. Moorman interview with Hall, 38.
23. Jimmy Carter, Presidential Directive 37, "National Space Policy" (Washington, DC: The White House, May 11, 1978), https://www.jimmycarterlibrary.gov/assets/documents/directives/pd37.pdf.
24. Moltz, 181–86. The Carter administration viewed Soviet actions holistically and would not separate space arms control from issues like human rights policy.
25. Moorman interview with Hall, 36.
26. Moorman interview with Hall, 41–42.
27. Moorman interview with Hall, 42.
28. David N. Spires, *Beyond Horizons: A Half Century of Air Force Space Leadership* (Maxwell AFB, AL: Air University Press, 2011), 194; James E. Hill, "Letter from Commander in Chief, Headquarters, Aerospace Defense Command, to Gen. Lew Allen Jr., Chief of Staff, United States Air Force," February 9, 1979 in David N. Spires, *Orbital Futures:*

Selected Documents in Air Force Space History, 2 vols. (Peterson AFB, CO: Air Force Space Command, n.d.), 1:565.
29. Robert S. Dickman, Moorman memorial service.
30. Dickman.
31. Moorman interview with Hall, 42; Thomas S. Moorman Jr., interview with Robert M. Kipp and Thomas Fuller, transcript of oral history interview, July 27, 1988, United States Air Force Historical Research Center, call no. K239.0512-1839 C.1, 5.
32. Moorman interview with Hall, 56.
33. Moorman interview with Hall, 56.
34. Moorman interview with Hall, 57.
35. Dickman.
36. Moorman interview with Kipp and Fuller, 10.
37. Spires, *Beyond Horizons*, 194.
38. Moorman interview with Kipp and Fuller, 8.
39. David N. Spires, *Assured Access: A History of the United States Air Force Space Launch Enterprise, 1945–2020* (Maxwell AFB, AL: Air University Press, 2021), 187.
40. Carter, "National Space Policy," 7; Spires, *Assured Access*, 187.
41. Spires, *Assured Access*, 187.
42. Moorman interview with Kipp and Fuller, 9–10.
43. Moorman interview with Kipp and Fuller, 10.
44. Moorman interview with Kipp and Fuller, 12.
45. Moorman interview with Kipp and Fuller, 13.
46. Moorman interview with Kipp and Fuller, 13.
47. Moorman interview with Kipp and Fuller, 13.
48. James V. Hartinger, *General Jim Hartinger: From One Stripe to Four Stars*, ed. John Pasarro (Colorado Springs, CO: Phantom Press, 1996), 219–20.
49. As quoted in Spires, *Beyond Horizons*, 203.
50. Moorman interview with Kipp and Fuller, 15–19. The space division was dual-hatted under the Aerospace Defense Center and NORAD, not to be confused with the Space Division of Air Force Systems Command.
51. Moorman interview with Kipp and Fuller, 19.
52. Moorman interview with Kipp and Fuller, 20; Moorman interview with Hall, 68.
53. Moorman interview with Kipp and Fuller, 20.
54. Spires, *Beyond Horizons*, 199.
55. Spires, *Beyond Horizons*, 205.
56. Moorman interview with Hall, 79.
57. Moorman interview with Hall, 81.
58. Spires, *Assured Access*, 223.
59. Spires, *Assured Access*, 223.
60. Charles P. Datema et al., *4C-1000: The Untold Story of the NRO Headquarters Staff (1962–1990)* (Chantilly, VA: Center for the Study of National Reconnaissance, 2021), 67.

61. Moorman interview with Hall, 90–92.
62. Joanne Isham, Moorman memorial service.
63. "White Paper on Air Force Space Policy," June 16, 1987, included in Maj. Gen. Albert L. Logan, director of plans, DCS/P&O, Headquarters, USAF, "Air Force Space Policy Letter," August 28, 1987, in Spires, *Orbital Futures*, 1:79.
64. "White Paper," in Spires, *Orbital Futures*, 1:79–80.
65. "White Paper," in Spires, *Orbital Futures*, 1:80.
66. Moorman interview with Hall, 95.
67. Moorman interview with Hall, 98.
68. Spires et al., *Beyond Horizons*, 234–36.
69. Larry D. Welch, chief of staff, United States Air Force, and E. C. Aldridge Jr., secretary of the Air Force, "Air Force Space Policy," December 2, 1988, in Spires, *Orbital Futures*, 1:92.
70. Welch and Aldridge, 1:92–93.
71. Michael A. Hamel, Moorman memorial service.
72. Thomas S. Moorman Jr., interview with George W. Bradley III, unpublished transcript of oral history interview, part 1, October 7, 1994, 3; Spires, *Beyond Horizons*, 238.
73. Spires, *Assured Access*, 247.
74. Spires, *Assured Access*, 248.
75. Moorman interview with Bradley, part 1, October 7, 1994, 3.
76. Spires, *Assured Access*, 248.
77. Moorman interview with Bradley, part 1, 2.
78. Moorman interview with Bradley, part 1, 2.
79. Moorman interview with Bradley, part 1, 3.
80. Moorman interview with Bradley, part 1, 3.
81. As quoted in Spires, *Assured Access*, 249.
82. Thomas S. Moorman Jr., "Space: A New Strategic Frontier," in *The Future of Air Power in the Aftermath of the Gulf War*, ed. Richard H. Shultz Jr. and Robert L. Pfaltzgraff Jr. (Maxwell AFB, AL: Air University Press, 1992), 241.
83. Moorman, "Space: A New Strategic Frontier," 242.
84. Moorman, "Space: A New Strategic Frontier," 242.
85. Moorman interview with Bradley, part I, 12.
86. Moorman interview with Hall, 124.
87. Moorman, "Space: A New Strategic Frontier," 243.
88. Spires, *Beyond Horizons*, 254.
89. Spires, *Beyond Horizons*, 259.
90. Moorman interview with Bradley, part I, 10.
91. United States Air Force, "General Thomas S. Moorman Jr.," n.d., https://www.af.mil/About-Us/Biographies/Display/Article/106158/thomas-s-moorman-jr/.
92. Moorman interview with Bradley, part I, 11.
93. Moorman interview with Hall, 142.
94. Moorman interview with Hall, 142.

95. Moorman interview with Hall, 142.
96. Moorman interview with Hall, 143.
97. Moorman interview with Hall, 135.
98. Moorman interview with Hall, 136.
99. William Shelton, Moorman memorial service.
100. John E. Hyten, Moorman memorial service.
101. Hyten, Moorman memorial service.
102. Hyten, Moorman memorial service.
103. Office of the Chief of Staff of the Air Force, "Blue Ribbon Panel of the Air Force in Space in the 21st Century," in Spires, *Orbital Futures*, 1:414.
104. George W. Bradley III, "A Brief History of the Air Force in Space," *High Frontier* 1, no. 2 (Fall 2004): 7.
105. Joseph W. Ashy, "Putting Space in the USAF Weapons School," *USAF Weapons Review* (Summer 1996): 2–4.
106. Office of the Chief of Staff, 1:401.
107. Spires, *Beyond Horizons*, 279.
108. Office of the Chief of Staff, 1:416.
109. Merrill A. McPeak, "Does the Air Force Have a Mission?" *Selected Works: 1990–1994* (Maxwell AFB, AL: Air University Press, August 1995), 155.
110. Office of the Chief of Staff, 1:408.
111. Spires, *Beyond Horizons*, xii.
112. Spires, *Assured Access*, 253.
113. Moorman, as quoted in Spires, *Assured Access*, 257.
114. Moorman, as quoted in Spires, *Assured Access*, 257.
115. Victor Hillard, correspondence with the author, December 2022.
116. Hillard correspondence.
117. Spires, *Assured Access*, 258.
118. Spires, *Assured Access*, 267.
119. McPeak as quoted by Martin C. Faga, Moorman memorial service.
120. Merrill A. McPeak, Moorman memorial service.
121. Merrill A. McPeak, presentation to the Commission on Roles and Missions of the Armed Forces, September 14, 1994, 185.
122. Merrill A. McPeak, "The Future of America in Space," speech at SPACETALK '94, Salt Lake City, Utah, September 16, 1994, in *Selected Works*, 320–21.
123. John P. White, "Directions for Defense: Report of the Commission on Roles and Missions of the Armed Forces," Washington, DC, May 24, 1995, ES-5.
124. Spires, *Beyond Horizons*, 283.
125. Ronald R. Fogleman, interview with Richard Kohn, December 11, 1997, transcript in Richard H. Kohn, ed., "The Early Retirement of Gen. Ronald R. Fogleman, Chief of Staff, United States Air Force," *Aerospace Power Journal* (Spring 2001): 10.
126. Fogleman interview with Kohn, 10.
127. Kohn, "The Early Retirement of Gen. Ronald R. Fogleman," 18.

128. Representative Steny Hoyer, 105th Cong., 1st sess., Congressional Record–House, June 26, 1997, H4825.
129. Donald G. Hard, Moorman memorial service.
130. Spires, *Orbital Futures*, 2:1208.
131. Richard W. McKinney, "Reconsidering the Space Commission 10 Years Later," *High Frontier* 7, no. 4 (August 2011): 12. The Air Force was estimated to "include 90 percent of space personnel, 85 percent of the military space budget, 86 percent of space assets, and 90 percent of space infrastructure."
132. Donald H. Rumsfeld et al., "Report to the Commission to Assess United States National Security Space Management and Organization," Washington, DC, January 11, 2001, 57.
133. Rumsfeld et al., xxii–xxiii.
134. Rumsfeld et al., 42.
135. Donald H. Rumsfeld and Stephen A. Cambone, "Enduring Issues: The Space Commission 10 Years Later," *High Frontier* 7, no. 4 (August 2011): 5; Rumsfeld et al., xxxiii.
136. Rumsfeld et al., vii.
137. William L. Shelton, introduction, *High Frontier* 7, no. 4 (August 2011): 2.
138. Dean J. Miller, "Space Education and Training Center Named for Space Pioneer," *Space Observer* 56, no. 38 (September 20, 2012), 1, https://www.yumpu.com/en/document/read/7601883/9-11-the-high-price-of-freedom-colorado-springs-military-.
139. D. T. Thompson, Moorman memorial service.
140. Hyten, Moorman memorial service.
141. Moorman interview with Kipp and Fuller, 51.
142. Spires, *Beyond Horizons*, xi–xii, xix.
143. Hyten, Moorman memorial service. At the time, General Raymond was the chief of space operations, General Thompson was the vice chief of space operations, and General Shelton was a former commander of Air Force Space Command.
144. Faga, Moorman memorial service.

8 STARS AMONG THE STARS

Charles F. Bolden, Kevin P. Chilton, and Susan J. Helms

STEPHEN J. GARBER AND JENNIFER M. ROSS-NAZZAL

Before the National Aeronautics and Space Administration (NASA) selected its first class of astronauts, the agency contemplated what the best skills, training, or background might be for those chosen to join Project Mercury. They looked at a variety of occupations that might prepare one for spaceflight: race car drivers, scuba divers, test pilots, and even mountain climbers. Eventually, President Dwight D. Eisenhower decided that the new spacefarers would be selected from the nation's military test pilots because being a "qualified jet test pilot appeared to be best suited" for spaceflight missions and the skills that might be needed in zero gravity.[1] Until 1978, when NASA named its first class of space shuttle astronauts, nearly all astronaut classes consisted solely of graduates from one of the two U.S. test pilot schools. Charles F. Bolden Jr., Kevin P. "Chili" Chilton, and Susan J. Helms come from a long line of highly qualified military astronauts from these prestigious institutions. These three stand out among their military astronaut colleagues as well as most career military personnel; they became astronauts, returned to their service branch, and rose through the ranks to become two-, three-, and four-star generals. Becoming even a one-star general officer is quite difficult, as approximately only 0.4 percent of U.S. military officers reach flag officer rank.[2]

The nation's test pilot schools at Edwards Air Force Base (AFB) and Naval Air Station Patuxent River are highly competitive programs that focus not only on mastering flight skills but also on understanding flight hardware as well as basic scientific and engineering concepts. Test pilots and flight test engineers are highly competent and some of the best in their chosen fields. They are leaders with outstanding judgment, the skills to manage highly technical missions, and the ability to perform calmly and admirably while under tremendous pressure. While perhaps contrary to the stereotype of such literally and figuratively high-flying officers as being iconoclastic individuals with the "right stuff," these three clearly are highly team-oriented leaders who support their people, work well with others on an individual level, and simultaneously are inspirational to broader groups.[3]

Charles F. Bolden Jr.

Charles F. Bolden Jr. was born on August 19, 1946, in Columbia, South Carolina. As a young black man growing up in the segregated South, he encountered significant racial discrimination in reaching his goals. He credits his parents, both educators, with instilling in him a strong sense of ethics and service to others. Bolden became "infatuated with the Navy" at the age of twelve because of several television programs based on the experiences of military academy cadets and midshipmen, and he set his sights on becoming a Navy "frogman" but at that time was not particularly interested in flying.[4]

Bolden secured an appointment to the United States Naval Academy through his character and persistence. He wrote to his two U.S. senators and congressional representative, as well as Vice President Lyndon B. Johnson, throughout high school, letting them know early on of his interest in securing a nomination to Annapolis. In addition to being too soon for them to act, he also knew that his congressional delegation would not support his nomination because of his skin color. As he later told National Public Radio, "It was clear why they were not supporting me, and it was because of the times. They were just not about to appoint a Black to the Naval Academy."[5] When President John F. Kennedy was assassinated and Johnson became president in November 1963, Bolden's senior year of high school, he again wrote to Johnson. Shortly afterward, a Navy recruiter knocked on his door, saying he had heard of the young man's interest in Annapolis. Around the same time, Johnson sent a former federal judge to Bolden's high school to recruit young men of color to the service academies. Yet because of his race, Bolden's appointment to Annapolis came from a Chicago congressman, Rep. William L. Dawson, a practice at the time to utilize unclaimed appointments for minority cadets from the South.[6] As hard as Bolden fought for his appointment, he almost quit during his first year there: "I cried all the time. I wanted to go home, and my father kept me there. Every time I'd call on the weekends, he'd say, 'Stay one more week and then we'll talk about it.'" Bolden persisted and later became president of his class.[7]

Shortly after graduation in 1968 with a degree in electrical science, he married Alexis (Jackie) Walker and decided he wanted to be a Marine Corps infantry officer like his company officer from his plebe year, Maj. John Riley Love, whom he admired. But he qualified for an aviation option, and both his wife and his Marine Corps basic school company officer convinced him to become a Marine aviator rather than an infantryman in Vietnam.[8] He turned out to be an excellent aviator, graduating first in his class at Naval Air Station Pensacola, giving him his choice of aircraft. Bolden chose to fly jets and ended up flying sorties over North and South Vietnam, Cambodia, and Laos in the A-6A Intruder. Parts of his experience there were positive, but he also came to believe that "war is not a good way to solve problems" and that the United States "failed" to accomplish its mission.[9]

In 1973 Bolden returned to the United States as a Marine recruiter in Los Angeles, California, where he had great success, unlike many of his fellow officer candidate recruiters facing antiwar protests on college campuses. He was so successful that his office was the Marine Corps' primary recruiting station for the nation.[10] Bolden understood the sentiments of the protestors, talked with them about his Vietnam experiences, and tried to understand their perspectives on the Vietnam War. He was the ideal choice for this assignment because he, like all great leaders, was a good listener and empathized with students and their beliefs about the war. Throughout his life, he made a point of trying to understand where people were coming from and to be honest with them in return.[11]

As soon as he started flying at Pensacola, Bolden decided he wanted to be a test pilot. He was so interested he applied eight times: "I started applying long before I even had the requisite flight time. I just wanted to go [to test pilot school]. It was sort of like my desire to go to the Naval Academy. I wanted people to know who I was, so that when I became eligible, people would be acquainted with me."[12]

While serving as a test pilot at Patuxent River, Maryland, he met Ronald E. McNair, one of the first three African American astronauts selected in 1978. Meeting McNair, who had also grown up in the segregated South, was a formative experience. McNair attended an event at Patuxent River and, after meeting Bolden, encouraged him to become an astronaut.[13] Bolden replied, "They'll never pick me." Even though NASA had just selected the first class of space shuttle astronauts, which included minorities, he still had doubts. Astronauts had typically been white Anglo-Saxon test pilots.[14] McNair told Bolden, "That's the dumbest thing I ever heard. How do you know if you don't apply?"[15] So he did.

NASA selected Bolden in 1980, and he went on to fly in space four times, twice as a pilot and twice as a shuttle commander. Bolden often exhibited grace under pressure, whether there was a technical problem or a high-profile mission. During his first spaceflight, as pilot on STS-61C, he quickly and calmly dealt with a potential leak in the space shuttle main engine on ascent and then during reentry handled another malfunction when the cooling system for a hydraulic power unit failed. Bolden, however, played the first incident down, saying, "We didn't have a real problem.... [it was only] an instrumentation problem," and the commander concurred that it was just a faulty sensor reading.[16]

Instructors respected Bolden's desire to participate in particularly challenging situations during training in case there was an in-flight failure. Typically, before their first space flight, shuttle pilots (and commanders) were required to have at least 750 landings in the shuttle training aircraft (STA), in addition to thousands of simulated landings in the shuttle mission simulator. Bolden commented, "I think I had 2,500 [to] 3,000 simulated landings in the STA, prior to the first time I flew the shuttle."[17] These experiences, as well as his military training, allowed him to be calm under pressure when landing the orbiter. As another astronaut and STA instructor noted, "I'd put Charlie number one if something went wrong... because he trained for the worst case."[18]

As any leader knows, teamwork is vital and results in better outcomes, no matter the field. When Bolden commanded shuttle crews, he followed the advice of his STS-61C commander, Robert L. "Hoot" Gibson, that no astronaut should act alone. He instructed his crew members that even in abnormal situations, "Nobody does anything. Don't you touch a switch until at least one other person has verified that what you think is wrong, is really wrong.... Let's at least make sure there are two of us that agree on the procedure, and then ... we're going to work it as a team." He explained that shuttle trainers would often be able to "divide and conquer ... with people who were typical pilots, arrogant, cocky, very confident.... You know, 'I got this one. You get that one.' And before you knew it, all hell broke loose, because [we] weren't communicating and you're working something over here that was in opposition to what was being worked over there. So eventually you learn ... this is a team effort and ... [the time to] be an all-star is over."[19]

As his son Ché Bolden, also a Marine, noted, leadership is mostly about inspiring people through mutual trust and established by demonstrating how to follow. "Every really good leader that I came across, starting with my father, knew how to follow first," he explained.[20] This is especially true for astronauts, when "everybody knows that they could be the leader but they aren't, and they're just helping the whole team go along."[21] Astronaut C. Michael Foale remembered that Bolden was always ready to give the reins of leadership to another crewmember and on STS-45 gave payload commander Kathryn D. Sullivan the authority to make important decisions in that role. "Kathy loved the title of being commander," Bolden remembered, and he called it a "blessing" because she kept up with the science on that mission.[22] Foale noted that "Charlie's strength was enabling followership.... The key to followership is having some empathy for what's going on around you," and he was very empathetic.[23] Ultimately, as Bolden learned, "Good leaders command respect; they don't demand it." Every person, not just a mission commander, a NASA leader, or a Marine general, had "something to contribute."[24]

On his fourth and final spaceflight, he commanded STS-60, the historic first joint American-Russian space shuttle mission that included Russian cosmonaut Sergei Krikalev as a mission specialist.[25] While Bolden was initially opposed to cooperating with a former Cold War adversary in this way, he made time to have an informal dinner with visiting cosmonauts, agreed to the mission, and soon became close friends with Krikalev and his family.[26]

Bolden's ability to develop a strong friendship quickly with Krikalev is indicative of his personal approach to work and life. One astronaut summed it up simply by saying that "he's just always friendly and cheerful . . . [and] will say hello to everybody and anybody."[27] Another astronaut noted that Bolden is a "'people person' in that he made every effort to know details about those he trained with—their backgrounds, family members and names, their interests, and their strong points in our crew environments. He made friends easily."[28]

He frequently invited not just his crewmates but also their extended families to his family's home. He would warmly welcome everybody and mention his wife, Jackie, every time. He also urged his crewmates to talk with him privately: "Please, please, please, if you have any difficulties, any issues in your families, outside of NASA, in the crew, or whatever, I am here to help you. Please ask me."[29]

Tellingly, astronaut trainer Lisa Reed's first impression of Bolden (before she even knew he was an astronaut) was of his low-key style coaching a NASA intramural softball team game in 1986. Bolden's team was doing poorly (many of them clearly hadn't played before), but he was very upbeat and cheerful. "Charlie was all about the camaraderie. I do also think that comes a little bit from the military upbringing. They're in close quarters, and they go through good and difficult times together, and it forges relationships."[30]

Said Bolden, "One of the cardinal rules of being a Marine is that you take care of your people, and they will take care of you."[31] Astronaut Terrence W. Wilcutt related a story that demonstrated Bolden's support for people in his care. NASA had signed agreements to fly a cosmonaut on the space shuttle as part of the Shuttle-*Mir* program, and in 1992 the Russian Space Agency sent two cosmonauts, Sergei Krikalev and Vladimir Titov, to train in Houston. The Russians had rented an apartment for them and filled it with cheap furniture. Bolden thought this was not the way to greet the two, so he stood up in the Monday morning astronaut meeting and asked his colleagues if they could redecorate and donate quality furniture that they were not using in order to help cement the bonds between the astronauts and cosmonauts. Wilcutt dubbed that call to action "the best leadership speech I had heard anyone give in the [Astronaut] Office."[32] Bolden described the preparation for this flight as his most memorable experience: "It was the opportunity to bring people from a really foreign culture to the United States and introduce them to our way of life and help them adjust and adapt and getting to know them as true friends and establishing a lifelong bond with them."[33]

With four successful spaceflights complete, he returned to the operating forces of the Marine Corps in 1994 and became deputy commandant of midshipmen at his alma mater. Coming back to Annapolis, Bolden noticed that racial discrimination remained an issue and urged the academy to expand its diversity and inclusion programs.[34] Three years later, the Marines recognized Bolden's outstanding leadership when they promoted him to major general, making Bolden "the highest-ranking African American ... serving in the corps."[35]

After several engineers with less admirable people skills served as NASA administrator, Bolden became NASA's top leader in 2009. Upon hearing that President Barack Obama nominated Bolden for administrator, former astronaut Kathy Sullivan described him as "a superb package overall for the job" because of his character, integrity, and leadership style.[36]

Less than a year after Bolden accepted the position, the Obama administration rolled out a budget in 2010 that terminated the Constellation program, an effort to take America back to the moon and on to Mars. With this decision, the Office of Management and Budget reimagined NASA as "more of a research and development agency," one no longer involved in human spaceflight operations for which it was widely known. Bolden, who was surprised by the cancellation, was deeply disappointed that "in one fell swoop ... we alienated the entire NASA family and the entire NASA workforce."[37] Cancelling Constellation meant there were no follow-on human spaceflight programs, since the thirty-year space shuttle program was nearly at an end. Many employees and contractors were anxious about NASA's future, but Bolden refused to give up on the work already completed on next-generation spacecraft. Even though Constellation was technically no longer funded, Bolden gave former astronaut and Johnson Space Center director Michael L. Coats permission to continue working on the vehicle, and Coats credits Bolden with saving human spaceflight because he convinced the administration of the need for the Orion multipurpose space vehicle and a new launch vehicle.[38] Nevertheless, Bolden credited the "whole leadership team," including Coats, for championing Orion and the new space launch system.[39]

Under his visionary leadership, NASA developed a commercial crew program to take astronauts to the International Space Station (ISS). Securing support for the initial concept proved to be difficult, and he recalled Congress initially redirecting commercial crew funding to other NASA projects.[40] Eventually, Bolden and his team secured support for the concept, which the agency relies on today to ferry crews to and from the ISS.

Bolden was very proud that NASA was first rated the "best place to work in the federal government" during his tenure as administrator. He tells the story of how Jeri Buchholz became NASA's human capital leader after holding a similar position at the Nuclear Regulatory Commission, which had previously won this award. She told him, "You have to tell people every single day that they are doing good stuff . . . and that you value what they do." Bolden naturally embraced this philosophy, setting NASA on a path to receive this award a record ten years straight.[41]

Ellen Ochoa, former Johnson Space Center director, credited Bolden with shifting the mindset of NASA managers and leaders away from quantifiable results when it came to guiding others. Instead, he encouraged them to mentor employees, offer career development, and consider how to create a more diverse and inclusive staff. As a result, Bolden "made a big contribution to the overall workforce at NASA" by directing management to consider less traditional approaches.[42]

Not surprisingly, Bolden's parents instilled in him the "importance of treating people right, and trying to build teams, because nobody does anything by themselves," particularly in the military where you must deal with "complex problems that you cannot solve by yourself."[43] He is a consensus-builder who sometimes takes his time to gather information, making sure he listens to all relevant voices before making a decision.[44] Fostering diverse opinions is important to him so people will know their voices have been heard. Bolden was an early proponent of diversity training for NASA managers, which was influential in practical ways for astronauts and other leaders.[45]

Similarly, Bolden is a strong believer in the value of delegation, and he "like[s] to push things down to the lowest level possible for decisions."[46] As a shuttle commander, he was supposed to fly the orbiter back home, but "that didn't make any sense" to him, so he let the pilot take the controls for a while on reentry before taking the controls again for the last part and the landing.[47] When he took the helm at NASA, the administrator traditionally wrote the personnel evaluations for all ten center directors and all the mission directorate associate administrators, so Bolden delegated those reporting relationships to Robert M. Lightfoot, the associate administrator.[48]

Charles Bolden—an aviator, a Marine, an astronaut, and the former administrator of NASA—is highly accomplished. But when asked to describe his leadership style, his colleagues all agree that it is how he made them and others feel—about themselves and their own abilities—that made him an outstanding leader. Astronaut Loren J. Shriver, who flew with Bolden on STS-31 (the deployment of the Hubble space telescope), said Bolden's leadership style "brings out the best in people." He had a calm demeanor and the unique ability to make people feel comfortable, without ever being intimidating even when people brought him bad news.[49] He fostered a cooperative spirit and friendship between American and Russian explorers that "opened the door to our future cooperation on the International Space Station," transitioned the United States from its longest-running human spaceflight program to the commercialization of missions to and from the ISS, and oversaw the first test of the Orion crew vehicle, designed to take astronauts out of earth orbit for the first time since 1972.[50] By consistently taking care of his people, he enabled his teams to achieve great successes.

Kevin P. Chilton

Kevin P. "Chili" Chilton was born November 3, 1954, in Los Angeles, California, the eldest of four children. His father was an aeronautical engineer for Douglas, and his mother once worked as a flight attendant for American Airlines. As a child, Kevin wanted to become a professional baseball player, but growing up in the shadows of Los Angeles International Airport, he soon became fascinated with flying and dreamed of becoming an airline pilot like a close family friend who flew for United Airlines. Even though he did not express any interest in flying in space, his parents woke him up to watch the early Mercury flights of

Alan B. Shepard, Virgil "Gus" Grissom, and John H. Glenn on television. "One day," his dad said, "you'll thank us for this."[51]

While reading did not interest Chilton then in the way airplanes or baseball did, he recalled the impact one book had on his life. He was "really touched" by a theme in Harper Lee's *To Kill a Mockingbird*—"you cannot understand another person until you've walked in their shoes"—and her novel influenced the way he has treated people throughout his life. Chilton found that learning more about other people's families, backgrounds, challenges, and experiences "gives you opportunities to be a better leader." He found the same concept applied to raising children. Listening to and understanding challenges faced by his own daughters helped him to understand their perspective and to be a better father.[52]

After high school, Chilton attended the Air Force Academy, primarily to achieve his goal of becoming an airline pilot, graduating in 1976 with a degree in engineering science. Chilton later regretted not participating in more cadet leadership training opportunities. "It was a mistake," he conceded, "as I look back on it in hindsight. Whether I was going to be an airline pilot or not, the academy is a leadership laboratory." Witnessing some classmates attempting to intimidate and ridicule subordinates, he quickly realized these techniques were ineffective in motivating people. Instead, Chilton preferred to correct subordinates gently when they made mistakes, which he continued to do as an operational instructor pilot and as an astronaut.[53] Years later his colleagues in the Astronaut Office reported that he remained "cool under pressure," calm, and softspoken.[54]

After completing his master's degree in engineering at Columbia University and undergraduate pilot training, Chilton flew RF-4C airplanes, a formative experience for him because the two-person crews needed to work closely together. While at Kadena Air Base, Japan, Chilton served under Lt. Col. Otto K. "Ken" Habedank, who mentored him early in his career. Habedank's leadership philosophy, which became Chilton's, rested on the simple yet fundamental tenet that he "truly in his heart believed no one in his command woke up in the morning, put on their uniform, and came to work saying to themselves, 'How can I screw up today?'" Thus, people's intentions are usually good, but if the results were not what was expected, then the leader needed to consider their communication style or the subordinate's learning style, personality, or home issues. The junior person should never be blamed first for failure.[55] This worked both ways. As another officer noted, "General Chilton had great confidence in his people, and we knew it. When you know how much the boss believes in you, you never want to disappoint."[56]

While he had studied D-Day and Gen. George Custer's last stand as a boy, Chilton developed a deeper interest in military leadership while stationed in Japan, and he visited the battlefields where Americans had fought in World War II. He became a voracious reader, picking up anything he could on the war in the Pacific. It was an evolution on many levels, he admitted: "I changed from this kid who wasn't really interested in being in the military to someone who said, 'I'm truly a member of the profession of arms ... [so I] better understand [my] profession.' I took to reading about it and loved it."[57]

When he returned to the United States, Chilton continued to study military history and visited other battlefields from Little Bighorn in Montana to Gettysburg in Pennsylvania. Walking those fields with docents and reading about events and biographies of generals from those campaigns, he learned more about leadership. While he often emulated the positive traits of people he worked with, he recognized that he had to be true to himself, and he often tells younger audiences, "God only made one George Patton.... [If] you try to be George Patton, you'll fail. You have to be yourself."[58]

Chilton also learned the importance of trusting teammates while flying Air Force missions in the F-15C Eagle fighter. Leadership for these missions was shared, with lower-ranking officers regularly given the opportunity to lead training scenarios, including large formations of fighters, bombers, tankers, and reconnaissance aircraft. Although he might not have known every pilot in the formation, he had to trust each team member's expertise. As a commander, it was vital to recognize that he could not accomplish the mission without their help. He explained, "You can't be in every cockpit. You have to trust people." Having faith and confidence in others is the basis for teamwork.[59]

Before each of Chilton's Air Force flights, teams held weather, threat, and safety briefings; each person flying that day had a role to play. But briefing three other airmen proved to be an initial challenge for Chilton: "My knees would knock together. I would stutter. I would stumble. I just was so afraid of standing in front of a group of people and talking." He grew more comfortable doing these briefings and then realized there was no difference between speaking in front of three people or three thousand, assuming he understood the issues. With practice, he became an outstanding speaker.[60]

Following the flights, there were debriefs with the flight crews. "That's where learning happened. That's where leadership skills were developed," Chilton recalled. The flight leader for that mission always ran the debrief—no matter their position, even if someone else outranked them on the team. Rank was not important; in these discussions, a lower-ranking captain could openly tell a colonel that the latter had made a mistake. The intent was to improve and learn from every flight but not to humiliate others. Airmen talked openly about the mistakes they made and their performance in the cockpit. If their team lost, they discussed what they might have done differently. These scenarios tested Chilton and reinforced his willingness to listen carefully to different perspectives.[61]

Chilton, who humbly called himself "eminently trainable," excelled in other learning environments. He was a distinguished graduate from the Air Force Academy in 1976 and first in his pilot training class. In 1982 he won the secretary of the Air Force leadership award as the top graduate of squadron officer school, an award that recognizes "the best of the best" who exemplify the traits valued by the U.S. Air Force.[62] Two years later, he was the top pilot to graduate from the Air Force test pilot school. Chilton's credentials and abilities were so impressive that he was regularly promoted "below the zone," meaning he received his promotions earlier than scheduled, an opportunity only offered to exceptional candidates.[63]

By the time he applied to be an astronaut in the mid-1980s, Chilton had the right mix of skills that the Astronaut Office desired. They looked for people with operational skills as well as the ability to lead and follow. Astronaut classmate William F. Readdy said Chili "had a reputation as 'team player,'" which was important to the selection board, a comment echoed by Duane L. Ross, former head of the astronaut selection office: "If you're not a team player, an astronaut's job is definitely not where you want to be."[64]

Chilton put those skills to work on his first space shuttle mission, aboard *Endeavour* on STS-49 in 1992. The crew's objective was to capture a stranded satellite and attach a new motor so the satellite could be boosted into geosynchronous orbit. After two failed attempts to capture the satellite, Chilton tried to cheer up crewmate Pierre J. Thuot, who had been unable to grapple the satellite on two extravehicular activities (spacewalks) due to equipment issues.[65]

Later that night, Chilton wanted to brainstorm with fellow crewmate Richard J. Hieb, who was tired but didn't want to disappoint him. "Soon most of the crew was back out of bed, engaged in the discussion, gaining enthusiasm and energy, and ultimately one of the crewmembers came up" with the then-radical idea of conducting a three-person extravehicular activity. The crew conveyed this idea to mission control, got some sleep, and then successfully retrieved the malfunctioning satellite. Hieb credits Chilton for initiating the brainstorming that led to this creative solution.[66]

Upon landing, Chilton returned home to his family and his pregnant wife, Cathy. Astronaut Kevin R. Kregel, who served as the family escort, remembered how Chilton quickly readjusted to life on the ground and began taking care of some daily chores that his wife could not do at the time because of her pregnancy. Chilton wryly commented to his friend, "Yes, Kevin, twelve hours ago, I was a space hero ... [but] right now I'm picking up dog poop." Kregel was impressed because it showed the type of person and leader Chili was, "a leader who's willing to do any job" to accomplish the task at hand.[67]

Following his final flight in 1996, Chilton volunteered to return to the Air Force, but those orders were canceled, and he was told to report to NASA's ISS program office to serve as the deputy program manager. At the time, the ISS was behind schedule, over budget, and facing technical issues. Looking back, Chilton believed that the decision to move him into the ISS program later served him well because of what he learned about program management and leadership as well as negotiating and working with international partners, skills he employed when he returned to the Air Force in 1998.[68]

Chilton was selected for the ISS position because he worked well with others and had flown to the *Mir* space station on STS-76, giving him valuable experience working with the Russians. Further, his experiences as a space shuttle commander and test pilot gave him credibility with the international partners. While he was not fluent, Linda M. Godwin remembered how well he could converse in Russian.[69] Chilton negotiated by building consensus and using a congenial, not dictatorial, approach. Astronaut Thomas D. Jones explained Chilton "called on the professionalism of the Russians—after all, they thought of themselves as world class experts on space stations." Recognizing their accomplishments was vital to reaching a consensus. "Then he praised them for their cooperation, grudging as it was, and used that as a foundation for greater progress."[70] He had a sense of Russian culture and used that to his advantage when negotiating. He found that the Russians would "ease up on their position" as the workday wound down. They preferred to wrap up business at five, but the Americans would "sit in a room, talk, lock the door, [and] slide pizzas under the door" until they reached a solution.[71] ISS program manager Randy Brinkley praised Chilton for his negotiating skills, recalling that "it took the wisdom of Solomon and the patience of Job and courage/vision of Moses to sort it out and get buy-in from the disparate parties."[72]

When Chilton joined NASA, he thought it would be a temporary assignment and that he would eventually return to the Air Force but soon learned that that was not Air Force policy. Attitudes began changing in the mid-1990s, however. Lt. Gen. John P. Jumper, deputy chief of staff for air and space operations, wanted to bring military astronauts back to the Air Force to bring operational rigor to space, just as the air service had brought pilots back on active duty before World War II. NASA and its human spaceflight program, he said, were "not contributing to the combat capability that I wanted [military] space to be a part of," but military astronauts had the credentials, experience, and knowledge that they could apply in command positions within the Air Force.[73]

On his first day back with the Air Force in 1998 as deputy director of operations for Air Force Space Command, Chilton was left in charge because his boss was leaving to oversee an accident investigation at Cape Canaveral Air Force Station. Having been at NASA since 1987 and with no job description, Chilton fell back on one of his basic tenets of leadership: "listen, not talk" when you are new to an organization. He gathered his colonels together to learn more about the command and the issues they were facing, quickly realizing they were some of the same challenges the ISS had faced. His ability to empathize and value everyone's input made his leadership stand out.[74] Soon after, the Air Force promoted Chilton to brigadier general. Gen. Richard B. Myers, who was then commander of Air Force Space Command, appreciated Chilton's skills, which would benefit the command: "We were so, so lucky to have him on our team.... Now we had somebody that had actually been to space."[75]

Chilton demonstrated his value the month he arrived. When North Korea attempted to launch a satellite that summer, the intelligence community concluded that they had tried to launch an intercontinental ballistic missile. Myers recalled that within minutes of looking at the data, Chilton concluded that the

North Koreans had not launched a missile but were trying to place an object into orbit. But, Myers recalled, "it took the folks in Washington . . . a lot longer than that."[76] Chilton, who had flown on the space shuttle, immediately recognized the familiar trajectory, but he humbly admitted he figured out the answer "not because I was brilliant, just because I was exposed to [spaceflight trajectories]."[77] Myers recalled there was an effort to rebuild Chilton's "chops" in the Air Force. "It's not very often you get somebody with these kind[s] of credentials back in the Air Force," he said. Senior leaders intentionally worked to ensure Chilton had the opportunity to succeed. He needed "operational credibility," so soon after his promotion to brigadier general, he was offered command of the 9th Reconnaissance Wing at Beale AFB, California.[78]

In California, while achieving the Air Force's mission was "the number one thing" for Chilton, he also prioritized the care and well-being of his personnel: "If you don't take care of your mind, body, spirit, and your family you're not going to be effective at doing the mission." He learned that families were struggling with the demands placed on the U-2 pilots, who were frequently gone two hundred days a year. He supported his people and focused on their welfare and how best to decrease the demands on pilots and families. He insisted on improving base housing, increasing recruiting efforts for the U-2 program, and improving access to training equipment.[79] He was a true champion for the wing, with one squadron commander remembering Chilton's time at Beale as "our Camelot."[80]

Subsequently assigned to the Pentagon and soon after the September 11 attacks, Brigadier General Chilton found himself negotiating in Pakistan and Uzbekistan to determine if the Air Force could use their airspace and military facilities to support strikes against the Taliban for harboring members of al Qaeda. Myers, now chairman of the joint chiefs of staff, remembered how effective Chilton, then head of the joint staff's office of politico-military affairs for the Asia Pacific and Middle East, was during that effort. Negotiators trusted Chilton because he was culturally sensitive.[81] He believed it was important to understand the backgrounds of those you were negotiating with and be sure not to insult someone, even unintentionally. "If you do," he said, "you're not going to make the progress you want to make in your discussions, dialogue, or the work you want to do with them."[82] General Jumper recognized Chilton's other strengths, including an understanding of the global issues and the players in the post-9/11 world. Soon after this mission, in April 2002, Chilton received a second star, becoming a major general while working under Jumper as director of programs at Headquarters U.S. Air Force. In that role, Chilton was responsible for ensuring that the Air Force would be able to "organize, train, and equip" its forces in the future and played a vital role in prioritizing the programs to be included in the congressional budget each year. When members of Congress questioned the need for the F-22 to fight against the Taliban, for example, he maintained a forward-looking vision and secured their support.[83]

In August 2005 Chilton was promoted to lieutenant general.[84] Less than a year later, he became a four-star general and given command of Air Force Space Command. In the fall of 2007, he became the commander of U.S. Strategic Command. Chilton's fourth star did not surprise Jumper: "I thought Chili Chilton could have been chairman of the joint chiefs of staff or chief of staff of the Air Force."[85] Chilton's outstanding leadership skills and abilities to learn new material and work with different kinds of people propelled his meteoric rise from one-star to four-star rank in a little over seven years.[86]

Chilton's concerns about the value of human life and safety were demonstrated again in 2007 when he was approached by the head of the National Reconnaissance Office about a nonfunctioning satellite set to reenter the atmosphere carrying a tank of frozen hydrazine that experts predicted would survive reentry to the ground. Hydrazine is a highly toxic substance, and even exposure to small amounts can be dangerous to humans. Chilton briefed President George W. Bush on the risks and asked for authorization to move forward on a plan called Operation Burnt Frost to intercept and destroy the satellite before it reentered the atmosphere. In the end, sixteen different agencies worked together in a compressed time frame to ensure the safety of citizens around the world. Ultimately, the satellite was successfully destroyed, and its debris, including the hydrazine tank, burned up on reentry over northern Canada without reaching the ground or causing harm. There were critics of the action—namely China, which called out the United States for shooting down the satellite, saying that human safety was merely a cover story. NASA's Nicholas L. Johnson, an orbital debris expert involved in Operation Burnt Frost, countered that assertion: "Not once did I ever hear of a reason to consider taking mitigation measures against USA-193 other than to protect human life."[87]

Gen. Kevin Chilton, former test pilot, astronaut, and four-star general, is a man of many accomplishments, as well as great character. His former colleagues have likened him to Gen. George C. Marshall and Gen. Omar N. Bradley for his selflessness and ability to navigate politics "without compromising his principles."[88] Chilton's humble belief in the value of all people, no matter their position, consistently shone through. As astronaut Kevin Kregel put it, "He treats the janitor and the president of the United States the exact same way."[89] He always focused on the mission, be it at NASA or the Air Force, but he was not all work and no play. He enjoyed playing guitar in the astronaut band Max Q, and even as a general he was not above being in on the fun, covertly putting plastic flamingos on a neighbor's lawn to select who would host the next base flamingo party and jamming with friends at the get-together.[90]

With all his achievements, he is most proud of helping to raise his four daughters, and his colleagues praise his devotion to his family. "He is the best example there is of what it means to be a good husband and father," fellow astronaut Robert D. Cabana said, and "those traits that make him such a good man also make him a great leader, for that's how he cares for his team."[91]

Susan J. Helms

Susan J. Helms was born on February 26, 1958, to Pat and Dori Helms in Charlotte, North Carolina. Pat was an Air Force officer, and Dori was a schoolteacher. The couple went on to have three more daughters. Her parents had high expectations for their children, and Susan excelled in school at a young age. She was an early reader and later demonstrated the ability to tackle complex math problems with her "algebra mind." Susan showed such promise that her parents chose to wake her, but not her sisters, to witness the Apollo 11 lunar landing, and she later recalled "that this would turn into . . . something more important in my future than anyone realized at the moment."[92]

Wherever they were stationed, the Helmses chose to live in the local community rather than on the military base, which gave young Susan the opportunity to have a more traditional middle-class suburban life and exposed her to people and life outside of the military. The family regularly took weekend road trips where they visited national parks and saw wide swaths of the country. When Susan entered seventh grade, the family settled in Portland, Oregon, where she noticed a trend: women in leadership roles. The students had even elected a female student body president. This was a visible change as more women entered the workforce, accepting positions in fields once closed to them. Seeing women in positions with great authority and outside of traditionally female fields was not a foreign idea to her—she had never been told she could not do or be something because of her sex. Instead, she felt like she could achieve anything she set her mind to.[93]

While Helms was an introverted child, she was mature for her age and began coming out of her shell in high school. She was president of the band, active in choir and glee club, and a youth leader at her church. Beverly B. Pratt, her junior high school counselor and an influential mentor who became a lifelong friend, recalled that Helms was always "wise beyond her years" and that other students looked up to her. Pratt also praised her outstanding listening skills and how she continued to stay in touch with friends, sending birthday and holiday cards. She and others even highlighted her loyalty to friends who had made serious mistakes.[94]

By eighth grade, Helms knew she wanted to be an officer in the Air Force, where she could earn equal pay for equal work, and she had a memorable conversation about her decision with her father a few years later. "The best way" to become an officer would be to attend the Air Force Academy, "but you can't do that," he told her. "Well, why not?" she asked. "Because you are a female." "That is not right!" she exclaimed. "How could that be the norm?"[95] Helms had never encountered such blatant sexism before. Things changed in 1975 when her mom stumbled across an article in *Parade* magazine—women could now apply to the military academies.[96] She asked her daughter, "Do you still want to do this?" If so, "you better start practicing your pullups." On March 19, 1976, the principal called Helms to his office, where she learned that Sen. Mark O. Hatfield had nominated her to the academy.[97] Helms was one of 157 women selected that first year that women were eligible.

Future astronaut Kevin Kregel was a junior at the academy when the female cadets arrived, and he remembered the barriers they faced. He attributed

Helms' success to her "quiet competence" approach: "If you show that you know what you're doing and you're competent, people forget what your sex is, they forget what color you are, what religion you are. They just accept you."[98] Helms and the other women initially shunned the media attention because "we just wanted to fit in and do the job and not keep being reminded that we were different."[99] Instead of consciously developing her leadership style at the academy, she was focused on performing in the cadet drum and bugle corps. While comfortable on the field in halftime shows, she remained an introvert and found she "had to work at" leading men in her element.[100] Being the only remaining woman in her squadron her senior year was a challenge. (Out of 157 women who started in 1976, ninety-seven graduated four years later.)

Upon graduation with a degree in aeronautical engineering, Helms moved to Eglin AFB, where she became a flight test engineer—which became a metaphor for how she led from the back, literally and figuratively—and had empathy for those under her command.[101] She sought guidance and emulated colleagues whose abilities she admired, such as Jeff Corey, another academy alumnus who emphasized competence more than drawing attention to his work. Thanks to his inspiration, she came to recognize that the "office was more than just a workplace. The office was a place that was important to everybody's wellbeing," a philosophy she embraced throughout her career.[102]

In 1984 Helms moved to California for graduate studies in aeronautics and astronautics at Stanford University. Stanford graduate Sally K. Ride visited the campus to talk about her second spaceflight. Helms initially did not believe she could be an astronaut but came to realize America's first woman in space was just like her. "Right then and there," Helms realized she wanted to be a spacefarer.[103]

Helms learned more about the application process from space shuttle astronaut Richard O. Covey. He first met Helms at the academy in 1985, where she taught aeronautics after graduate school, and told her about his path to the Astronaut Office. He had attended Air Force test pilot school, as had many other military astronauts, and Covey encouraged her to apply to their flight test engineer course, which she did. Covey believed Helms was an exceptional astronaut candidate and encouraged her to apply for the next selection. He was so confident she would be selected for an interview that she remembered him saying at her test pilot graduation in 1988, "I'll see you in Houston."[104]

Competition to become an astronaut was fierce, and when she learned she had been selected in 1990, Helms felt like she won "the lottery."[105]

For the next twelve years Helms dedicated her life to the space program. Within the Astronaut Office she impressed her colleagues and office chief Daniel C. Brandenstein with her work ethic. Years later, classmate Eileen M. Collins identified Helms as "one of the hardest workers" in their class.[106] She was intensely focused and demonstrated in-depth knowledge of shuttle systems, which led to her being one of the first three in that class to receive a flight assignment.[107] She went on to fly four shuttle flights: STS-54, STS-64, STS-78, and STS-101, with all of her commanders noting her attention to detail. John H. Casper, commander of STS-54, believed Helms "was so competent in all assigned tasks that she seemed a seasoned space veteran rather than a first-time space flyer."[108] STS-64 commander Richard N. Richards remembered how confident he was in her abilities because she was always prepared.[109]

Helms' expertise was vital to her next assignment as the STS-78 payload commander, a life science mission. Previous payload commanders for those flights held medical degrees, but the ability to pull together an international payload crew to ensure that they captured the scientific data proved more important.[110] Helms played an important role in drafting their flight plan. Payload specialist Robert B. Thirsk said that was "a significant contribution from Susan" because she provided a "practical big picture perspective" on eliminating redundancies by suggesting ways of sharing data between investigators.[111] This contribution was an eye-opening moment for her. "That's when I realized," Helms said, "I had a gift of operational leadership, which ended up being my forte" for the rest of her career.[112] The best operational leaders have a clear understanding of how to achieve a mission's objectives, and Helms used STS-78's goals to draft a plan that worked for the scientists.

In training, Helms looked after the payload crew, who affectionately called her "Queen Mum" because she took them "under her wing." Her leadership style included sharing with and delegating to the team. Although all the payload specialists were rookies, she included them in science matters and asked them to monitor mission hardware. Helms even asked Thirsk to redesign a piece of flight hardware, an opportunity he greatly appreciated. She further built camaraderie by inviting the scientists to group dinners and to movies after work.[113] Training to fly in space was not just about getting the job done and putting in long hours; it was also about enjoying the experiences with friends and colleagues. Helms enjoyed another outlet and used her musical background to become the keyboard player in the all-astronaut band Max Q.[114]

When NASA assigned crews for the ISS, the Astronaut Office struggled to find volunteers. Most wanted to fly on the space shuttle, not the ISS, because training times were longer, and crewmembers had to spend many months away from friends and family. Astronaut Wilcutt recalled that the "toughest duty" in the Astronaut Office was to be assigned to one of those early ISS missions.[115] All the fliers for Expedition 1 through Expedition 4 were military astronauts because they agreed to serve at the request of their commanding officer.[116] As Cabana, then chief of the Astronaut Office, explained, "I knew if I asked Susan, she wouldn't shun the task, and I wasn't disappointed."[117] Recalled astronaut and ISS veteran Carl Walz, "Even though it was at substantial cost to her personally," she trained for almost four years, then packed up her apartment, sold her car, canceled her credit cards, and "moved off the planet" for the duration of her five-month flight.[118]

Within NASA, Helms was known for her organizational abilities and leadership, but she was an unknown in Russia. When it came to flight assignments in Russia, their male-dominated space program relied heavily on credibility and trustworthiness. "You have to be trusted to do your job," Helms explained. "Your word has to be trusted. You have to be an honorable person, or they simply don't fly you in space." Seeing her dedication and abilities, the Russians recognized she was more than prepared to fly on Expedition 2 in 2001.[119]

Afterward, Helms decided to return to the Air Force. Retired Maj. Gen. Suzanne M. Vautrinot said Helms did not come back wanting "to become a three-star or a four-star" general; rather, "it was a determination to serve."[120] As Helms had been gone for twelve years, not everyone agreed with the decision to name her to a leadership position. Even though she was an astronaut and had attended test pilot school, she did not fit the "worldview of what military leadership" was, and there was an overall sense that she had to be "reblued."[121] In Air Force culture at the time, its leaders were pilots, not astronauts or flight test engineers.[122] As chief of the space superiority division at Air Force Space Command, she worked to cultivate space situational (later called domain) awareness in space, which involves tracking space objects and inferring satellite operators' intentions within the context of national security. "That fundamental concept was not [fully mature] when I showed up" because space was not thought of as a battlefield at that time, she said.[123] Helms envisioned future military space operations and situations that might occur in that domain and was open to hearing other ideas on how the Air Force could move forward in this arena. She favored collaboration and informal discussions among personnel in divisions with different missions and organizational viewpoints.

Helms was thinking about the future of space deterrence instead of advocating for any new programs or systems upgrades per se, and she pushed her colleagues to better understand what might happen in the future and how best to prepare for those challenges. She encouraged change through "quiet professionalism," not "pounding her fist on the table or calling anyone out." Instead, she relied on "intellect and research and understanding and vision" and patiently waited "for the next opportunity to make those changes."[124] So that they could see what skills they needed to manage the space battlefield and be aware of potential threats, she took her team to visit the mission control center in Houston.[125]

Transformational change takes many years to implement, and Helms had the requisite patience. When she later served as vice commander of the 45th Space Wing at Patrick AFB, Florida, issues involving the head of launch safety at Cape Canaveral Air Force Station became paramount. Retired Brig. Gen. Mark H. Owen, then commander, believed that the Air Force launch program had failed to evolve with the growth and demands of commercial space launch start-ups. SpaceX's Elon Musk and other interested commercial entities demanded changes to launch safety protocols, but the Air Force was slow to adapt.[126]

After her promotion to brigadier general and taking command of Patrick AFB and the 45th Space Wing in 2006, Helms pushed back when engineers and technicians explained why they could not accommodate any changes to their processes. She listened to their concerns, and eventually they came to see that she "cared about safety as much as they did." Using techniques she had learned at Harvard University's Kennedy School of Government, she appointed a more forward-thinking employee as the new civilian head of safety. By the spring of 2007, the Air Force reached an agreement with SpaceX to launch from Cape Canaveral.[127] Recalling that achievement, Helms humbly said, "I feel like I can take a little bit of credit" for this transformation. Helms came to see herself as a "change agent" who subtly but firmly pushed people beyond their comfort zones in order to move organizations in new directions.[128]

Throughout her career, her colleagues noticed that Helms was willing to "stand up" for others even "when it's hard."[129] She publicly demonstrated this when the Supreme Court found the Defense of Marriage Act unconstitutional in 2013. Helms, then commander of the Fourteenth Air Force and commander of the joint functional component command for space, recognized that the decision might be an issue for some. Helms asked her staff if anyone had an issue with the new policy and actively listened to their concerns. She relayed how women were forced upon men at the Air Force Academy and recognized that members of the lesbian, gay, bisexual, and transgender communities might be similarly harassed or treated poorly. Helms told her colonels to watch for signs of prejudice, to make sure every servicemember felt included and welcome on base. A decision had been made, and there would be no turning back.[130]

Though Helms was known for her principled leadership, some members of Congress questioned her judgment when a personnel decision she made became national news. In 2012 as a three-star general, she overturned a sexual assault conviction. Upon receiving the military tribunal's decision, Helms spent hours poring over the pages of evidence, looking at the facts, and studying every detail, and she eventually concluded that justice had not been served. She downgraded the charge to indecent behavior and spent sixteen hours drafting a memo that explained her reasoning.[131] Her decision reflected her habit of delving into the details of a situation and listening to multiple perspectives to understand a subject or situation as fully as possible.

When President Barack Obama nominated Helms for another three-star position in Colorado, Missouri Sen. Claire McCaskill put a hold on her nomination, saying, "Lt. Gen. Helms sent a damaging message to survivors of sexual assault" that a sentence can be overturned "with the stroke of a pen."[132] Air Force Space Command Gen. William L. Shelton commented, "She refused—steadfastly refused—to compromise her integrity by refusing to be used as a political cause."[133] Obama withdrew the nomination and, eventually, Helms retired. Chilton and many others who were fully informed about the inner workings of the military justice system admired her decision not to bow to political pressure, which reflected her values.[134]

Before retiring, Helms spent time mentoring young women in various ways. Owen remembers her holding a "Space Station Day" with the junior and high school daughters of the senior staff at Patrick AFB. She rented all the *Lord of the Rings* movies and spent the day eating pizza and making cookies with the girls. They had the chance to ask Helms questions in a casual setting, which Owen believed inspired "a young person for life. Not to become an astronaut but just to be a better person, to do better in school. To be excited about science."[135] Vice wing commander and retired Col. Keith Balts recalled her writing a "very heartful entry in an 'advice journal'" for his daughter.[136] After retiring, Helms, along with other former female astronauts, formed AstraFemina to inspire young girls to study science, technology, engineering, and math and explore careers in the field.

As an outstanding student and hard worker from an early age, along with her career achievements, Helms typified the kind of female role model needed to inspire young girls to do their best. She was one of the first women to graduate from the Air Force Academy. She later became the first military woman to fly in space and the first woman to live on board the ISS. She steadily rose through the ranks to become a lieutenant general, and she left her mark on military space, cultivating the important concepts of space situational awareness and space commercialization. Perhaps even more notable than her achievements is her quiet but effective leadership style. Helms has long known that competence—understanding military and space systems—and building organizational relationships are important skills for leaders in the Air Force and at NASA, but the real key to leadership is understanding and supporting people. Her ability to listen and to be present, even when she was juggling many issues, meant so much to those she served with, as did her joy of sharing her favorite pastimes with colleagues and their families, who felt cared for.[137] As Lt. Gen. Nina M. Armagno, USSF, said, "It felt like we were always in good hands under Helms' leadership."[138]

Conclusions

Overall, in discussing these astronauts who became true leaders, one observer noted that they "should be willing to do what any of their people, their subordinates are going to do. You've got to be willing to step in and roll up your sleeves and help." They never felt that they were too good to help out that way, which is what made them so well liked. They were "servant leaders" who had empathy and knew that they weren't always leaders so they "lead by example."[139]

Notes

1. Robert Gilruth, "From Wallops Island to Project Mercury, 1945–1958: A Memoir," in *Essays on the History of Rocketry and Astronautics: Proceedings of the Third through the Sixth History Symposia of the International Academy of Astronautics*, 2 vols., ed. R. Cargill Hill (Washington, DC: NASA Scientific and Technical Information Program, 1977), 2:471; George Low, "Status Report No. 6–Project Mercury," February 3, 1959, cited in Loyd S. Swenson Jr., James M. Grimwood, and Charles C. Alexander, *This New Ocean: A History of Project Mercury* (Washington, DC: NASA Historical Series SP-4201, 1966), 131.
2. Officers represent only about 18 percent of the U.S. military; the other 82 percent are enlisted personnel. Lawrence Kapp, "Defense Primer: Military Officers," Congressional Research Service report, November 23, 2022, https://crsreports.congress.gov/product/pdf/IF/IF10685.
3. Tom Wolfe, *The Right Stuff* (New York: Farrar Straus Giroux, 1979).
4. Charles F. Bolden, interview by Sandra Johnson, January 6, 2004, transcript, Johnson Space Center Oral History Project (hereafter JSC OHP).
5. Laura Wagner, "First Black NASA Administrator Charles Bolden 'Pleaded' to Get into Naval Academy," February 9, 2016, National Public Radio, https://www.npr.org/sections/thetwo-way/2016/02/09/466191748/first-black-nasa-administrator-charles-bolden-pl.eaded-to-get-into-naval-academy.
6. Charles F. Bolden, interview by Cathleen Lewis and Ian Cooke, February 13, 1995, transcript, United States Naval Academy, Annapolis, Maryland.
7. Bolden interview by Johnson, January 6, 2004.
8. Bolden interview by Johnson, January 6, 2004.
9. Bolden interview by Lewis and Cooke, February 13, 1995.
10. Fred H. Allison and Kurtis P. Wheeler, eds. and comps., *Pathbreakers: U.S. Marine African American Officers in Their Own Words* (Washington, DC: History Division, United States Marine Corps, 2013), 105.
11. Allison and Wheeler; Franklin R. Chang-Diaz, email to Jennifer Ross-Nazzal, August 11, 2022.
12. Bolden interview by Johnson, January 6, 2004.
13. McNair was one of the mission specialists selected to fly on the space shuttle in 1978, but he was not a test pilot.
14. Thomas M. Baughn, "Major General Bolden to Head NASA," *Fortitudine: Bulletin of the Marine Corps Historical Program* 34, no. 4 (2009): 27, https://www.marines.mil/Portals/1/Publications/Fortitudine%20Vol%2034%20No%204.pdf.
15. Tom Fox, "NASA Administrator Charles Bolden on Leadership: 'At NASA We Do Big Things,'" *Washington Post*, June 1, 2011, https://www.washingtonpost.com/blogs/ask-the-fedcoach/post/administrator-charles-bolden-on-leadership-at-nasa-we-do-big-things/2011/03/04/AGqBsQGH_blog.html; W. Henry Lambright, "Reflections on Leadership and Its Politics: Charles Bolden, NASA Administrator, 2009–17," *Public

Administration Review 77, no. 4 (July/August 2017): 616. Tragically, just ten days after Bolden's first shuttle flight landed in January 1986, the *Challenger* accident claimed McNair and the other crewmembers' lives.
16. Bill Nelson with Jamie Buckingham, *Mission: An American Congressman's Voyage to Space* (Orlando, FL: Harcourt Brace Jovanovich, 1988), 113, 177; Bolden Awards File, "Citation for Defense Superior Service Medal," JSC History Office, Houston, TX; Bolden interview by Johnson, January 6, 2004.
17. Bolden interview by Johnson, January 6, 2004.
18. Kevin R. Kregel, interview by Jennifer Ross-Nazzal, August 2, 2022, transcript, JSC OHP.
19. Bolden interview by Johnson, January 6, 2004.
20. "Daughters and Sons: Leadership through Generations: A Conversation with Charles F. Bolden Jr. and Ché Bolden," Council on Foreign Relations, June 3, 2021, https://www.cfr.org/event/daughters-and-sons-leadership-through-generations-conversation-charles-f-bolden-jr-and-che.
21. C. Michael Foale, interview by Jennifer Ross-Nazzal, August 17, 2022, transcript, JSC OHP.
22. Charles F. Bolden, interview by Rebecca Wright, March 18, 2016, transcript, NASA Headquarters Oral History Project (hereafter NASA HQ OHP).
23. Foale interview.
24. "Daughters and Sons."
25. See Bolden's official NASA biography, January 2017, https://www.nasa.gov/sites/default/files/atoms/files/bolden-cf.pdf.
26. Robert D. Cabana, email to Jennifer Ross-Nazzal, September 7, 2022. One of Bolden's crewmates commented that it was always fun going to his house for crew get-togethers. In particular, Sergei Krikalev liked dessert and ice cream specifically, so the Boldens always had lots of ice cream; Sergei once ate five bowls. (See Kenneth S. Reightler, notes from a phone conversation with Jennifer Ross-Nazzal, July 21, 2022.)
27. Foale interview.
28. Loren J. Shriver, email to Jennifer Ross-Nazzal, September 9, 2022.
29. Foale interview.
30. Lisa Reed, interview by Jennifer Ross-Nazzal, July 7, 2022, transcript, JSC OHP.
31. Fox.
32. Terrence W. Wilcutt, interview by Jennifer Ross-Nazzal, August 12, 2022, transcript, JSC OHP.
33. Charles F. Bolden, interview by Sandra Johnson, January 15, 2004, transcript, JSC OHP.
34. Allison and Wheeler, 184.
35. "Chronologies—1997," Marine Corps University, last updated 2009, https://www.usmcu.edu/Research/Marine-Corps-History-Division/Research-Tools-Facts-and-Figures/Chronologies-of-the-Marine-Corps/1997/.
36. Kathryn D. Sullivan, interview by Jennifer Ross-Nazzal, May 28, 2009, transcript, JSC OHP.

220 | Chapter 8

37. Bolden, interview by Rebecca Wright, March 31, 2014, transcript, NASA HQ OHP.
38. Michael L. Coats, interview by Jennifer Ross-Nazzal, August 5, 2015, transcript, JSC OHP.
39. Bolden interview by Wright, March 18, 2016.
40. Bolden interview by Wright, March 31, 2014.
41. Lambright, 620; "Decade of Excellence: NASA Named Best Place to Work 10th Year in a Row," July 13, 2022, NASA, https://www.nasa.gov/press-release/decade-of-excellence-nasa-named-best-place-to-work-10th-year-in-a-row; Bolden, interview by Rebecca Wright, June 5, 2017, transcript, NASA HQ OHP.
42. Ellen Ochoa, email to Jennifer Ross-Nazzal, July 24, 2022.
43. Bolden interview, June 5, 2017; "Daughters and Sons."
44. Reightler notes; Bolden interview by Wright, June 5, 2017.
45. Foale interview; Bolden interview by Wright, March 18, 2016.
46. Bolden interview by Wright, March 18, 2016.
47. Bolden interview by Johnson, January 15, 2004.
48. Bolden interview by Wright, June 5, 2017.
49. Shriver email; Eileen M. Collins, email to Jennifer Ross-Nazzal, August 2, 2022; Wilcutt interview.
50. Cabana email.
51. "Distinguished Graduate General Kevin P. Chilton Class of 1976," *Checkpoints*, March 2015, 52, https://aog-websites.s3.amazonaws.com/usafa-org/documents/heritage/distinguishedgrads/2014DG-Chilton.pdf.
52. Kevin P. Chilton, interview by authors, July 14, 2022, transcript, JSC OHP.
53. Kevin P. Chilton, interview by authors, July 13, 2022, transcript, JSC OHP.
54. Collins email; William Readdy, email to Jennifer Ross-Nazzal, July 24, 2022.
55. Chilton interview, July 13.
56. Jon Eric Stroberg, email to Jennifer Ross-Nazzal, August 28, 2022.
57. Chilton interview, July 13.
58. Chilton interview, July 13.
59. Chilton interview, July 14.
60. Chilton interview, July 14.
61. Chilton interview, July 14.
62. Chilton interview, July 13; Rhonda Smith, "Air University Airmen Receive the 2021 SecAF Leadership Award," Maxwell Air Force Base, May 18, 2021, https://www.maxwell.af.mil/News/Display/Article/2622003/air-university-airmen-receive-the-2021-secaf-leadership-award/.
63. Collins email.
64. Readdy email; Duane L. Ross, intern presentation, June 29, 2001, transcript, JSC History Office, Houston, Texas.
65. Daniel C. Brandenstein, interview by Jennifer Ross-Nazzal, August 10, 2022, transcript, JSC OHP.
66. Richard J. Hieb, email to Jennifer Ross-Nazzal, July 21, 2022.

67. Kregel interview.
68. Chilton interview, July 13.
69. Linda M. Godwin, email to Jennifer Ross-Nazzal, July 24, 2022.
70. Thomas D. Jones, email to Jennifer Ross-Nazzal, June 15, 2022.
71. Chilton interview, July 13.
72. Randy H. Brinkley, email to Jennifer Ross-Nazzal, September 12, 2022.
73. Richard B. Myers, interview by authors, August 24, 2022, transcript, JSC OHP; John P. Jumper, interview by authors, September 13, 2022, transcript, JSC OHP.
74. Chilton interview, July 13.
75. Myers interview.
76. Myers interview.
77. Chilton interview, July 14.
78. Myers interview.
79. Chilton interview, July 14.
80. Stroberg email.
81. Myers interview.
82. Chilton interview, July 13.
83. Jumper interview.
84. United States Air Force, "General Kevin Chilton," biography, August 2010, https://www.af.mil/About-Us/Biographies/Display/Article/104791/general-kevin-p-chilton/.
85. Jumper interview.
86. Chilton biography.
87. Chilton interview, July 14; Nicholas L. Johnson, "Operation Burnt Frost: A View from Inside," *Space Policy* 56 (May 2011), https://doi.org/10.1016/j.spacepol.2021.101411.
88. Jumper interview; Sidney M. Gutierrez, email to Jennifer Ross-Nazzal, June 16, 2022.
89. Kregel interview.
90. Mark H. Owen, interview by Jennifer Ross-Nazzal, August 4, 2022, transcript, JSC OHP.
91. Cabana email.
92. Susan J. Helms, interview by authors, July 11, 2022, transcript, JSC OHP.
93. "Preflight Interview: Susan Helms," NASA, last updated December 6, 2012, https://www.nasa.gov/mission_pages/station/expeditions/expedition02/helms_interview.html.
94. Beverly B. Pratt, interview by Jennifer Ross-Nazzal, August 22, 2022, transcript, JSC OHP; Kregel interview; Suzanne M. Vautrinot, interview by authors, July 20, 2022, transcript, JSC OHP.
95. Susan J. Helms, unidentified interviewer, February 28, 2003, transcript, United States Air Force Academy, Special Collections, Colorado Springs, Colorado.
96. President Gerald R. Ford signed Public Law 94–106 in 1975, which allowed women—for the first time—to be admitted into the nation's prestigious military academies.
97. Helms interview by authors, July 11.
98. Kregel interview.

99. Helms interview, February 28.
100. Helms interview by authors, July 11.
101. Owen interview.
102. Susan J. Helms, interview by authors, July 12, 2022, transcript, JSC OHP.
103. Samantha Saulsbury, "First U.S. Military Woman in Space Reveals Secrets of Success," March 27, 2014, https://www.af.mil/News/Article-Display/Article/475104/first-us-military-woman-in-space-reveals-secrets-of-success/; Lisa Sonne, "Walking on Air," *Stanford Magazine* May/June 2001, https://stanfordmag.org/contents/walking-on-air.
104. Richard O. Covey, email to Jennifer Ross-Nazzal, June 13, 2022.
105. Helms interview, February 28.
106. Collins email.
107. Brandenstein interview.
108. John H. Casper, *The Sky Above: An Astronaut's Memoir of Adventure, Persistence, and Faith* (West Lafayette, IN: Purdue University Press, 2022), 183.
109. Richard N. Richards, interview by Jennifer Ross-Nazzal, August 11, 2022, transcript, JSC OHP.
110. Carl E. Walz, interview by Jennifer Ross-Nazzal, August 8, 2022, transcript, JSC OHP.
111. Robert B. Thirsk, interview by Jennifer Ross-Nazzal, August 18, 2022, transcript, JSC OHP.
112. Helms interview by authors, July 11.
113. Thirsk interview.
114. The astronauts formed Max Q following the *Challenger* accident to build morale at the Johnson Space Center.
115. Wilcutt interview.
116. Walz interview.
117. Cabana email.
118. Walz interview.
119. Helms interview, February 28; Kregel interview.
120. Vautrinot interview.
121. Owen interview. "Rebluing" meant reindoctrinating her into Air Force culture and operations.
122. Helms interview by authors, July 11.
123. Helms interview by authors, July 12.
124. Vautrinot interview.
125. Helms interview by authors, July 12.
126. Owen interview.
127. SpaceX was given clearance to launch its Falcon rocket at the Cape. "SpaceX Cleared for Launches," *Florida Today*, April 25, 2007, https://web.archive.org/web/20070930030616/http://www.floridatoday.com/floridatoday/blogs/spaceteam/2007/04/spacex-cleared-for-cape-launches.html.
128. Helms interview by authors, July 12.
129. Vautrinot interview.

130. Helms interview by authors, July 11.
131. Helms interview by authors, July 12.
132. Craig Whitlock, "Senator to Put Permanent Hold on Promotion of Air Force General," *Washington Post*, June 7, 2013, https://www.washingtonpost.com/world/national-security/senator-continues-to-block-promotion-of-air-force-general/2013/06/06/bbf9ea0a-cee3-11e2-ac03-178510c9cc0a_story.html.
133. Janene Scully, "Helms Leaves Amid Controversy, to Retire April 1," *Santa Ynez Valley News*, February 6, 2014, https://syvnews.com/news/local/military/helms-leaves-amid-controversy-to-retire-april-1/article_4208dd8e-8e12-11e3-8ff2-0019bb2963f4.html.
134. Kevin P. Chilton, interview by Jennifer Ross-Nazzal, September 27, 2022, transcript, JSC OHP.
135. Owen interview.
136. Keith Balts, email to Jennifer Ross-Nazzal, July 11, 2022.
137. Balts email; Thirsk, interview.
138. Nina M. Armagno, interview by authors, notes, September 15, 2022.
139. Reed interview.

MATURING LEADERSHIP OF SPACE

Lance W. Lord, C. Robert Kehler, and William L. Shelton

HEATHER P. VENABLE

"**Leadership has been a defining** hallmark of the U.S. space effort since the beginning of the Space Age," remarked Gen. C. Robert Kehler, former commander of Air Force Space Command (AFSPC) in 2012.[1] He was not referring to leadership as a trait of military officers but rather to how the United States led the way in pioneering key space capabilities. This use of the term "leadership" referring to a nation rather than to individuals might be taken as a bleak commentary on either the limited leadership of U.S. space professionals, the historical tendency in the Air Force to overstress technology, or both. It could also be taken as an indication of the slow maturation of space leadership over decades, as space professionals finally began growing into seasoned leaders ready to lead their own community rather than being led by fighter pilots.

This change was something worth celebrating at the beginning of the twenty-first century. The space community struggled paradoxically as the twentieth century had come to a close. Although the U.S. military had increasingly recognized the importance of space capabilities after Operation Desert Storm, the space community now had to satisfy a huge increase in demand for a very expensive capability. Air Force leaders tried to support a broad base of joint customers, pursuing the development of a number of new, highly desired space capabilities that proved challenging and expensive to engineer. They did all of this in an era of budget cuts. As space capabilities filtered down from the strategic level where they had flourished during the Cold War to the tactical level after the Cold War, they no longer received the same budgetary handwaving.[2]

The three leaders of Air Force Space Command surveyed in this chapter—Lance W. Lord (2002–6), C. Robert Kehler (2007–11), and William Shelton (2011–14)—each commanded and propelled the command through deep organizational unrest while providing key space capabilities during the global war on terror. Although they did not urge an independent space force, pushing for an independent space organization should not be the measure of leadership effectiveness.

Of the three leaders of AFSPC discussed in this chapter, Lord probably left the greatest mark on AFSPC. Not only did he drive toward the increased operationalization of spacepower, he also stressed spacepower's importance in the space domain itself in terms of the need to secure space superiority. Despite not being a career space officer but rather a missileer, Lord sought a unique identity for space and intercontinental ballistic missile (ICBM) operators, which were combined in a single operational career field at the time.

Importantly, Lord was the first four-star nonrated (nonflying) officer to lead AFPSC. While some view Lance Lord's ascent to the position as representing a change in the Air Force's philosophy about who should run its space business, another way to view this departure from the past is that it reflected the nonrated community having matured enough to have grown nonpilots from second lieutenants into generals.[3] Selected in the wake of United States Space Command's disestablishment after the September 11 attacks in favor of creating United States Strategic Command (USSTRATCOM), Lord's four-year tenure offered some stability to the Air Force and to the command.

After Lord, the institution faced some uncertainty given the short tenure of its subsequent commanders. Frank G. Klotz replaced him as acting commander before pilot-astronaut Kevin Chilton assumed command in June 2006. However, Chilton held command only briefly, as he quickly received command of USSTRATCOM.

Some stability returned to AFSPC when Kehler assumed command. A nonpilot and nontechnical officer like Lord, Kehler came up through the ICBM ranks and transitioned to the space business as a colonel following the Commission to Assess United States National Security Space Management and Organization (space commission) report in 2001. After the report's release, Kehler became director of the national security space integration office for the undersecretary of the Air Force, which some saw as a staff-in-waiting for a future U.S. Space Force. Kehler followed Chilton to command USSTRATCOM.

Finally, the next major change occurred with the tenure of William Shelton, who held multiple degrees in astronautical engineering. Shelton's appointment represented the maturation of space leadership, as he became the first officer who began his career as a space officer to lead Air Force Space Command.

These AFSPC commanders will be assessed not by evaluating their charisma or how well they instituted change. It is important to eschew the assumption that since leaders accomplish significant things, significant things must have been accomplished by great leaders. Rather, their ability to navigate and negotiate what is far more difficult to control—institutional culture—will be used to evaluate them. In this vein, Dr. Edgar Schein, an expert in organizational culture and management, argues that the sole "important thing leaders do. . . . is create and manage culture."[4] Tom Karp and Thomas I. T. Helgo begin with a similar view of leadership but go one step further, contextualizing it within complexity theory to argue that "leaders do not always have choices and do not have the control that most leadership theory suggests."[5] Karp and Helgo offer a number of important insights. First, leaders are almost always followers, and thus opportunities for charting their own course are often highly constrained, which is particularly true of military organizations. Second, leadership is not a one-way street but, rather, fundamentally interactive. Finally, their approach is in keeping with ongoing trends to decenter the leader. Leadership studies have been "obsessed" with a leader's characteristics and have overemphasized a leader's role within an organization as a kind of "special super-heroic individual."[6] But that view lends to a more static view of leadership while also overstressing what leaders can do rather than being realistic about their limitations. In short, it is only in interactions that leadership emerges, which departs from the "dominant way of thinking about a leader as holding an assigned position to oversee activities and seeking to achieve a desired outcome."[7]

This approach also avoids evaluating these commanders' leadership by developments that have occurred since then, most notably the Space Force's establishment. The tendency among many, especially those tending toward zealotry, will be to critique these and other space leaders by one key measurement for leadership success: the extent to which they pushed against an air-centric view of space. This would be problematic for many reasons, including the fact that circumstances out of their control intervened, a point previously highlighted in Karp and Helgo's research. As will be shown, moreover, AFPSC commanders walked a fine line between advocating for spacepower within the Air Force and working to operationalize and normalize space capabilities in the air operations center and elsewhere.

Little has been written about space culture in this period. Air Force officer Carl M. Jones argued in 2011 that General Kehler had a unique opportunity to reshape space culture, which Jones maintained was functioning in its organizational midlife—a phrase borrowed from the work of Edgar Schein—as seen in the rise of subcultures. Jones identified three subcultures: space operations, ICBM operations, and acquisitions, with the first two sharing an Air Force specialty code and initial qualification training. Each of these communities had begun its existence outside of Space Command, thus developing its own unique culture. Once kluged into Space Command, the result became unworkable.[8] But the notion of an organizational midlife does not apply neatly to AFSPC, as most of its subculture resulted from organizational change imposed on it from the Air Force and thus outside much control of space leadership. Perhaps more important, though, is Schein's postulation of organizational maturity ending in destruction and rebirth. In other words, the process of cultural change is iterative.

Desert Storm and Spacepower

In 1991, Operation Desert Storm had launched spacepower into prominence in supporting what some have described as the "first information war," during which the space domain provided communication, navigation, intelligence, surveillance, and reconnaissance, and other key capabilities using about sixty military and civilian satellites.[9]

Born more at the strategic level of war during the early Cold War, the first generation of space capabilities helped detect nuclear launches as well as deter nuclear war, such as by building ICBMs that required operations through space. Satellites' importance in detecting nuclear war had led to intense secrecy, leading many to be ignorant of space's increasing capabilities.[10]

As spacepower developed, its effects increasingly became user-friendly at the operational and tactical levels of war.[11] Spacepower had provided information to warfighters as early as the Vietnam War, such as by transmitting weather information and enabling some satellite-based communication, although that information only rarely reached the tactical user.[12] Operation Urgent Fury in 1983 marked another step forward, with space capabilities increasingly contributing to command and control.[13]

The Air Force had organizationally recognized this increasing capability, establishing Air Force Space Command in 1982 to better operationalize space capabilities within the broader Air Force. After AFSPC's formation, the cascading down of many strategic space capabilities from Strategic Air Command, such as the space-based missile warning systems, followed. The question for some spacepower professionals increasingly became how to make the joint community aware of its increased operational and tactical capabilities.[14]

They found their answer in Operation Desert Storm. Then–Air Force Chief of Staff Merrill A. McPeak belatedly recognized the importance of space capabilities, quickly dubbing Desert Storm the first "space war" at a briefing held at the National War College in March 1991. It had, he explained, provided the necessary information to numb Iraq, analogous to a "general anesthetic inducing operational paralysis."[15] In a boost of confidence to spacepower professionals, he now sought to make airpower and spacepower "equals" in the Air Force's mission statement.[16] McPeak even recognized that the Air Force had a "spotty" past record for its treatment of space capabilities.[17]

In making up for its neglect, the Air Force established the Fourteenth Air Force, the first numbered air force for space, in July 1993. Now, space capabilities received an organization analogous to other Air Force units.[18] As an "operational space component" for USSPACECOM, the Air Force intended to normalize space capabilities alongside airpower to better train and use air and space together. Other new organizational changes included establishing the Air Force space warfare center.[19]

Even as the Air Force expanded its space organization in response to Desert Storm, the institution also reorganized itself in response to the Cold War's end. As a result of McPeak's emphasis on organizational change, many communities experienced significant institutional turmoil. In what McPeak has acknowledged as his most significant mistake while serving as chief of staff of the Air

Force, for example, he shifted ICBMs into Air Combat Command to integrate all kinetic capabilities into one command. Another subsequent change combining missileers and space operators into Space Command in July 1993 served, he hoped, to correct his mistake while "infus[ing]" the space community with "more operational thinking." McPeak's incorporation of ICBM officers into Space Command to fix his earlier decision did not fully resolve the problem.[20] Despite these missteps, given the huge organizational changes McPeak implemented due to the Cold War's end and ensuing budget cuts, space historian David Spires has argued that in "reinventing the Air Force," McPeak "made space a top priority." Convinced of the space domain's central role in enabling future power projection, the chief of staff sought to bring all of the Air Force's operational capabilities under AFSPC by 1993.[21]

Despite these organizational changes, civilian leadership outside the Air Force remained concerned about U.S. civilian and military space capabilities given anticipation regarding future competition in space. As such, in 2001, external agents intervened in the space community, recognizing that only the president had the authority necessary to reorganize it.[22]

Led by Secretary of Defense Donald H. Rumsfeld, the space commission consisted of twelve key members, including seven retired flag officers—one from the Navy, two from the Army, and four from the Air Force. Of the retired Air Force officers, three had been fighter pilots, with only one, Gen. Thomas S. Moorman Jr., being a seasoned space professional. Still, one of the fighter pilots, Gen. Howell M. Estes III, had commanded the North American Aerospace Defense Command and U.S. Space Command from 1996 to 1998.[23] The commission's most relevant finding for AFPSC centered on the Air Force needing to "take steps to create a culture within the Service dedicated to developing new space system concepts, doctrine, and operational capabilities."[24] The commission also wanted to increase the centralization of the Air Force's space assets into AFSPC. The Space and Missile Systems Center, for example, belonged to Air Force Material Command. The commission hoped that "consolidating space functions into a single organization would create a strong center of advocacy for space."[25] External actors—some of whom had once been internal actors at the Air Force—had given the service its mission statement. Now it was up to the Air Force writ large, and Air Force Space Command more narrowly, to consider how to operationalize that intent.

Gen. Lance Lord

Gen. Lance Lord assumed command shortly after the space commission report's release. In preparing for his interview for the position with Secretary Rumsfeld, Lord read the space commission report repeatedly. When Rumsfeld asked him his vision for running Space Command, he began reciting key points from the report. Rumsfeld subsequently interrupted, telling him to "hold on a minute," at which point Lord announced, "I'm going to go put my arms around those folks and lead them."[26] This account reveals the extent to which Lord may have needed some pointed coaching in how to lead an organization, with Rumsfeld clearly setting his expectations on leading people more than implementing bullet points.

Several key forces had primed Lord to lead. Growing up in Tennessee with a father who served in the Army in World War II and continued to serve as a reservist after the war, Lord looked to the prospect of military service from childhood. He ultimately sought to follow his father into the Army. However, the Army and then the Navy rejected him because of his eyesight; the Air Force, however, willingly overlooked his eyesight, although it kept him from being a pilot.[27] Similarly, working in his grandfather's store taught him to go above and beyond expectations, which Lord believed inculcated him with "creative" and "innovative" tendencies.[28] Finally, seeing his peers letter in football before he did intensified his determination to achieve goals.[29]

Graduating with a bachelor of science in education from Otterbein College, Lord commissioned into the Air Force in 1969. Unable to fly, Lord chose ICBMs, the next most operational job he could find.[30] He subsequently served four years on alert duty in support of the Minuteman II. A notable part of that experience consisted of serving alongside many members of Strategic Air Command who had been navigators on bombers, which he believed inculcated him deeply into the culture of that community.[31] He also began growing as a leader, realizing that ensuring change required more than one-time implementation. Observing women being integrated into ICBM crews, for example, he recognized how leaders must continue to disseminate clear messages to ingrain change. He also became sensitive to diversity's importance.[32]

Lord subsequently served on the air staff as part of an intern program for junior officers. There he learned about taking risk. He recalled being in a crowded room full of higher-ranking officers when a two-star general asked for recommendations regarding the testing of a new carousel guidance system on a Titan II missile. Lord recalled how no one had the gumption to support it until the general finally got to him, a captain, and he recommended trying it.[33]

Lord later commanded two ICBM wings before receiving an assignment to a space wing, where he oversaw ballistic missile launch tests and satellite launch. Due to the great secrecy surrounding space, Lord remained largely compartmentalized within the ICBM community despite being assigned to a space wing.[34]

Lord's responsibilities for space capabilities continued to increase as he served as director of plans and as vice commander for AFSPC headquarters. He drew on his bachelor's degree in education in a variety of positions at Air University, including commanding the squadron officer school. He then served as the assistant vice chief of staff for headquarters U.S. Air Force, where he began considering how to implement the space commission's recommendations.[35]

Assuming command of AFSPC, Lord found himself running an organization of almost 40,000 individuals who had been buffeted by repeated organizational change since Desert Storm. Whereas space professionals had spent the previous decade increasingly working to operationalize space capabilities into the fabric of joint warfighting, Lord more actively sought to ensure space superiority to preserve U.S. asymmetric advantages, which rested on a more quietly weaponized notion of space.[36] In terms of the leadership of AFSPC, he also wanted to create a more holistic culture in which both missileers and space operators believed they belonged to the same community while feeling like they had as much value to the Air Force as pilots.[37]

Lord subsequently translated the commission's intent into internal guidance and collaboration to fuel the pursuit of seven areas for organizational improvement: (1) commanding the future, (2) enterprise, (3) partner, (4) unleash human talent, (5) warfighters, (6) wizards, and (7) rapidly moving technology to warfighters. Taken together, AFSPC intended for the organization to transition from providing force support into a full-fledged "Space Combat Command" capable of employing spacepower more holistically.[38] On the flip side, these areas were not entirely self-evident and overlapped in some places. The essence of his plan, however, was to "maintain" support to the joint force while "increas[ing] its focus on producing warfighting effects with space superiority and strike capabilities—in short, to become a full spectrum space-combat command."[39]

The relative importance of this dual focus for spacepower did not always emerge clearly. Releasing the first issue of *High Frontier*—a new quarterly journal for space professionals—in the summer of 2004, Lord wrote an introduction that began with a boldfaced quote stressing the need to increase "the intellectual properties of our space professionals" to "harvest more decisive, innovative and integrated effects on the battlefield."[40] This wording accorded with the space community's focus since at least Desert Storm. In the penultimate paragraph, though, Lord made clear that the efforts of space professionals should not stop there. Spacepower could do more than improve the nation's "ability to fight an enemy today; ultimately, it will provide full spectrum combat effects on the battlefields of the future."[41] Thus, Lord may not have communicated his message as clearly as possible.

Still, *High Frontier*'s establishment notably provided an important new forum for space professionals to share their ideas.[42] Stressing a different theme each issue, the publication first focused on the importance of space professionals in accordance with the space commission's emphasis on this topic. The commission had found that less than 20 percent of 150 individuals in key space positions had significant space experience. Most professionals had previously operated ICBMs, air defense artillery, or aircraft, having about 2.5 years of space experience to draw on. Notably, the Air Force had placed ICBM operators in space positions to "broaden" their "career opportunities," but this decision came at AFSPC's expense.[43]

Lord did not always clearly articulate how AFSPC would develop a space professional strategy. As he explained in the introductory issue of *High Frontier*, he intended to ignite a "quest for continual learning" in space professionals.[44] Lord's early interest in education, as demonstrated by his college degree and his postings to Air Force professional military education, bore fruit on the pages of *High Frontier*. He shared, for example, his vision to create a National Space University in Colorado Springs that would bring together civilian and military space professionals as well as academics.[45] But the space professional strategy also held a deeper intent aligned with increased advocacy. As Michael Stumborg argues, this strategy not only sought broad-based improvements to space professionals but also looked for powerful officers in other tribes to win them to APSC's cause.[46]

In what could be considered a moment of zealousness in *High Frontier*'s inaugural edition, Lord claimed that spacepower has "matured quicker and provided more substantive contributions to the American way of fighting wars" since 1945 compared to airpower's first fifty years of existence. Not only did Lord make a claim impossible to substantiate but he may also have unnecessarily provoked parochial antagonism; this could only undermine support from the airpower community, especially given the emphasis on jointness in the U.S. military.[47]

Lord did not always take the same loud and proud approach to advocacy, however. Externally, for example, he penned a somewhat staid 2002 article for *Joint Force Quarterly* that more objectively and dispassionately made a case for the range of capabilities AFSPC provided, concluding by hewing to the party line of the last decade that AFSPC would "ensure unparalleled space capabilities for joint forces when and where they are needed."[48]

Beyond his outward advocacy, Lord shepherded internal changes at AFSPC to improve support for joint warfighting.[49] The Fourteenth Air Force joint operations center, for example, normalized incorporating space planning into joining planning, as did in-theater directors of space forces by investing a single person with the space coordinating authority required to provide unity of effort of disparate civilian, military, and other space assets.[50] As hungry joint users eagerly "pull[ed]" these capabilities in Operation Iraqi Freedom with space professionals no longer having to "push" them, improvements continued, with Lord claiming that sensor-to-shooter times had been reduced to "single-digit minutes" by providing satellite data links to the MQ-1 Predator remotely piloted aircraft.[51]

In a similar vein, to "normalize" these processes, Lord offered lower-ranking officers opportunities to promote the integration of space planning, such as holding the first conference of space weapons officers in 2005 at Air University regarding how best to incorporate space capabilities into the operational level of warfare.[52] Some observers believed that such a conference would have been impossible prior to Operations Enduring Freedom and Iraqi Freedom because space professionals lacked real-world experience integrating air and space capabilities.[53] Indeed, then-commander of Air Force Material Command and fighter pilot Gen. Gregory Martin remarked that air and space capabilities had been truly integrated in a significant way for the first time in Operation Iraqi Freedom.[54]

Despite these improvements, Lord had learned early in his Air Force career that he needed to convey his vision in a systematic and powerful way. Viewing himself as one of many leaders within AFSPC, he drew on his earlier lessons in an ICBM squadron regarding the importance of consistency of message. Believing that the "hardest job you have as a leader is communication," he explained that "keeping people on messages and understanding the goals and objectives of the organization was a very . . . important part of leading."[55] In this way, Lord recognized that leadership was not a one-way street in which he could simply issue an edict from above.

For AFSPC, this advocacy meant educating others as to the threats likely to emerge in the space domain as it became increasingly militarized.[56] In his article "Commanding the Future," Lord issued his battle cry for AFSPC: "If you're not in space, you're not in the race."[57] His struggle to achieve success in high school football provided the foundation for this battle cry, or what he considered his means of advertising the value of space for national security.[58]

Lord also loudly advocated for spacepower within the Air Force behind closed doors, worried that the service might even lose the space mission due to neglect.[59] Exaggerating for dramatic effect, he described how he had all but thrown himself on the floor kicking and screaming to convince other Air Force generals to give as much attention to space capabilities as they had lovingly lavished on the F-22 and F-35. He anticipated Air Combat Command increasingly "atrophy[ing]," with Langley Air Force Base becoming a museum by 2049.[60]

Internally, Lord represented the first officer to seek to distinguish space professionals in important ways culturally from the air side of the house. In doing so, his top-down approach seeking buy-in from powerful nonspace officers was complemented by a bottom-up approach to get buy-in from his own space professionals, who increasingly pushed the boundaries of the possible, letting theory run ahead of technology.[61] One space weapons officer, for example, envisioned space weapons at some unknown point in the future being able to decisively target centers of gravity from space, either directly or indirectly.[62] Perhaps most controversially, an Air Force report in 2003 suggested using hypervelocity rod bundles—what then became dubbed "rods from gods"—a solution Congress found unpalatable, placing limits on developing these and similar weapons systems.[63]

But how exactly he intended to reshape space culture more broadly is not clear. In *High Frontier*'s first issue, Lord did explain that "cultural shifts and change are sometimes met with apprehension and skepticism; however, these initiatives are needed to face the asymmetric challenges of the 21st century."[64] In a similar vein, Col. James C. Hutto proclaimed AFSPC to have a "unique space culture," although he provided no subsequent explanatory details.[65] AFSPC vice commander and fighter pilot Lt. Gen. Dan P. Leaf gave something more specific in the third issue of *High Frontier*, explaining the need to establish a true "combat culture" by "meld[ing] the combat sense and discipline of the missile culture with the creativity of the space operations community."[66] Approaching the end of his two-year tour at AFSPC, Leaf seemingly struggled to identify the cultural traits of his command, perhaps still a fish out of water in his first space assignment.[67] After all, did missileers really have more combat "sense" given the increased integration of space capabilities into air planning?

Still, Leaf's emphasis on a "combat culture" overlapped with Lord's desire to impart a warrior culture into AFSPC that reflected his personal goal of emulating his father's combat service in the Army during World War II and his own subsequent decision to become a missileer. A warrior culture, however, can be problematic in moving away from the traditional notion of those in uniform as defenders toward one of warriors on the offense.[68] As such, the embrace of "guardians" by the Space Force in 2020 reflects a move back in the correct direction while embracing its earlier heritage.[69]

More authentically, though, Lord reinforced this intensified identity by introducing important "artifacts," or what organizational scholar Schein contends is the most visible form of culture.[70] In this case, Lord directed the development of a new space badge for space and ICBM operations, even presenting the first space badge to Gen. Bernard A. Schriever, who had contributed so much to space and missile capabilities in the early Cold War in a timely and classy tribute, given Schriever died the following month. A short article Lord penned heralded Schriever for being a "lone voice advocating the space and missile capabilities many now take for granted." He had been "chastised for his outspokenness. He talked openly of Space Supremacy and Space Superiority well before the launch of Sputnik. Following one notable speech the Secretary of Defense admonished [Schriever], "Do not use 'Space' in any of your speeches in the future."[71] Lord viewed himself, perhaps somewhat egotistically, in some ways as a successor to Schriever, even though notions of space superiority had become increasingly palatable to the national security community by this point in time. He also implemented the badge as part of a larger effort to increase the sense of self-worth among missile and space operators given their and Lord's resentment at being something akin to "orphans" in the Air Force compared to the pilots.[72] It did not help that AFSPC acquired so much of its senior leadership from the fighter pilot community; however, it took time for the space community to grow its own high-ranking leadership.[73] This trend would subsequently be replicated for cyber operators as well under Lord's successors.

Lord retired on March 3, 2006. In terms of supporting the joint warfighter in the global war on terror, he pointed to how space capabilities served as a force multiplier for the joint community, contending that "we have traded the traditional necessity for massed forces by using space capabilities for precision, speed, and the ability to quickly maneuver on the battlefield."[74] Internally, he promoted the same emphasis on jointness within the space community, including by mentioning other services prominently on *High Frontier*'s pages and seeking to integrate more Army and Navy space professionals into the space and missile systems center.[75]

Lord had also called for a greater emphasis on space superiority and increasing the weaponization of space-based systems. However, external events intervened with implementing the space commission's guidance to transform space. Whereas Operation Desert Storm had been a boon for the space community in making it a household name among the joint community, the global war on terror inhibited more investment in future capabilities. The joint communities' hunger for space capabilities paradoxically had become an albatross around AFSPC's neck.[76]

As such, Lord worried more about resources than organization, which he considered a "fourth order" issue."[77] While it can be tempting in the wake of the Space Force's 2019 establishment to look for early ideas and advocates of this notion, it is clear that Lord "did not see a need to create a separate space service in the military." That does not mean that he did not have a vision for AFSPC's future. Rather, his vision of the future of the Air Force entailed bifurcating capabilities, placing kinetic ones in Air Combat Command and nonkinetic ones in the Air Force, for which he pushed for bringing together cyber and space under the same roof.[78]

Lord's retirement greatly concerned supporters of the space community as the Air Force considered how to trim numbers across its service. Rumors abounded, such as possibly moving the Space and Missile Systems Center back into Air Force Material Command.[79] Senator Wayne Allard, for example, worried so much that he demanded the Air Force convey its intent in writing to continue to fill AFSPC with a four-star commander, ignoring Lord's reassurances.[80]

Upon Lord's retirement, AFSPC underwent some leadership uncertainty, with Lt. Gen. Frank G. Klotz succeeding Lord as the acting commander from April to June 26, 2006. Klotz sought to assure anyone who would listen that the Air Force valued a space command enough to keep AFSPC intact.[81] Next, Gen. Kevin P. Chilton served as commander of AFSPC for just over a year until October 3, 2007, when he received a new assignment as the commander of USSTRATCOM.[82] Interestingly, Chilton believed that space advocates best served the space community not by separating themselves from the Air Force, as had recently been tried with acquisitions, but more often than not "by having space as an integrated part of Air Force planning and operations."[83] Chilton had a broad background as a fighter pilot, reconnaissance pilot, test pilot, astronaut, and AFSPC deputy director of operations that made him a particularly compelling choice to better integrate air and space.[84] Another acting commander served in the position for only nine days until finally AFSPC received a leader who would serve for more than three years, Gen. C. Robert Kehler, who had been temporarily appointed to lead USSTRATCOM while Congress considered his nomination for a fourth star.[85]

Gen. C. Robert "Bob" Kehler

Gen. Bob Kehler assumed command of AFSPC in 2007 in the wake of two critical events: the Chinese antisatellite weapon test earlier that year and two scandals that had resulted from improperly moving nuclear material.[86] Of those two, the latter proved to be a more important priority for him, with China dropping largely off AFSPC's radar due to ongoing efforts in the global war on terror. His third priority centered on developing a sound strategy for space acquisitions.[87] Kehler viewed himself as perhaps "the first of the really insider commanders. Gen. Lord [Gen. Lance W. Lord] was pretty good. . . . [H]e was about 75% ICBM guy and about 25% space guy. I'm about 50/50 probably. Shelton's probably about 75% or 80/20 space."[88]

Ironically, then, the first "insider" spent more of his tenure at AFSPC fixing the ICBM community rather than being focused on space, by no fault of his own. As he had stated early in his tenure, he had hoped to make strengthening space situational awareness his primary focus.[89]

But in noted contrast to Lord, who showed more interest in the command's internal workings, Kehler looked to improve AFSPC's status far more externally within the Air Force writ large. Kehler believed that the problem was not that AFSPC was an orphan, as Lord had thought, but that it lacked the key connections to give it its rightful clout. Kehler explained, "Some describe it as the 'Rodney Dangerfield' command. 'We don't get no respect.'. . . . It isn't that no one likes space. It isn't that you're the stepchildren. It isn't that the Air Force doesn't want you or like you. It's mostly that the command has never grown the attachment points that it needs to be [e]ffective."[90] As a result, Kehler viewed his role as commander not to focus on capabilities but rather to advocate for AFSPC in Washington, DC, where he did much of his leading upwardly. Internally, he brought fighter pilots to AFSPC precisely for the same reason—to use their connections in hopes of improving AFSPC's ability to advocate for itself.[91]

In some respects, however, Kehler and Lord had some important overlaps. Kehler graduated from Pennsylvania State University in 1975; like Lord, he also obtained a bachelor of science degree in education. Also, like Lord, he became a missileer, qualifying on both the Minuteman II and Minuteman III. As a legislative liaison for the Air Force, he worked on issues regarding the ICBM modernization program. Twenty years after commissioning, he received his first assignment to a space billet, becoming the inspector general of AFSPC before serving as its deputy director of operations. He then commanded the 30th Space Wing at Vandenberg Air Force Base, where he gained experience with the Titan II, Titan IV, and Delta II launch systems. Before assuming command of AFSPC, he filled several other positions and, ultimately, served as the deputy commander for USSTRATCOM.

Early on, Kehler took the traditional stance of space as supporting the terrestrial warfighter; in a speech he gave in February 2008, a few months after taking command, he stated, "Space power has helped shape the American approach to warfare; it gives our warfighters a precise advantage. . . . Without space, military operations would be far less precise, focused, timely, coordinated and efficient, and far more costly."[92] To support this approach, AFSPC considered how best to contribute to joint counterinsurgency efforts, such as moving global positioning system (GPS) satellites to better support troops in mountainous Afghanistan and sending more AFSPC airmen on longer deployments to the Middle East to increase their firsthand understanding of the ground forces' requirements.[93] AFSPC also continued to interweave air and space professionals together, incorporating more than seventeen hundred intelligence and communications airmen into its training and education beginning in September 2008.[94]

Meanwhile, Kehler and Secretary of the Air Force Michael Wynne, who had extensive experience in space technology, remained vague on needing to acquire offensive space capabilities, which previous AFSPC commander General Lord had found to be a "lightning rod" before Congress.[95] If Kehler had openly advocated for preparing for warfare in space, against guidance from the secretary of the Air Force, it would have detracted from civil-military relationships.[96] AFSPC tried to be more forward-leaning in other areas to preempt congressional intervention. On March 31, for example, it established a space protection program to determine "how best to protect our space systems and stay ahead of the threat."

Externally, problems in controlling nuclear material led to an explosion in organizational change sweeping AFSPC and the Air Force as a whole. In the wake of AFSPC's mishandling of nuclear material, the secretary of defense task force on nuclear weapons management urged the inclusion of the bomber community into AFSPC for two reasons. First, such a change would "broaden" AFSPC's vision into an Air Force Strategic Command. Second, and more practically, it would help to increase promotion opportunities for nuclear-related career fields, which the task force identified as deficient compared to those afforded to pilots in Air Combat Command.[97] Instead, though, the Air Force chose to establish Global Strike Command the following year, which suggests that the Air Force's top leadership (primarily pilots) was not ready to have space professionals in leadership positions over pilots.

Rather than incorporating bombers into AFSPC, the Air Force decided to make it the major command for cyberspace rather than establishing cyberspace as its own major command. This may have been a compromise decision, especially after the Air Force provoked the other services by awkwardly setting up a provisional cyber command in what some perceived as a power grab.[98] Although the move gave more clout to AFSPC in the short term, it may have hurt the Air Force in the long term in its possible goal to become the service most responsible for the cyber domain. One outsider thought that this move was the right one given his belief that the Air Force needed to get its nuclear capabilities in order, whereas cyber capabilities were "less urgent."[99] Ultimately, Kehler supported the idea of bringing together the two domains given their interdependence.[100]

In 2009 AFSPC lost its ICBMs to the newly formed Global Strike Command, and Kehler found himself in a space community newly augmented by cyber capability. Chief of Staff Gen. Michael T. Mosely had told Kehler's predecessor General Chilton that his experience with air and space power would be critical to solving the "challenges ahead of us as an air and space force."[101] Under Kehler's watch, AFSPC would now help transform the Air Force into an air, space, and cyber force.

In terms of space capabilities, Kehler acknowledged that the problem AFSPC faced in the space domain was nothing new. The few satellites the United States had were highly vulnerable and increasingly aged, something he pointed out had been presciently anticipated in the 1970s by Sir John Hackett in his book, *World War III, August 1985*.[102] Kehler, then, did not have much to propose in the way of new solutions. Rather, he sought a new strategic approach to pursuing the required capabilities, arguing he had done so since taking command.[103] Under his command, AFSPC also nurtured Twenty-Fourth Air Force and its cyber capabilities into operational readiness.[104]

One expert space analyst views Kehler as too "committed to space as a supporting capability" and as being "all sound bites."[105] This is perhaps overstated. First, if Kehler seemed to consist of sound bites, that is because he saw his leadership role as externally focused on improving AFSPC's position with relation to the Air Force. Second, his upfront commitment to supporting terrestrial capabilities was matched behind the scenes with an intense dislike of the current Air Force's tendency to make everything from the B-2 to the F-22 into an existential "Billy Mitchell" moment. Kehler felt like the Air Force had lost its mind and its credibility.[106]

Gen. William L. Shelton

Next to take the helm was Gen. William L. Shelton, who became hooked on space as a child while watching President John F. Kennedy address the nation about putting a man on the moon and subsequently watching space launches.[107] Shelton graduated from the Air Force Academy with a degree in astronautical engineering in 1976 before obtaining a master's degree in the same field at the U.S. Air Force Institute of Technology.

Unlike Lord and Kehler, Shelton began his military career in the space field rather than as a missileer, first serving as a launch facilities manager at Vandenberg Air Force Base's space and missile test center. He then became a space shuttle flight controller for four years at the Johnson Space Center before returning to the military space world, serving as a staff officer at Peterson Air Force Base as the deputy chief of staff for operations and the office of space plans and policy in the office of the secretary of the Air Force.

Subsequently he commanded the 2nd Space Operations Squadron, responsible for GPS satellites, and later the 50th Operations Group, which operated numerous satellite types, in addition to various other positions and schools. He served in a vast array of assignments at USSTRATCOM and elsewhere, including commanding Fourteenth Air Force in 2008. Like Lord, Shelton also served as the assistant vice chief of staff of the Air Force before assuming command of AFSPC.[108] In short, he arguably had more space experience than any other previous AFSPC commander. The year after assuming command, he received a vice commander who was also a space professional, Lt. Gen. John Hyten. Having two space professionals in the top two positions at AFSPC represented the maturation of the space leadership community as it shifted away from its reliance on fighter pilots as leaders.

Hyten also complemented Shelton in being more of a people-focused leader.[109] In many ways, Shelton's leadership style reflected his engineering background. He tended to be straightforward, even blunt, seen in his terse commentary citing historical and present space capabilities. He had little patience with buzzwords and bullet points. Speaking to an audience at the space symposium, for example, he explained that the process of developing space capabilities "faster, better, cheaper" would always fail when the "Air Force treated it as nothing more than a PowerPoint slogan."[110]

Shelton had three overarching priorities that evolved naturally from Kehler's tenure: supporting terrestrial warfighting capabilities, reducing the cost of space capabilities, and "normaliz[ing] and operationaliz[ing]" cyber capabilities as had occurred with space over the previous twenty years.[111]

Less than a month after assuming command, the space professional made it clear that he was serious about cyber, putting Air Force commanders on notice that they needed to take the cyber domain as seriously as they took the air domain. At an Air Force Association conference, Shelton insisted that commanders respond to cyber tasking orders upon receipt to ensure network defenses were updated as necessary.[112] Under his tenure, the community of cyber professionals also grew, in some years, up to 15 percent.[113]

By contrast, with *High Frontier* shutting down, one important forum for discussion went away. Shelton justified this change by claiming it was time for space and cyber professionals to "move into the mainstream academic discussion in other professional publications."[114] Perhaps heightening budgetary constraints also drove this decision, making it difficult to fund highly expensive and difficult-to-defend satellites packed with multiple capabilities.

Calling this moment in time a "strategic crisis" given the convergence of both a "radically different operating environment and a declining budget," Shelton called for change.[115] Possible solutions included "disaggregation," increased partnership with more affordable civilian spacepower solutions, and commercial off-the-shelf buses for spacecraft, which furnish propulsion, power, and other satellite essentials.[116]

Disaggregation subsequently received the most focus, with AFSPC releasing a white paper entitled "Resiliency and Disaggregated Space Architectures" in August 2013. By building a few smaller satellites and spreading capabilities throughout satellite constellations, AFSPC hoped to make U.S. space capabilities not only more resilient to space debris but also work as a deterrent against other nations' deliberate aggression. Early in his tenure, Shelton stressed the likelihood of accidental debris over adversarial actions, although his emphasis on the contested nature of space increased over the course of his command.[117] What made disaggregation even possible was the fact that space launch was becoming more affordable.[118] From today's vantage point, it is difficult not to read the huge increases in civilian launch capability into these statements. However, it is important to note that Shelton doubted in 2012 that launch costs would "ever get cheaper" than they currently were.[119] That is not to critique him for being shortsighted; rather, it is to point out that we must conduct our historical analysis from the past forward rather than read the present back into the past.[120]

Although Shelton faced budgetary cuts from the beginning of his tenure, he received an elevated challenge in March 2013 when sequestration went into effect. As a result, he lost $508 million in AFSPC's budget. AFSPC responded in a variety of ways, such as shutting down one-third of space fence radar coverage, which had the effect of no longer covering the eastern portion of the United States and reducing civilian contractors by 50 percent.[121] Shelton wryly commented publicly that what Congress was doing to AFSPC was far more dangerous than adversaries' plans, considering his decisions akin to "choosing among your children."[122] In this case, though, the space fence, which tracked debris and satellites, had already been determined to be outdated.

In seeking to save money, the engineer made several important changes to AFSPC's cultural fabric. First, he introduced a clearer distinction between pilots, missileers, and space operators by prohibiting the wearing of flight suits by AFSPC personnel unless actively "engaged in flying operations." While he largely justified the decision as a practical one that saved money and made sense given that satellite operators did not actually fly, it had cultural implications as well by dividing the perceived "haves" and "have nots." The idea spread outside AFSPC through some other quarters of the Air Force, where it caused some consternation.[123] While moving away from associating with pilots, Shelton also narrowed down who could wear the space badge, seeking to associate it solely with space operators.[124]

Looking forward, he hoped to move beyond thinking about space or cyber capabilities to a more seamless emphasis on "information" to end a stovepiped view of what a single domain could provide. Seeking that holistic integration of capabilities may be one reason he, like his predecessors, opposed AFSPC morphing organizationally into something more akin to U.S. Cyber Command.[125]

Meanwhile, space professionals' concern for increasing adversarial threats ratcheted up. In response, perhaps, in his final year in command in 2014, Shelton appeared to casually—no doubt for theatrical effect—reveal the existence of a formerly classified satellite: the Geosynchronous Space Situational Awareness Program, which AFSPC intended to use to observe Russian and Chinese satellites more closely.[126] Some outsiders viewed military space professionals as increasingly demonstrating an overhyped anxiety about conflict in space.[127] Others, however, saw developments like this one as a natural response to China's launch of a satellite into geosynchronous orbit in 2013, which subsequently resulted in the 2014 space strategic portfolio review.

Conclusions

General Lord, General Kehler, and General Shelton each took the helm of AFSPC during a period of vast organizational change in the wake of the normalization of space capabilities in the Department of Defense after seemingly being "discovered" by the joint force in Operation Desert Storm in 1991. The next decade saw enormously bumpy organizational changes for space professionals, many of which were out of their control as the Air Force instituted them. Meanwhile, AFSPC commanders helped more fully integrate spacepower capabilities into joint planning.

From today's perspective, it is tempting to look backward at these three leaders and evaluate them based on the extent to which they loudly proclaimed and anticipated the birth of the Space Force, pushing hard for more recognition and equality. This would be a mistake, given that we have the benefit of knowing how the story turned out.

Evaluating leaders is difficult; there are multiple factors that are outside many leaders' control because most leaders are also simultaneously followers. Importantly, though, each AFSPC commander discussed in this chapter not only advocated for the operational specialty in which he first served but also increasingly brought together capabilities from different domains. Even as the first generation of space leadership matured during this era, as seen by Shelton and Hyten filling the top two positions of AFSPC, they continued to evolve, appreciating the need to integrate other emergent domains.

Notes

1. C. Robert Kehler, "Implementing the National Security Space Strategy," *Strategic Studies Quarterly* 6, no. 1 (Spring 2012): 18–26. I am grateful to Molly Schmidt for her thoughtful comments.
2. Adam J. Hebert, "Turning a Corner on Space," *Air and Space Forces Magazine*, January 1, 2007; https://www.airandspaceforces.com/article/0107space/; David N. Spires, *Beyond Horizons: A Half Century of Air Force Space Leadership* (Maxwell Air Force Base, AL: Air University Press, 1997), 266–67.
3. Email conversation with Everett Dolman. He argued this represented a change in the Air Force approach. I suggest the rationale for the change was simply that nonrated officers had matured into this position.
4. Quoted in Carl M. Jones, "E Pluribus Unum: Strengthening the Air Force Space Command Culture," Strategy Research Project, U.S. Army War College, 2011, https://apps.dtic.mil/sti/citations/ADA543654.
5. Tom Karp and Thomas I. T. Helgo, "Leadership as Identity Construction: The Act of Leading People in Organisations: A Perspective from the Complexity Sciences," *Journal of Management Development* 28, no. 10 (2009): 882.
6. Karp and Helgo, 883.
7. Karp and Helgo, 884.
8. Jones, 2, 8, 10. Jones discusses the role of ICBMs in a cohesive Strategic Air Command but never addresses that Strategic Air Command similarly had subcultures that separated the bomber and ICBM communities, which undercuts his analysis to some extent. He does note a key common ingredient linking the two being an emphasis on "perfection."

9. Quoted in Frank Gallegos, "After the Gulf War: Balancing Space Power's Development," in *Beyond the Paths of Heaven: The Emergence of Space Power Thought*, ed. Bruce M. DeBlois (Maxwell Air Force Base, AL: Air University Press, 1999), 63; Spires, 245.
10. Gallegos, 63.
11. Tyler M. Evans, "Space Coordinating Authority Information Services from Space," in *Space Power Integration: Perspectives from Space Weapons Operators*, ed. Kendall K. Brown (Maxwell Air Force Base, AL: Air University Press, 2006), 1–2; Spires, 244.
12. Evans, 15.
13. Gallegos, 64.
14. Spires, 275.
15. Merrill A. McPeak, *Selected Works 1990–1994* (Maxwell Air Force Base, AL: Air University Press, 1995), 207.
16. Spires, 275.
17. McPeak, 222, 289, 350–51.
18. Gallegos, 83.
19. Gallegos, 65.
20. Gallegos, 80.
21. Spires, 283–84; Jones, 15.
22. Donald H. Rumsfeld, chairman, *Report to the Commission to Assess United States National Security Space Management and Organization* (Washington, DC: U.S. Government Printing Office, 2001), ix, xxi.
23. General Estes, for example, commanded North American Aerospace Defense Command and United States Space Command from 1996 to 1998, but this represented a departure from the bulk of his career as a fighter pilot. U.S. Air Force, Gen. Howell M. Estes III, biography, https://www.af.mil/About-Us/Biographies/Display/Article/107122/general-howell-m-estes-iii/.
24. Rumsfeld, xxiii.
25. Rumsfeld, 90.
26. Air Force Space Command Directorate of History, oral history program, interview of Gen. Lance W. Lord by George W. Bradley III, Dr. Rick W. Sturdevant, and Maj. Donald Holloway, January 29, 2007, 3–4, 8.
27. Lord interview, 34.
28. Lord interview, 5.
29. Lord interview, 7–8.
30. Lord interview, 10.
31. Lord interview, 11.
32. Lord interview, 9.
33. Lord interview, 17–18.
34. Lord interview, 12.
35. U.S. Air Force, Gen. Lance W. Lord, biography, n.d., https://www.af.mil/About-Us/Biographies/Display/Article/105049/general-lance-w-lord/.

36. Gen. Lance W. Lord, "Commanding the Future: The Transformation of Air Force Space Command," *Air and Space Power Journal* 18, no. 2 (Summer 2004): 12.
37. Lord interview, 30.
38. Michael F. Stumborg, "Air Force Space Command: A Transformation Case Study," *Airpower Journal* (Summer 2006): 82, www.airuniversity.af.edu/Portals/10/ASPJ/journals/Volume-20_Issue-1-4/2006_Vol20_No2.pdf.
39. Lord, "Commanding the Future," 13.
40. Gen. Lance W. Lord, "Welcome to *High Frontier!*" *High Frontier* 1, no. 1 (Summer 2004): 3.
41. Lord, "Welcome," 4.
42. Stumborg, 82.
43. Rumsfeld, 43.
44. Lord, "Welcome," 4.
45. Lord, "Developing Space Professionals," 7.
46. Stumborg, 83.
47. Lord, "Welcome," 4.
48. As will be discussed later, Lord also worked internally to strengthen support to the joint warfighter. Lance W. Lord, "Forging Space Warriors," *Joint Force Quarterly* 33 (Winter 2002–2003): 43. The title, ironically, is not indicative of the article's content.
49. Kendall K. Brown, "Acknowledgments," in *Space Power Integration*, ix.
50. Stumborg, 84, 85; Evans, 1–2, 16. This authority had formally been the responsibility of the joint Air Force component commander, having entered Air Force doctrine in this format in 2001.
51. U.S. Congress, 108-1, "Hearings: Department of Defense Authorization for Appropriations for Fiscal Year 2004, S. Hrg. 108–241, Part 7, March 12–April 2, 8, 2003," 2004, 35–36.
52. Kendall K. Brown, "Space Power Integration: Perspectives from Space Weapons Operators," *Air and Space Power Journal* (Summer 2006): 50–51, https://apps.dtic.mil/sti/citations/ADA463471.
53. Brown, "Space Power Integration," 53.
54. Brown, "Space Power Integration," 53; U.S. Air Force, Gen. Gregory Martin biography, n.d., https://www.af.mil/About-Us/Biographies/Display/Article/104705/general-gregory-s-martin/.
55. Lord interview, 19. For the importance of that continuing approach in the subsequent AFSPC commander General Shelton, see end of tour interview, tape two, General Shelton, Commander, Air Force Space Command, 10.
56. Stumborg, 82; Lord, "Commanding the Future," 13.
57. Stumborg, 84.
58. Lord interview, 7–8, 29.
59. Lord interview, 22.
60. Lord interview, 42.
61. Email correspondence with Dr. Everett Dolman, March 4, 2022.
62. Evans, 10–11.

63. Ben Iannotta, "Space Protection: How Far Will America Go to Protect Its Satellites?" *C4ISR—The Journal of Net-Centric Warfare* (June 1, 2008): 18.
64. Lord, "Developing Space Professionals," 7. For other bland uses of "culture," see Col. John E. Hyten, "The First Line of Defense," *High Frontier* 2, no. 3 (n.d.): 29. In a different issue, a major claimed more precisely: "The service culture of the United States Air Force has always been characterized by a hunger for new technology that will aid in its mastery of the high ground. Nowhere is this service culture more apparent than in Air Force Space Command (AFSPC)," with all the large models of platforms hanging in the headquarters lobby. Maj. Patrick H. Donley, "Space Systems—More than Hardware and Hope," *High Frontier* 2, no. 1 (n.d.): 30.
65. Col. James C. Hutto, "Developing Space Professionals Crucial to Critical Wartime Roles," *High Frontier* 1, no. 1 (Summer 2004): 10.
66. Lt. Gen. Dan P. Leaf, "Providing Combat Effects to the Battlefield," *High Frontier* 1, no. 4 (n.d.): 4.
67. U.S. Air Force, Lt. Gen. Dan P. Leaf, biography, n.d., https://www.af.mil/About-Us/Biographies/Display/Article/104990/lieutenant-general-daniel-p-leaf/#:~:text=Lt.%20Gen.%20Daniel%20P.%20Leaf%20is%20Deputy%20Commander%2C,of%20Wisconsin-Madison%20Air%20Force%20ROTC%20program%20in%201974.
68. Lord, "Commanding the Future," 13. Also see Michael Howard, "Rendezvous in Space: Looking in on Military Space Power," *Army Space Journal* (Summer 2010): 31, https://apps.dtic.mil/sti/pdfs/ADA559420.pdf, in which he compared airpower's trajectory, noting that although "space power emerged as a critical enabler of combat power during the Iraq and Afghanistan wars, it has not proven that it can deliver the 'game-changer' while getting its nose bloodied in combat on par with airpower's combat test in WWII"; William Tresder, "'Warfighter' Is Not the Best Way to Define Service Members," *Task and Purpose*, March 3, 2015, https://taskandpurpose.com/news/warfighter-wrong-way-define-american-service-member/.
69. Secretary of the Air Force Public Affairs, "U.S. Space Force Unveils Name of Space Professionals," December 18, 2020, https://www.spaceforce.mil/News/Article/2452593/us-space-force-unveils-name-of-space-professionals/.
70. Edgar Schein, *Organizational Culture and Leadership*, 4th ed. (Hoboken, NJ: Wiley, 2010), 23–33.
71. Gen. Lance W. Lord, "We Walked with a Legend: General Bernard A. Schriever, 1910–2005," *High Frontier* 1, no. 4 (n.d.).
72. Lord interview, 21; also see similar comments on 13–14, 16, 23. This point is one of the dominant themes in Lord's oral history.
73. Lord interview, 32.
74. Quoted in Evans, 21.
75. "Lord Opens Doors to Army, Navy Spacemen," *Air and Space Forces Magazine*, February 3, 2006, https://www.airandspaceforces.com/1517lord. This had been a recommendation of the 2001 Space Commission. See Rumsfeld, 92. The first issue of *High Frontier*, for example, contained a "Total Force Space Development Section" with representatives

from the Army, Navy, and Marine Corps each contributing an article. See *High Frontier* 1, no. 1 (Summer 2004): 15–21.
76. Jeremy Singer, "War on Terror Supersedes 2001 Space Commission Vision," *Space News*, January 23, 2006, https://spacenews.com/war-terror-supersedes-2001-space-commission-vision/.
77. Howard, 31.
78. Lord interview, 22.
79. "Allard Vows to Fight Space Command Downgrade," *Air and Space Forces Magazine*, March 20, 2006, https://www.airandspaceforces.com/1070allard/.
80. "Take It to the Bank?" *Air Force*, April 21, 2006, https://www.airandspaceforces.com/1040bank/; "All Quiet on the High Frontier," *Air and Space Forces Magazine*, February 23, 2006, https://www.airandspaceforces.com/1040quiet/.
81. "Space Command Lives," *Air and Space Forces Magazine*, April 7, 2006, https://www.airandspaceforces.com/1060space-2/.
82. "Air Force's Chilton Named to Take Over STRATCOM," *Defense Daily*, July 25, 2007.
83. Jeremy Singer, "Vacancies Offer Pentagon Opportunity to Integrate Space Efforts, Experts Say," *Federal Times*, October 15, 2007, 18.
84. "Chilton Tapped for AFSPC," *Air and Space Forces Magazine*, May 1, 2006, https://www.airandspaceforces.com/1030chilton/.
85. "Chilton Takes Over U.S. Strategic Command," *Defense Daily*, October 4, 2007.
86. However, space leaders did not necessarily see this as a Sputnik-type event in terms of consequence given the United States had already demonstrated this capability. See Lord interview, 25. Also, several leaders anticipated it, and felt that now they no longer had to convince Congress that adversaries had the capability of targeting U.S. space capabilities. Lord interview, 38.
87. Air Force Space Command, directorate of oral history program, interview of Gen. C. Robert Kehler on end of tour by Mr. George W. Bradley III, December 22, 2010, 1.
88. Kehler interview, 26.
89. Michael Hoffman, "Space Maneuvers Satellite Shootdown Brings Security Concerns to the Fore," *Air Force Times*, March 3, 2008, 16.
90. Kehler interview, 2.
91. Kehler interview, 2–4.
92. Monique Randolph, "AF Leaders Discuss Future Challenges at Symposium— Commanders Drive Home the Need to Recapitalize Aging Aircraft and Space Systems," *Hansconian* (February 29, 2008), 8.
93. John Reed, "U.S. Space Worries: Budget Cuts, Cyber Threats," *Defense News* (April 19, 2010), 29.
94. Erik Holmes, "More Airmen to Be Included in Space Training," *Air Force Times*, September 29, 2008, 21.
95. Michael Hoffman, "Space Command, NRO Plan Satellite Protection Strategy," *Air Force Times*, May 19, 2008, 25; Erik Holmes and Michael Hoffman, "Radical Change—Proposal to Move Bombers and Their Nuclear Weapons into Space Command Would Likely

Benefit Airmen's Careers," *Air Force Times*, September 29, 2008, 16; Gen. C. Robert Kehler, Global Warfare Symposium, November 21, 2008, 8, https://secure.afa.org/events/natlsymp/2008/scripts/Kehler.pdf. Wynne resigned after the Air Force mistakenly transported nuclear weapon fuses to Taiwan in 2006, and then in 2007 a B-52 with six nuclear weapons flew across the United States. For Lord's comment, see Lord interview, 31.

96. Kehler's successor also, when asked about a future "space war," stressed defensive capabilities, at least early on in his tenure. Gen. William L. Shelton, speech, Air Force Association's Air and Space Conference and Technology Exposition, September 20, 2011, https://www.afspc.af.mil/About-Us/Leadership-Speeches/Speeches/Display/Article/252873/general-sheltons-afa-speech-2011/.

97. Jason Simpson, "Task Force: Put Overall Nuke Mission under Air Force Space Command," *Inside Missile Defense* 14, no. 20 (2008): 16–17, http://www.jstor.org/stable/24785875; Marcus Weisgerber and Jason Simpson, "In Wake of Schlesinger Report ...: USAF Exploring Several Options for Nuclear Bomber Re-Structuring," *Inside Missile Defense* 14, no. 20 (2008): 18–19, http://www.jstor.org/stable/24785877; Holmes and Hoffman, "Radical Change," 16.

98. Kehler interview, 19; Erik Holmes and Michael Hoffman, "Questions Remain about Cyber Force," *Air Force Times*, October 28, 2008, 12. Interdependence may not be the correct word. Rather, it may be the space domain's reliance on the cyber domain. See, for example, Sean Gallagher, "Cyber-Overhaul," *C4ISR*, August 1, 2009, 30. Air Force rhetoric, however, may have spouted this interdependence. See, for example, Carlo Muñoz, "Decisions Announced: Air Force Creates Nuclear Command, Re-Assigns Cyber Command," *Inside the Air Force* 19, no. 41 (2008): 14, http://www.jstor.org/stable/24795094.

99. Bob Brewin, "Force to Place Cyber Operations Under Space," NextGov.com, October 8, 2008, 1.

100. Holmes and Hoffman, "Questions Remain."

101. "An Astronaut Leads Space Command," *Air and Space Forces Magazine*, June 29, 2006, https://www.airandspaceforces.com/1070afspc/.

102. Kehler, Global Warfare Symposium, 2008, 13.

103. Kehler, Global Warfare Symposium, 2008, 11.

104. Gen. C. Robert Kehler, speech, Armed Forces Association Global Warfare Symposium, November 2010, https://www.afspc.af.mil/About-Us/Leadership-Speeches/Speeches/Display/Article/252877/afa-global-warfare-symposium-speech/.

105. Email correspondence with Dolman.

106. Kehler interview, 12, 16. After leaving AFSPC, he became the commander for U.S. Strategic Command.

107. Gen. William L. Shelton, speech, "National Security Space: Then. Now. Tomorrow," May 20, 2014, https://www.afspc.af.mil/About-Us/Leadership-Speeches/Speeches/Display/Article/731713/national-security-space-thennowtomorrow/.

108. Everett Dolman contends that the Air Force was "not sending its best" to AFSPC, stating that Shelton believed in space being subordinate to airpower. In an alternative perspective, one would assume the position of vice assistant chief of staff of the Air Force is hardly filled by subpar officers. Email with the author, March 4, 2022.
109. "CV Earns 3rd Star, Brings Wealth of Experience," *Targeted News Service*, June 22, 2012.
110. Dan Ward, "Acquisition Reform . . . for Real," *Armed Forces Journal*, April 1, 2012, 16. For other blunt comments, see, for example, "Shelton Talks Space, Cyberspace at AFA Air and Space Conference," *Targeted News Service*, September 19, 2013. In one case, his style did not help him when he belittled a Government Accountability Office (GAO) report as conducted by a person with a "bean-counter perspective," that subsequently the GAO auditor fired back regarding costs overruns, "We were right." Titus Ledbetter III, "GAO Auditor Works to Keep U.S. Space Activities within Budget," *Army Times*, July 16, 2012, 12.
111. End of tour interview session, session one, Gen. William L. Shelton, commander, Air Force Space Command, September 4, 2014, 1. These differed somewhat from the ones he gave before Congress. See Department of Defense Authorization for Appropriations for Fiscal Year 2014 and the Future Years Defense Program, part 7, S. Hrg. 113–108, 2014, 58. There he stated he wanted to "provide assured full-spectrum space capabilities," "develop highly skilled and innovative space professionals," and "provide resilient, integrated systems that preserve operational advantage."
112. Scott Fontaine, "Respect Cyber Orders, AFSPC's Shelton Warns," *Air Force Times*, February 28, 2011, 19.
113. "Airmen Must Understand Business of Cyber, General Says," *Targeted News Service*, June 6, 2013.
114. Gen. William L. Shelton, "Introduction," *High Frontier* 7, no. 4 (Spring 2011): 2.
115. Shelton, "Strategic Crossroad," 4–10.
116. Shelton, "Strategic Crossroad," 8–9. One could argue, on the other hand, that this was not a new development. The operational responsive space program had demonstrated a similar interest in quickly launching less expensive satellites. That program was stood up as a joint effort at Kirtland Air Base in May 2007, in part because of Congress' urging. See Michael Hoffman, "Wynne Calls for More Investment in Space Program," *Air Force Times*, April 21, 2008, 22.
117. See, for example, Alex Brown, "Military to Congress: Help Us Avoid 'Gravity II,'" *National Journal: Web Edition Articles*, March 12, 2014. For the increased emphasis on space's contested characteristics, see, for example, "Space Superiority Remains Vital to National Security," *Targeted News Service*, April 8, 2014.
118. Aaron Mehta, "AF: Pentagon Must Update Space Policy," *Air Force Times*, September 9, 2013, 20; Mike Gruss, "Air Force Report Cites Benefits of Disaggregation," *Space News*, August 28, 2013, https://spacenews.com/36982air-force-report-cites-benefits-of-disaggregation/. For some of the changes in launch, see Colin Clark, "ULA, SpaceX Rumble Shaping Up to Rival Tanker Wars," *Breaking Defense*, May 5, 2014,

https://breakingdefense.com/2014/05/ula-spacex-rumble-shaping-up-to-rival-tanker-wars/.
119. Don Jewell, "A Conversation with General William Shelton, Commander, Air Force Space Command," September 13, 2012, https://www.gpsworld.com/a-conversation-with-general-william-shelton-commander-air-force-space-command/.
120. A possible counterpoint is increasing interest in commercial spaceflight as early as 2008. See, for example, see Dave Ahearn, "Commercial Space Sector Poised For Liftoff: Experts," *Defense Daily*, February 8, 2008, https://www.defensedaily.com/commercial-space-sector-poised-for-liftoff-experts/space/.
121. Colin Clark, "Space Command Juggles Budget in Face of North Korean Threat, Sequestration," *Breaking Defense*, April 9, 2013, https://breakingdefense.com/2013/04/space-command-juggles-budget-in-face-of-north-korean-threat-spa/; Aaron Mehta, "'Tough Choices for Space Command with Budget Cuts," *Air Force Times*, April 22, 2013, 15. However, the space fence was an obvious choice because it had already been "aging" and due to be replaced. Bob Brewin, "Sequester Shutters System for Tracking Threats from Outer Space," NextGov.com, August 13, 2013, https://www.nextgov.com/cxo-briefing/2013/08/sequester-shutters-system-tracking-threats-outer-space/68651/. AFSPC provided *Space News* a detailed response of its decision in Mike Gruss, "Gen. Shelton on Space Fence Closure and the Road Ahead," *Space News*, August 28, 2013, https://spacenews.com/36974sn-blog-gen-shelton-on-space-fence-closure-and-the-road-ahead/.
122. Mike Gruss, "Shelton: Sequestration Could Break Military Space Program," *Space News*, September 23, 2013, https://spacenews.com/37270shelton-sequestration-could-break-military-space-program/; Mehta, "Tough Choices," 15. Whereas Shelton believed that these cuts would "break" every space program, some believed that AFSPC remained in far better shape than other areas of the military. Quoted in Joan Johnson-Freese, *Space Warfare in the 21st Century: Arming the Heavens* (New York: Taylor & Francis, 2016), 17. For a competing perspective, see Barbara Opall-Rome, "U.S. Seeks Space Hotline with China," *Defense News*, February 6, 2012, 6.
123. AFSPC Public Affairs, "Space Command Officials Eliminate Flight Suit, Jackets," April 16, 2012, https://www.af.mil/News/Article-Display/Article/111361/space-command-officials-eliminate-flight-suit-jackets/#:~:text=O#112-1105023-4890641n%20April%2010%2C%20a%20policy%20letter%20released%20from,to%20hang%20them%20up%20for%20good%20May%201; Jeff Schogol, "Flight Suit Furor," *Air Force Times*, June 18, 2012, 16.
124. Kevin Williams, Air Force Space Command Public Affairs, "Space Badge Renamed, New Guidance Issued," December 10, 2013, https://www.afspc.af.mil/News/Article-Display/Article/731417/space-badge-renamed-new-guidance-issued/.
125. Tom Budzyna, Air Force News Service, "'Teammates Wanted' to Deliver Future," September 10, 2012, https://www.af.mil/News/Article-Display/Article/110431/teammates-wanted-to-deliver-future/; "Gen. William Shelton," *Defense News*, September 17, 2012, 30.

126. Colin Clark, "New Spy Satellites Revealed by Air Force; Will Watch Other Sats," *Breaking Defense*, February 21, 2014, https://breakingdefense.com/2014/02/new-spy-satellites-revealed-by-air-force-will-watch-other-sats/.
127. Joan Johnson-Freese and Theresa Hitchens, "Stop the Fearmongering Over War in Space: The Sky's Not Falling," part 1, *Breaking Defense*, December 27, 2016, https://breakingdefense.com/2016/12/stop-the-fearmongering-over-war-in-space-the-skys-not-falling-part-1/.

10 THE BLIND KID FROM ALABAMA

John E. Hyten

GREGORY W. BALL AND WADE A. SCROGHAM

If Gen. Bernard Schriever led the first generation of military space pioneers, and Gen. Thomas Moorman perhaps led the second generation, it is safe to say that Gen. John Hyten stands at the forefront of space leaders in the first two decades of the twenty-first century. From humble beginnings to the pinnacle of his profession as the vice chairman of the joint chiefs of staff, Hyten demonstrated an enduring passion for space that led him to think deeply about space as an instrument of national power and then to articulate those issues to his fellow space professionals, defense leaders, and industry partners. Without a doubt, his leadership and mentorship of the officers who would one day lead the U.S. Space Force were significant.

In a career spanning four decades, Hyten served in Air Force space operations during a time of major transition. As he recalled as a four-star general in 2017, "I've been involved in something that is truly remarkable, because I've been involved in a fundamental change in warfare, a fundamental revolution in military affairs. An overused term, but it really is a revolution. . . . [T]hey created an environment where information became the key to the battlespace."[1] General Hyten's role in grappling with those changes, that revolution in military affairs that space operations became, is a story worth telling, because Hyten not only witnessed significant changes in how the United States understood and then operated in space but also gleaned many leadership lessons that future space professionals could apply in their careers.

Early Years

John Hyten was born in Torrance, California, in 1959, but his parents moved to Huntsville, Alabama, in 1965. For Hyten in the 1960s, Huntsville was an "amazing place" because it was the heyday of "Wernher von Braun and the German scientists and NASA [National Aeronautics and Space Administration]," and it fired and fed his imagination. His father, Sherwyn Hyten, worked for NASA as an engineer and was assigned to the Saturn V rocket program. During his childhood, Hyten saw the "Saturn V built up. I got to see the F-1 main engine test. . . . Some of my science projects in school were about rocket engines because we were building a rocket that would go to the moon."[2]

The development of the F-1 rocket engine also affected him another way: through its deep-throated roar as it was tested. Since his father worked on the project, Hyten got a closeup view on several occasions, but what he really remembered was,

> Every time it would test the whole town would shake right to its core. Dishes would go bouncing off our countertops, pictures would fall off the wall. . . . I asked my mom . . . I asked her do you remember? Because my remembrance is that nobody ever complained. She said that's right, nobody ever complained because that was us going to the moon and we were just, that's just the price we had to pay. And we all knew we were building the rocket that was going to go to the moon.[3]

Hyten attended schools that were filled with teachers with advanced degrees, once recalling that he had a fifth-grade teacher with a doctorate in mathematics, and a "physics teacher in high school that had a PhD from

Georgia Tech," while his chemistry teacher had a doctorate from Purdue.[4] For Hyten, education made him enthusiastic about science and math; that excitement about the sciences opened the world to him in a number of ways, and he recalled that "I got to do things that when I'd say them out loud they'd sound almost impossible." One of those things that would have been impossible for most fifth graders at the time was meeting Wernher von Braun, the legendary NASA engineer: "I got to walk up and talk to Wernher von Braun in 5th grade, and you know what? That was a cool thing for a 5th grader, but when I look back on it, I'm going, man, I wish I'd asked him this, this and this. And all I could think of was holy cow, I'm meeting Wernher von Braun."[5]

The space race pervaded Huntsville. Hyten met von Braun because the city was opening three new schools, named in honor of the Apollo astronauts who were killed in January 1967 on the launch pad at Cape Canaveral while testing their capsule. The elementary school was named for Roger Chaffee, the junior high for Ed White, and the new high school was named in honor of Virgil "Gus" Grissom. Being selected by his fifth-grade teacher to go to the opening ceremony of the school was how he got to meet von Braun.[6]

Growing up in Huntsville instilled in Hyten two loves: a love for space and a love of education.[7] But it was more than just learning: it was the impact that his teachers had on him personally. As he later recalled, it was his math teacher who "saw something in me. She's the one that got me to meet von Braun. And then she saw something in me that said there's something in this person and if I just encourage him. So, she would actually set aside math problems that were different than the rest of the kids were doing and have me do them." Hyten was never averse to giving credit to his teachers, some of whom even helped teach him calculus so he could get into the college that he had set his sights on: Harvard University.[8]

Although he may have loved the space program, he realized he would never go to space. As he recalled, "I love space but that wasn't going to be my future." It was not in his future because of his vision: "I'm basically blind. I have 20/900 vision, which means without my glasses I can't see anybody. I can't see anything.... it's just all a blur and all colors."[9] Thus, because of the love of learning instilled in him by his teachers and the encouragement they gave him to do things he didn't know he could, he was inspired to study engineering at Harvard.[10]

But the support of his parents and teachers was simply not enough to pay the costs of attending Harvard, as the stark reality of tuition appeared to be an unbreakable barrier that his parents could not shatter. However, his guidance counselor found him another option and told him the U.S. Air Force offered a scholarship through its Reserve Officers Training Corps (ROTC) program that would pay for his education and provide him with a monthly stipend of one hundred dollars. But because of the protests against the war in Vietnam that were rampant in the late 1960s, Air Force ROTC had been pushed off Harvard's campus in 1969, and it had not yet returned for Hyten's freshman year, 1976. He did enroll in the Air Force ROTC program at the nearby Massachusetts Institute of Technology, which allowed him to take his ROTC classes there while pursuing his academic degree at Harvard. As Hyten later said, the path to Harvard and the Air Force opened "all because of my teachers. All because of the fascination that I had with science and math."[11]

Still, his vision problems haunted him when he applied to the Air Force ROTC program and caused him to fail his physical the first time. Fortunately for Hyten, his eye doctor was an Air Force Reserve officer and encouraged Hyten to stay the course and helped him seek a waiver for his vision.[12] Although Hyten was thrilled that the Air Force would pay his way to college, he did not consider the Air Force as a long-term career, and as he said, "When I came out of Harvard my whole plan was to be in the Air Force four years and get out. Because the Air Force paid for my college education, I only owed them four years. And that was the plan all along." Like any new college graduate, he just wanted to travel and see the world.[13]

But Hyten was surprised to find out that instead of travelling the world courtesy of the Air Force, his first assignment in 1981 took him right back to Alabama, to Gunter Air Force Base (AFB) in Montgomery. As he later said, "I told the Air Force on my dream sheet, I don't care what I do, where I go," and so they sent him back to Alabama.[14] His first duty assignment was as a software engineer, where he focused on base level data automation and the interservice/agency automated message processing exchange program, both of which were considered at the time as "two of the largest data automation and telecommunications programs ever undertaken."[15] Although only a lieutenant, Hyten was soon the "recognized authority in the area of configuration management," all without "prior training or experience." He also demonstrated an "aggressive approach to resolving seemingly insurmountable problems between the various services and agencies," which proved invaluable to the success of those programs.[16]

While there, he also learned one of his first leadership lessons, one that he would take with him all the way through his career, and that was about respect. As he recalled,

> What I found when I got to Montgomery was that the Air Force had a lot of overt racism in . . . in 1981, especially in Montgomery, Alabama. And it wasn't long after George Wallace had just been Governor. George Wallace was the Governor when I was growing up. And somehow that had pervaded our United States Air Force. So, I made the commitment that that would not stand in my Air Force. And every time I saw it, and I probably didn't handle it as well, but I became aggressive, I guess, whenever I saw that, trying to eliminate that from my Air Force. From the time I was a lieutenant.[17]

By 1985, although Hyten enjoyed his four-year stint at Gunter, he still planned to leave the Air Force and pursue a civilian career. To that end, he completed a master of business administration degree at the University of Alabama–Montgomery. At the same time, he realized had grown more attached to the Air Force and his military career. In discussions about his future, his supervisor asked him what would keep him in the service. Hyten was ready with his answer: get him into the Air Force "space business." And he got his wish. Hyten agreed that he would give the Air Force one more assignment before leaving. In 2017 he quipped, "That was 32 years ago now, so I think the Air Force got their money back out of me."[18]

During these early years of his career, Hyten learned several more leadership lessons. As he recalled, "When I was a captain, my bosses made me come up with a plan. The plan was, what is your career going to be like? What are you going to do each step?" Hyten came to believe that everyone should write down a plan simply because it "gives you goals and things to work for." Hyten kept that plan he wrote down as a captain and decades later recalled: "And I looked at that plan I wrote down, and I realized that absolutely nothing I'd wrote down in my plan happened. Not a thing. The end state of that plan was I was going to be a program director of a major space program someday. That's what I wanted to be, because I was going to be an engineer."[19] Indeed, he tended to think that life never goes according to plan, and to him, that meant anything could happen, but it was also why life was "such a wonderful thing, because that's what makes it such an amazing ride. A plan is important because it gives you goals to work toward, but it's also true that if you plan early, you better plan often because it just never turns out the way you think it will. You should definitely have a plan, but you have to be prepared and excited for what comes at you out of left field."[20]

Space Acquisitions

At the time, getting an assignment in the Air Force "space business" meant sending Captain Hyten to the space defense programs office at Los Angeles AFB, where he found himself part of the software development branch and then serving as the chief of the engineering and acquisition division. The space defense office was at the time part of Air Force Systems Command. In that role, he participated in a major antisatellite (ASAT) program that developed the procedures and then successfully tested an air-launched ASAT in September 1985 when Maj. Doug Pearson destroyed the defunct Solwind satellite in low earth orbit. Hyten participated in two of the antisatellite tests—the medium altitude probe and the low altitude probe—that took place in August and September 1986. Although the program was eventually shelved, Hyten impressed his superiors and demonstrated "exceptional leadership, sound technical skill and superb judgment."[21] For someone who had just come into the space business, Hyten had found a home. Thinking back on that part of his career, he was amazed at the responsibility he was given:

> I become a chief engineer on the F-15A antisatellite program, [where] we actually shot down the satellite back in the 1980s. I was the chief engineer on the Air Force Strategic Defense Initiative, the initial Star Wars program as a captain. It was an amazing thing and I kept getting these amazing jobs, so I think I'll stay for this next job. And then I get a call one day and the caller says, hey John, this is so and so, I'm calling from Systems Command Headquarters. You grew up in Huntsville, didn't you? I said yes, I did. Well, this won't be so bad, click. Next thing I knew, I had orders to Huntsville, Alabama, to work for the U.S. Army. What the heck is that? Join the Air Force and see the world.[22]

That job turned out to be his first joint position as the chief of the Army/ Air Force integration in surveillance and battle management/command, control, and communications division, kinetic energy antisatellite joint program office of the Army's Strategic Defense Command in Huntsville. That was a long title for what was popularly known as the Strategic Defense Initiative, or Star Wars.[23]

As an engineer for the Strategic Defense Initiative, Hyten worked on both the Brilliant Eyes and Brilliant Pebbles programs. Brilliant Eyes characterized targets, while Brilliant Pebbles were space-based interceptors. As Hyten recalled, "The reason you needed space-based interceptors was because of the size of the threat that was going to be there. But the challenge with that is that it really is focused on large-scale attacks, because the thing about space that you have to realize is that space is everywhere. Just the earth, the geosynchronous orbit is 73 trillion cubic miles around the surface of the earth. So, it's very difficult to be everywhere all the time."[24]

Even though Hyten's first two assignments in the space business saw him participating in significant programs, he was still not certain that he wanted to make the Air Force a career and mulled getting out and pursuing life as an engineer to "make my fortune." In fact, Hyten later recalled, "I got this great job at Omaha. I interviewed with companies, and I found a job, triple my salary, signing bonus with stock options, member of the country club, all of the things I thought I wanted." But when he went home and told his wife that he wanted to leave the Air Force for a new career, she urged him to stay the course. After an hour of discussion, he realized that he did love the Air Force. And if he loved it, why would he leave? But he told himself that if he was going to stay in the Air Force, he was going to stay as an "operator."[25]

After Huntsville, Hyten returned to Los Angeles as the deputy for engineering in the space defense programs office at Los Angeles. In the summer of 1990, Iraqi dictator Saddam Hussein overran Kuwait and stood poised to invade Saudi Arabia, a key U.S. ally. The immediate response to the invasion of Kuwait was Operation Desert Shield, and one of the key threats that emerged during the buildup of coalition forces in the region was the use of Scud missiles by Iraq to attack coalition forces. In response to those attacks, the Air Force studied ways to detect Scud launches with enough time to give coalition forces warning. As part of this effort, Hyten briefed the commander of Strategic Air

Command, Gen. John T. Chain Jr., on whether the Defense Support Program missile warning satellites would be able to detect Iraqi Scud missile launches. Much later, he recalled that briefing to General Chain:[26]

> That was probably the most scared I've been. And I've been shot at now, and I still don't think I was as scared as I was when I walked into that headquarters the first time to go see the commander-in-chief of [Strategic Air Command]. I was just a captain. And my job was to explain to the commander-in-chief of Strategic Air Command whether the Defense Support Program satellite, the old missile warning satellite, would see Scuds when they launched. And to be honest, we didn't know. We had built that satellite to see [intercontinental ballistic missiles] coming out of the Soviet Union. We hadn't built it to see Scuds. So as an engineer, you can do the math. But we never tested it, so I didn't really know. That's why I was scared. I didn't know whether it could. Because I was going to tell him it could, and we were going to base a lot of what we did with Israel, a lot of what we did in the Middle East, a lot of what we did with force protection based on whether it could see it or not. I told him it could. And he just closed the book and looked at me and said, you guarantee that it will see Scuds. And I was an idiot captain, so I just looked at him and said yes, sir, it will see it. I came out five pounds lighter, because I had just you know, sweat bullets. But that was space in the Gulf War.... We were neophytes in that, but we saw the potential.[27]

After Desert Storm, Hyten finally broke the pattern of assignments in Alabama or California when he got an assignment as a speechwriter for the assistant secretary of the Air Force for acquisition at the Pentagon in 1991. From there he transitioned to working budget issues for the advanced technology programs office. At the Pentagon, Hyten continued to demonstrate his leadership skills including "keen technical knowledge," "in-depth management expertise," and "tremendous policy breadth," and displayed what was described as "a cooperative synergistic spirit that resulted in successful development and fielding of the most advanced space technology worldwide." Furthermore, Hyten was credited with the publication of "significant policy direction to formulate the future of the Air Force in space for years to come." Soon, he was promoted to major and selected to attend the Air Command and Staff College at Maxwell AFB in 1993, which would send him once more back to Alabama.[28]

Perhaps just as importantly, before heading off to the Air Command and Staff College, Hyten got to meet the father of Air Force space and missiles, Gen. Bernard Schriever. As Hyten recalled, "One of the honors of my life was to be an idiot major in the Pentagon when the Chief of Staff, General [Merrill] McPeak, looked for a space guy and he said who's that idiot major that does space down in AQ [acquisition]? I was in the room at the time. I did not raise my hand. But I was that idiot major, and I got to go pick up General Schriever in a C-21 and take him to some places and show him what we were doing. Because General Schriever, even at 80 years old, was criticizing what the Air Force was doing. And rightly so because we were not moving fast enough."[29]

Many years later, Hyten recalled reading an article in *Air Power Journal* published in 1998 by Lt. Col. Bruce DeBlois that stuck with him. Indeed, Hyten discussed DeBlois' article in a piece he penned for the *Air and Space Power Journal* in 2004. DeBlois had argued that the United States should pursue a "sanctuary" strategy for space and "that a strategy of weaponization in space was flawed in a number of ways." Although DeBlois argued for space as a sanctuary, he understood that the development of space weapons was a necessity to ensure space remained a sanctuary. Such themes of deterrence would echo through Hyten's own thinking and writing and would come full circle when he took command of U.S. Strategic Command (USSTRATCOM).[30]

In the early 1990s, this changing attitude toward space capabilities was demonstrated in another *Air Power Journal* article, this one penned by General Moorman. He highlighted the growing importance of space, writing that "space operations are finally coming into their own—specifically the application of space assets to support Air Force missions," particularly in light of their success in the Gulf War.[31] But Moorman also recognized the changing character of the space domain, noting that "space technology for the range of military functions will become available to many nations." Concluding that the Gulf War might have had a very different outcome if Saddam Hussein possessed his own space assets, Moorman argued that "an operational ASAT capability designed to eliminate an adversary's space capabilities must be considered an integral part of this country's force structure." Once again, the theme of deterrence loomed large.[32]

And while Hyten was at the Air Command and Staff College, the strategic thinker and writer Colin Gray weighed in on the survivability of space systems and echoed the sentiments of Gen. Colin Powell that the "United States learned from Operation Desert Storm that it had 'to achieve total control of space if [it is] to succeed on the modern battlefield.'" Furthermore, Gray noted two key points that surely influenced Hyten as his own thinking about the space domain matured while at the college. First, Gray observed that "more and more defense professionals and commentators are coming to appreciate that the right to use space will need to be fought for, no more and no less than the right to use the sea and the right to fly." Second, Gray believed that "space will witness a full transition from being a convenient place to perform useful force-enhancement tasks to being key to mission accomplishment." Thus, Hyten was studying and thinking about space in a cauldron of new ideas and new paradigms about how space might be used in the future, and whether the space domain might be characterized as a sanctuary or a future battleground.[33]

After graduating from the Air Command and Staff College, Hyten spent two years at United States Space Command in Cheyenne Mountain Air Force Station near Colorado Springs, where he learned another important lesson about the space domain and space warfare:

> Gen. Roger DeKok was a mentor of mine, boss of mine. He taught me one of the most important lessons I ever learned as a young space officer and that was, he looked at me and he said, "Major, I want you to always remember this about space. Satellites don't have mothers." To be honest, as a major I had no clue what he was talking about. But I've learned over the years that it's true. Satellites don't have mothers. Therefore, if a satellite is attacked, the country really doesn't care.[34]

Space Command

After Cheyenne Mountain, Hyten received his first command opportunity, assuming command of the 6th Space Operations Squadron (6 SOPS) at Offutt AFB, Nebraska. The squadron was a geographically separate component of its parent 50th Space Wing near Colorado Springs. The 6 SOPS's mission was to manage the defense meteorological satellite program constellation. At the time, he believed that commanding the 6 SOPS "would be the only time I'd be given the opportunity to command in my career and it was a special time" for him and his family. He wanted to make the most of it.[35]

As it turned out, Hyten did an outstanding job as a squadron commander despite having aging equipment and lacking a dedicated ground antenna for the first time in the squadron's thirty-five-year history. Nevertheless, the squadron earned two major awards, including the 1997 Gen. Richard C. Henry award for the best space operations squadron in Air Force Space Command and the Guardian Challenge space operations competition, while he was its commander.[36]

Still, Hyten's first command opportunity offered him leadership lessons that he would not forget, one of which he later called his "most important leadership story." For the first year of his command tenure, he recalled working on average sixteen hours a day and spending more time with his airmen than with his family. As he recalled, "You have an organization where 40 percent of your organization is under 21. You're going to have some challenges." To connect with the younger airmen, Hyten visited them frequently in their dorms and played on the squadron's intramural sports teams with them. As he said, "I made it a point of knowing everybody's story in the organization. It was my job to know what they needed. It was my job of being the boss. So, I was literally killing myself, and I'm not being a good husband and a good father. I'm not being a bad husband or a bad father, but I'm not there. So, since I'm not there, I'm not being a good husband or a good father."[37]

Then one day, after a meeting, Hyten returned to his office frustrated and tired. Suddenly, three of the squadron's senior noncommissioned officers (NCOs) came in and asked him why he was so frustrated.[38] Hyten bemoaned the fact that he was spending all his time with the troops and that the mission was going great, but the squadron continued to have behavioral problems. As he remembered, "I can't seem to get to the airmen, and it's just frustrating the heck out of me." What the three NCOs then told Hyten changed his perspective completely. They said, "Why don't you just tell us. It's our job to take care of the airmen, not your job. Your job is to be our commander. Your job is not to take care of every airman. If you would tell the NCOs what you want them to do, we will go do that and we'll do it spectacularly well."[39] And with that plea from his senior NCOs, Hyten learned the power of delegation: "And so, I made a commitment to myself right there on the spot that I was going to tell the NCOs what I wanted; I was going to tell the company grade officers what I wanted; I was going to tell the field grade officers what I wanted; I was going to tell the airmen what I wanted, all the way through. And then I would let them do their jobs and I would not do their jobs for them any longer." As he put it, "Leadership was about doing your job and letting others do theirs."[40]

As it turned out, Hyten was the last active duty commander of the 6 SOPS, as the squadron was transferred to the Air Force Reserve, where it continued to carry out its mission. From Offutt, Hyten was selected as a national defense fellow at the University of Illinois, a prestigious Air Force program to place high-performing field grade officers in various institutions to study national security policy issues. To complete the program, fellows were expected to write a five-thousand-word paper on a key policy issue. Hyten recalled that period as a "great year," and he quickly realized that the program allowed him to discuss his thoughts and ideas with people who did not necessarily think the same way he did. And since people in the military frequently thought along similar lines, having the opportunity to discuss ideas with people of different and varied mindsets helped him refine his own thinking on space issues: "It turned out to be that was spectacular because I think I helped move their thinking a little bit, but they helped move my thinking a lot."[41]

Thinking about Space

During this time, the cauldron of new ideas and new thinking about space continued to swirl. First, in 1999 the administration of President Bill Clinton published its own national defense strategy, which defined space as a vital national interest of the nation. In 1996 the Clinton administration had published a national space policy, although Hyten argued that it failed to provide a "coherent, long term space vision." Another formative event during this period was the creation of the Commission to Assess United States National Security Space Management and Organization, more commonly known as the Rumsfeld commission or the space commission. It was an inflection point in the history of military space operations, and Hyten was ready to be a major participant.[42]

The culmination of Hyten's time as a national defense fellow was a paper titled "A Sea of Peace or a Theater of War? Dealing with the Inevitable Conflict in Space." This paper allowed Hyten to crystallize his thinking about the nature of the space domain and space warfare. Hyten also sought publication of his paper, and it appeared in 2002 in *Air and Space Power Journal*. His article became a landmark in understanding space as a military domain and marked a shift in his career—far from thinking that he would end his career as a squadron commander or program manager, if he had not realized it by now, his selection as a national defense fellow showed that his supervisors believed in him and that he was destined for bigger things. His paper added more evidence that they had judged him correctly.[43]

In his article Hyten underscored how essential space had become to "all aspects of our existence," an idea that had also been explored in the Rumsfeld commission report. Hyten also looked more deeply at a question posed by the commission: Would the United States recognize the dangers posed by adversaries in space and take steps to protect national space capabilities? Or would it wait until a "space Pearl Harbor" occurred "to galvanize the nation and cause the U.S. government to act?" As he wrote, "We are on notice, but we have not noticed." According to Hyten, the time for discussion was now, and he lamented the lack of debate in "any significant way in public."[44]

Hyten also believed that the lack of debate stemmed from two key ideas that were in opposition to each other: the need for space weapons versus the need to maintain space as a peaceful sanctuary, harkening back to the DeBlois argument. Hyten also clearly recognized that space systems could "become a significant part of any future military conflict involving the United States."[45]

Hyten's paper was a visionary call to arms for the future of military space; it recognized that the U.S. military was proceeding into space "unprotected" and that the nation had to develop a clear, cohesive national space strategy. He made several recommendations that subsequently bore fruit over the next decades. It is no stretch of the imagination to say that Hyten's paper influenced the thinking of those officers who would later establish the U.S. Space Force. Hyten later said of his writings, "I don't think there was any magic involved in that study. I think you just look at what the Chinese were saying and what they were writing, and you projected it out and you could see exactly where they were going to end up and that's exactly where they did end up."[46]

After leaving the University of Illinois, Hyten spent the next four years at the Pentagon, split between the joint staff and the air staff but working space programs at both jobs. This gave him time to refine and publish his thesis from Illinois as well as to work on a second article, published in 2004 and cowritten by Dr. Robert Uy. That article explored the moral and ethical dimensions of space warfare. More specifically, the two authors argued that the greater good might be better served by the development of certain space weapons as deterrent factors. Once again, the idea of deterrence appeared in Hyten's thinking.[47]

Hyten clearly believed that earlier thinkers such as DeBlois and members of the Rumsfeld commission had set the precedent that space weapons were a legitimate area of focus, as was the distinction between the militarization of space and the weaponization of space. As he and Uy concluded, "The appropriate, measured development and use of certain space weapons will allow the United States, in circumstances where the nation is forced into war, to conduct warfare in ways that increase combat effectiveness while at the same time limiting collateral damage here on Earth—a more moral and ethical decision."[48]

Prior to leaving the Pentagon in 2004, Hyten was promoted to colonel and returned to Air Force Space Command for a year before moving onto his second and third command opportunities, first at the 595th Space Group and then as commander of the 50th Space Wing in April 2005, both at Schriever AFB.[49]

Hyten also had the opportunity to fill the space advisor role when he deployed to Pacific Air Forces at Hickam AFB, Hawaii, where he served under then-Maj. Gen. David Deptula, who was Pacific Air Forces director of air and space operations and was serving as the joint force air component commander for Operation Unified Assistance, providing disaster relief operations in South Asia. Hyten recalled the effect that working for Deptula had on him:

> I think the first time I met him I was his director of Space Forces in the Pacific when he was the [joint force air component commander] at Hickam. I remember going in, and the first time I really met him one on one, and he tasked me with a very interesting task. And the task was to build a space campaign plan. And this is 12, 13 years ago. And so, building a space campaign plan that would integrate with the other campaigns that were going on in the Pacific was an interesting challenge. But I can tell you, most of the time when I brought back the answer to him, it was not quite right. You can imagine how those meetings went. I was a colonel at the time, but I never expected to be a general, so I probably wasn't as respectful as I should have been, and I apologize. But it's interesting. When we put the campaign plan

[together], we put everything in place. The problem with the plan was that [when] we went to source the intelligence that we needed to execute the plan, there was no intelligence. Because we had taken our eyes off of the ball that was the strategic problem for years and years.... And it took us a long time, and we're still on that path. So, it's been a challenge.[50]

Space and War

During his tenure as a wing commander, Hyten also served as an early director of space forces when he deployed to Southwest Asia as the U.S. Central Command director of space forces. The position was a key operational assignment in a theater where active combat operations were taking place. In that job he traveled extensively across the U.S. Central Command area of responsibility and recalled, "I was in Iraq, Afghanistan, all up through the Middle East. I was in Fallujah, Ramadi, Balad, Bagram, and Kandahar. Just pick it." While the Marine Corps was heavily engaged in the battle of Fallujah, Hyten worked directly with a joint space support team in the city. As he noted, "There were Marines in Fallujah and there was a joint space support team in Fallujah run by a Marine lieutenant colonel with an Air Force captain and a bunch of soldiers whose job was to bring space capabilities at the time and tempo of the battle to the Marines in Fallujah." Supporting active combat operations gave Hyten a sharper perspective on how space capabilities were a key component of modern warfare. When it was time for him to redeploy home, he was replaced in theater by Col. John W. "Jay" Raymond. During his deployment, Hyten also was promoted to brigadier general.[51]

After completing his time as wing commander, Hyten moved to Air Force Space Command (AFSPC) headquarters as the director of requirements for two years, learning some difficult lessons about the requirements process and acquisitions—particularly the length of time it took to acquire new systems like the next-generation intercontinental ballistic missile and the replacement for the UH-1 helicopter. While also selected for his second star, perhaps most importantly, he learned the importance of keeping interest in and support for projects from lapsing because, as he quickly learned, when there was no interest in the Pentagon, support dried up and programs died quickly.[52]

Hyten returned to the Pentagon in 2009, working on the air staff and then in the secretariat, the civilian side of the department. In 2012 he was confirmed as a lieutenant general and as the new vice commander of Air Force Space Command, where he recognized the critical role between cyber and the space domain. He also continued to advocate for maturing space capabilities quickly.[53]

Finally, in 2014 the "blind kid from Alabama" who never thought he would serve more than four years in the Air Force or be more than a program manager pinned on his fourth star and assumed command of Air Force Space Command. Of his selection to this position, he later ruminated, "When I came in the Air Force there was no such thing as Space Command. Then I grew up to become the commander of Space Command. That's crazy, that's impossible, but it happened."[54]

As the senior space officer within the Air Force, Hyten found himself in a position to be heard as the leader of the military space community. He used his position to change the military space enterprise. Two of the key initiatives he pushed were the space enterprise vision (SEV) and its associated space warfighting construct (SWC), both of which included deterrence as key components. In fact, the SEV focused on making space systems more resilient, protected, and defensible. Hyten and others had come to believe space systems had been developed and built in an era when it was not necessary to take into consideration threats an adversary might pose. Since space had largely been considered a peaceful sanctuary since the fall of the Soviet Union (and for much of Hyten's career), potential threats had mainly been dismissed. But the emergence of increasingly realistic threatening activities by other spacefaring nations, particularly Russia and China, caused Hyten to recognize that acquiring space systems based on "longevity" and "cost" was "no longer an adequate methodology to equip space forces." The SEV offered a new paradigm to build space capabilities responsive to twenty-first-century threats.

The SWC, described as the "framework for making the Space Enterprise Vision an operational reality," was a broader approach that included several component parts that, when taken together, aimed to mature the space enterprise and "provide space leaders with tools, decision aids and response options necessary to prevail if conflict extends into space." The SWC included the space enterprise vision, the space warfighting concept of operations, the space mission force, resilient architectures, enterprise agility, and partnerships. Hyten later testified about the SEV/SWC before the Senate Armed Services Committee, saying, "In 2016, Air Force Space Command and NRO [National

Reconnaissance Office] developed the joint SEV to advance their shared interest in designing, acquiring, and operating more agile and resilient space capabilities in response to emerging threats." Expounding on the SWC, he explained, "This construct supports the National Space Policy and focuses on the forces, operations, and systems needed to prevail in a conflict that extends into space. As an enterprise, we must normalize how we think of space, operate in it, and describe it to each other. It is unique for many reasons, but the concepts that govern other military operations such as intelligence, maneuver, fires, protection, logistics, and command and control apply just the same."

These ideas were evident in the space mission force, which focused on force presentation to combatant commanders by building a force that was properly trained and equipped to provide the required space capabilities to operations. The centerpiece of the space mission force was a stronger focus on training when not conducting real-world combat operations. The Space Force today ensures a trained and ready force is presented to combatant commanders, and that force presentation construct developed by the service was influenced by Hyten's space mission force.

As Hyten had lamented earlier, a lack of intelligence caused their space campaign plan to founder. Thus, his own personal experience likely played a role in his efforts while AFSPC commander to revamp the space intelligence community to foster a greater unity of effort and information flow among space operators, the Department of Defense, the broader intelligence community, and external space organizations. His efforts along those lines led to the establishment of the National Space Defense Center on October 1, 2015, which Hyten described as a "center that synergizes the National Reconnaissance Office, the National Intel Community, and the [Department of Defense]. It is focused on achieving interagency unity of effort while defending against space threats." Such efforts to mature space intelligence capabilities would be carried forward into the Space Force as well with the establishment of the national space intelligence center in 2022.[55]

USSTRATCOM

In October 2016 Hyten relinquished command at AFSPC to General Raymond and moved to command USSTRATCOM, where he continued to focus on deterrence. At the time he assumed command, USSTRATCOM also managed space capabilities at the unified command level, which it had done since the disestablishment of U.S. Space Command in 2002. In his change of command speech, Hyten was humble when reflecting how he had reached this point, observing that "there's no way I should be the commander of Strategic Command. That's the command of Curtis LeMay. It's the command of people that were legends in the Air Force. I'm a blind kid from Alabama. That can't happen, but it did. And it happened because I happened to be involved in the fundamental change in warfare that was created by space."[56]

Hyten's priorities and expectations as USSTRATCOM commander offered another lesson in leadership. They were based on what he called his "two red lines." As he said, "Some may disagree, but I think I'm actually pretty easy to get along with, unless you cross one of these lines." The first red line was communication: "I can handle any news that comes through my door, except old news. If something good happens, let me know and I'll celebrate it with you. If something bad happens, let me know and I'll help you fix it. But if something happens and you choose not to tell me, when I [find] out, and I will, I'm not going to worry about you, I'm just going to worry about fixing that problem. I will not worry about you at all."[57]

The second red line was respect: "Everybody that chooses to come to work in this command ... each and every one of you has made a special commitment to this nation and you deserve to be treated with respect. I have no patience for those who don't treat others with respect that they have earned, and I don't think those people have any place in this business. I don't even want them around, period."[58]

While Hyten focused on commanding the organization and getting out of the way and letting his subordinates do their jobs, he oversaw a much larger portfolio with deterrence as his top priority, which did not allow him to think about space as much as he wanted. To ensure that space received the attention it needed, he decided to establish a command within USSTRATCOM for space, known as the joint force space component command, in 2017. Its first commander was the AFSPC commander, General Raymond. As Hyten jokingly recalled, "The great thing about Gen. Raymond is he has a boss at U.S. Strategic Command that cares passionately about the space business. The bad news about Gen. Raymond is he has a boss at U.S. Strategic Command that cares passionately about the space business." Thus, Raymond enjoyed the support of Hyten, but Hyten also had step back and let General Raymond do his job.[59] This allowed Hyten to serve as a key advocate for space who was considered by some to be a "successful insider, working quietly and forcefully within the institution to advance the cause of space and cooperating smoothly with both his four-star peers and his civilian masters."[60]

Hyten also advocated for many of the concepts that the Space Force would bring to fruition just a few years later. For example, in a January 2017 speech, Hyten acknowledged that the military space enterprise was "at a place now where we fundamentally have to change our architectures. We have to think about space as a warfighting environment. And if we think about space as a warfighting environment, and we treat it as such and we prepare for it, my hope is that we never fight a war in space."[61] Likewise, he understood the space domain was changing and that "we must normalize how we think of space, how we operate in it, and how we describe it to each other. It is unique for many reasons, but the concepts that govern other military operations: intelligence, maneuver, fires, protection, logistics, and [command and control] apply just the same."[62] All of that converged in his efforts to advocate for reforming space acquisition. He was keenly aware that "it is so important that we have the right authorities in space and cyberspace to respond to a threat in a timely manner. It is so important that we go fast enough to keep pace with our adversaries who are going unbelievably fast in terms of developing new capabilities. And we have built a bureaucracy in our country that is very slow to react. We can't be slow to react to the world that we're in today."[63] Furthermore, he recognized the need for the United States to keep pace with peer adversaries such as China.

Vice Chairman

On April 4, 2019, the president nominated Hyten to become the vice chairman of the joint chiefs of staff, the second highest military position in the nation. The Senate confirmed General Hyten as the vice chairman by a vote of seventy-five to twenty-two on September 26, 2019. Just a week after being nominated for the vice chairmanship and while awaiting his confirmation hearing, Hyten was invited to testify before the Senate Armed Services Committee on the need for a Space Force. Hyten joined Acting Secretary of Defense Patrick Shanahan, Chairman of the Joint Chiefs of Staff Gen. Joseph Dunford, and Secretary of the Air Force Heather Wilson. In his opening statement, Hyten acknowledged that space was, at best, his third priority as the USSTRATCOM commander, behind deterrence and the nuclear mission, and so he supported both a new unified command for space and a Space Force within the Department of the Air Force. The hearing went well and was another step toward the creation of the U.S. Space Force later that year.[64]

After his confirmation as vice chairman, Hyten focused on new priorities. First, he wanted to ensure that he provided his bosses the best possible military advice he could. Second, he wanted to "put speed into everything that we do." Finally, he wanted to always "take care of the people that actually get the job done."[65] As his second priority indicated, Hyten still did not believe the United States had done enough to reform the space enterprise. In a 2021 interview, he expressed frustration with the continued slow pace of change:

> We've been talking about the challenges we have with vulnerable space architectures for a decade now. Because a decade ago, when we started looking at the advancing threats we were going to face ... I said [our satellites are] just a bunch of fat, juicy targets. And that's what they are, because everybody knows where they are, everybody knows what goes through them, everybody knows how important they are. And so, we've told the entire world. And then we've said at the same time, we recognize that and so we're going to build a more resilient architecture.

This attitude spurred the general to continue to advocate for space as vice chairman.[66]

Conclusions

On November 19, 2021, Gen. John Hyten retired from the U.S. Air Force after forty years of service. While he never served in the U.S. Space Force, he should surely be considered one of its forefathers. He led the space community in the twenty-first century, both in recognizing that the space domain was changing and in serving as an advocate to reform space acquisition. Perhaps more importantly, he brought the conversation out into the open. Hyten's leadership and style might be summed up in just a few key principles. The first was to focus on the mission at hand. As he said, "It's funny how life turns out when the only thing you're really focused on is doing the job you're asked to do right now."

The second principle was to delegate responsibility—tell subordinates what the vision is, set their boundaries, and then turn them loose.[67] The third principle was to respect those who work for you.[68]

In sum, Gen. John Hyten should be viewed as a farsighted leader and thinker who played a key role in maturing national military space operations—certainly among the first rank of visionaries who were there at the right moment to not only recognize future challenges but also formulate solutions that would allow military organizations to meet those challenges head on when the time came. Toward the end of his career, Hyten summed it up: "There's two things that all four-star generals have in common. Number one, we're all circling the drain, which means we're all old and we're about done, we're all going down. And we know that. It's just the nature of the beast. And number two, we always wish we were you. We do. We wish that we could go back and do it all again because holy cow, it's an amazing life."[69]

Notes

1. John E. Hyten, speech at Center of International Security and Cooperation (CISC), Offutt AFB, NE, January 24, 2017, https://www.stratcom.mil/Media/Speeches/Article/1063244/center-for-international-security-and-cooperation-cisac/.
2. John E. Hyten, speech at Armed Forces Communications and Electronics Association (AFCEA) Greater Omaha Chapter Luncheon, Offutt AFB, NE, April 13, 2017, https://www.stratcom.mil/Media/Speeches/Article/1156937/armed-forces-communications-and-electronics-association-greater-omaha-chapter-l/.
3. Hyten, speech at CISC.
4. Hyten, speech at AFCEA.
5. Hyten, speech at AFCEA.
6. Hyten, speech at CISC.
7. Hyten, speech at CISC.
8. Hyten, speech at AFCEA.
9. John E. Hyten, commencement address to College of Nursing and Health Sciences, College of Public Policy and Justice, and College of Business graduates, Auburn University at Montgomery (Montgomery, AL, May 6, 2017), https://www.stratcom.mil/Media/Speeches/Article/1184147/auburn-university-at-montgomery-commencement-address-to-college-of-nursing-and/.
10. Hyten, speech at CISC.
11. Hyten, speech at AFCEA.
12. Hyten, commencement address.
13. Hyten, speech at CISC.

14. John E. Hyten, speech at Air Force Association—Aksarben chapter annual awards ceremony, Offutt AFB, NE, April 20, 2017, https://www.stratcom.mil/Media/Speeches/Article/1169704/air-force-association-aksarben-chapter-annual-awards-ceremony/.
15. United States Air Force, citation to accompany the award of the Meritorious Service Medal to John E. Hyten, 1986.
16. United States Air Force, "Biography–General John E. Hyten," April 2021.
17. John E. Hyten, speech at USSTRATCOM Fellows Leadership Conference, Offutt Air Force Base, NE, May 2, 2017, https://www.stratcom.mil/Media/Speeches/Article/1177286/usstratcom-fellows-leadership-conference/.
18. Hyten, speech at CISC.
19. Hyten, speech at Air Force Association.
20. Hyten, commencement address.
21. United States Air Force, citation to accompany the award of the Air Force Achievement Medal to Capt. John E. Hyten, August 25, 1987.
22. Hyten, speech at Air Force Association.
23. United States Army, citation to accompany the award of the Army Commendation Medal to Capt. John E. Hyten, July 27, 1990.
24. John E. Hyten, speech at Strategic Deterrent Coalition Symposium, Arlington, VA, May 9, 2017, https://www.stratcom.mil/Media/Speeches/Article/1177286/usstratcom-fellows-leadership-conference/.
25. Hyten, speech at Air Force Association.
26. Hyten, speech at Air Force Association.
27. Hyten, speech at CISC.
28. Hyten biography.
29. John E. Hyten, speech at Mitchell Institute Breakfast Series, Washington, DC, June 20, 2017, https://www.stratcom.mil/Media/Speeches/Article/1226883/mitchell-institute-breakfast-series/.
30. John E. Hyten and Robert Uy, "Moral and Ethical Decisions Regarding Space Warfare," *Air and Space Power Journal* 18, no. 2 (Summer 2004): 51–60; Bruce M. DeBlois, "Space Sanctuary: A Viable National Strategy," *Airpower Journal* 12, no. 4 (Winter 1998): 41–57.
31. Thomas S. Moorman Jr., "Space: A New Strategic Frontier," *Airpower Journal* 6, no. 1 (Spring 1992): 14–23.
32. Moorman.
33. Colin S. Gray, "Space Power Survivability," *Airpower Journal* 7, no. 4 (Winter 1993): 27–42.
34. Hyten, speech at Mitchell Institute Breakfast Series.
35. John E. Hyten, USSTRATCOM change of command remarks, Offutt AFB, NE, November 3, 2016, https://www.stratcom.mil/DesktopModules/ArticleCS/Print.aspx?PortalId=8&ModuleId=1541&Article=1028833.
36. United States Air Force, citation to accompany the award of the Meritorious Service Medal (Second Oak Leaf Cluster) to John E. Hyten, November 17, 1998.
37. Hyten, speech at USSTRATCOM Fellows Leadership Conference.

276 | Chapter 10

38. It was Master Sgt. Robby Robinson, Master Sgt. Jim Patton, and Master Sgt. Bill Roquel, respectively the maintenance superintendent, the operations superintendent, and the chief of plans, so the three leaders of the enlisted force.
39. Hyten, speech at USSTRATCOM Fellows Leadership Conference.
40. Hyten, speech at USSTRATCOM Fellows Leadership Conference.
41. Hyten, speech at CISC.
42. John E. Hyten, "A Sea of Peace or a Theater of War? Dealing with the Inevitable Conflict in Space," *Air and Space Power Journal* 16, no. 3 (Fall 2002): 78–92.
43. Interestingly, and as a nod to the growing importance of space, the same issue in which Hyten's article appeared also saw the title of the journal shift from *Air Power Journal* to *Air and Space Power Journal*. Air Force Chief of Staff Gen. John P. Jumper noted that the 2001 space commission (also known as the Rumsfeld commission) did not use the term aerospace because it did not "give the proper respect to the culture and to the physical differences that abide between the environment of air and the environment of space." Furthermore, Gen. Michael Ryan noted that "we will respect the fact that space is its own culture, and that space has its own principles." Jumper concluded his remarks by acknowledging the need to develop space warriors ("Those trained in the planning and execution of space-based operational concepts") and to transform space organizations.
44. Hyten, "Sea of Peace or a Theater of War?"
45. Hyten, "Sea of Peace or a Theater of War?"
46. Hyten, speech at CISC.
47. Hyten and Uy.
48. Hyten and Uy.
49. Article, 50 SW/PA, "50th SW Commander Shares Wing Vision," May 18, 2006.
50. Hyten, speech at Mitchell Institute Breakfast Series.
51. John E. Hyten, Military Reporters and Editors Association Conference, Arlington, VA, March 31, 2017, https://www.stratcom.mil/Media/Speeches/Article/1153029/military-reporters-and-editors-association-conference-keynote-speech/.
52. Hyten, speech at Strategic Deterrent Coalition Symposium.
53. Kenneth Kesner, "Huntsville Native Receives Third Star on Air Force Uniform and Post with Space Command," *Huntsville Times*, January 14, 2019.
54. Hyten, speech at CISC.
55. John E. Hyten, "Statement of John E. Hyten, Commander, United States Strategic Command, before the Senate Committee on Armed Services," April 4, 2017.
56. Hyten, speech at CISC.
57. Hyten, USSTRATCOM change of command remarks.
58. Hyten, USSTRATCOM change of command remarks; Hyten, speech at USSTRATCOM Fellows Leadership Conference.
59. John E. Hyten, "Integrating and Normalizing Space for the Warfighter," 33rd Space Symposium, Colorado Springs, CO, April 6, 2017, United States Space Command, https://www.stratcom.mil/DesktopModules/ArticleCS/Print.aspx?PortalId=8&ModuleId=1541&Article=1152751.

60. Theresa Hitchens and Sydney J. Freedberg Jr., "Hyten: Sexual Abuser or Billy Mitchell of Space?" *Breaking Defense*, July 29, 2019.
61. Hyten, speech at CISC.
62. Hyten, statement before the Senate Committee on Armed Services.
63. Hyten, keynote speech, Military Reporters and Editors Association Conference.
64. John E. Hyten, "Testimony to Senate Armed Services Committee on Proposal to Establish a United States Space Force," April 11, 2019.
65. Michael O'Hanlon, "A Conversation with Vice Chairman of the Joint Chiefs of Staff Gen. John E. Hyten," The Brookings Institution, Washington, DC, September 13, 2021.
66. O'Hanlon.
67. Hyten, speech at USSTRATCOM Fellows Leadership Conference.
68. Hyten, speech at Air Force Association.
69. Hyten, speech at AFCEA.

11 PRESENT AT THE CREATION

John W. "Jay" Raymond
LANCE JANDA

If the United States established a military aristocracy, Gen. John W. "Jay" Raymond would surely hold an honored place among its ranks. Members of his family have served in uniform since 1861, when his great-great-grandfather, Charles W. Raymond, entered the United States Military Academy at West Point. Charles graduated first in his class in 1865, and one of the highlights of his long and accomplished career as an engineer came when he led the American expedition to Tanzania in 1874 to observe the transit of Venus. Over the next seven generations, many of his descendants followed his example and joined the Army or one of the other military service branches.[1] His great-grandson, John A. Raymond, graduated from West Point in 1958 and had a son in 1962.[2] He named his son John as well, and to distinguish the boy from his father, the family began calling him "Jay" almost immediately, and the nickname stuck.[3] Jay watched Neil Armstrong walk on the moon in 1969 from the comfort of his living room at West Point, where his father taught in the social sciences department, and then he built a model of an Apollo rocket in the dining room.[4] Then his father was stationed in West Germany and the family lived near Wiesbaden Air Base (AFB). Perhaps those early exposures to space and the Air Force helped Jay resist the siren song of the fortress on the Hudson, for rather than join the family business and attend West Point, he joined the Air Force instead. It proved a fortuitous decision, for by the end of his military career in 2022, Raymond outranked everyone in his family, past and present. He retired as a four-star general, the first chief of space operations (CSO) for the U.S. Space Force, the first guardian, and the officer most responsible for shaping the creation of the sixth service branch of the U.S. armed forces, the only independent space force in the world.

Early Years

Raymond's thirty-nine-year career as a commissioned officer began when he became a second lieutenant in the U.S. Air Force after graduating from Clemson University in 1984 with a bachelor of science degree in administrative management.[5] Raymond chose Clemson for its "world class" Air Force Reserve Officers Training Corps program, because he "fell in love" with the campus, and because the university was relatively close to his family in Washington, DC.[6] He was commissioned in the midst of the Cold War and the massive American military buildup championed by President Ronald Reagan and initially chose to pursue a career as a pilot. His skill set proved a poor fit for the cockpit, however, and after washing out of pilot training, he found himself in California at Vandenberg AFB training on strategic missiles.[7] Raymond became a Minuteman intercontinental ballistic missile crew member, stationed at Grand Forks AFB in North Dakota in 1985, because he wanted to serve in an operational career field.[8] He stayed until 1989, serving in a variety of positions and rising to the rank of captain.[9]

Raymond then moved back to Vandenberg for duty with the 1st Strategic Aerospace Division and within a few years found himself seriously considering leaving the Air Force. The Cold War ended in 1991, and nuclear missile forces were on the verge of being drastically reduced. Raymond saw limited career options as a missileer and would have left the service had fate not intervened in the form of Brig. Gen. Sebastian F. Coglitore, who took an interest in Raymond and made him his executive officer while commanding the 30th Space Wing. Raymond described the experience as "life changing," saying that "the whole world opened up for opportunities in an area that I was really passionate about, and that was space."[10]

From that moment, Raymond's career accelerated dramatically, and he found himself promoted regularly and progressing through a series of increasingly challenging command positions. Highlights of the next twenty-three years include serving as chief of commercial space lift operations at Peterson AFB; as a student at the Air Command and Staff College at Maxwell AFB; as a space and missile force programmer at the Pentagon; as commander of the 5th Space Surveillance Squadron at Royal Air Force Base Feltwell in the United Kingdom; as a student at the Naval War College; as a transformation strategist in the office of force transformation in the office of the secretary of defense; as commander of the 30th Operations Group and director of space forces for the combined air operations center in Southwest Asia; as commander of the 21st Space Wing; dual-hatted as vice commander of Fifth Air Force and deputy commander of Thirteenth Air Force in Japan; and as commander of Fourteenth Air Force.[11] He then served as the only nonrated officer ever to be deputy chief of staff for operations. By late 2016 he was a four-star general with a remarkably diverse set of experiences within the Air Force, with every branch of the American military, and with many U.S. allies, and he possessed as much or more expertise in space operations as anyone in uniform. Those experiences provided him with a unique perspective on the long history of the Air Force in space and on the fragmented way the United States had historically managed space operations.

For decades, space operations were initiated and managed by an amalgam of civilian and military agencies and branches of the armed forces. The Air Force was by far the largest player, but no truly centralized oversight of space operations existed, and by the time the Cold War ended in 1991 many military and civilian leaders were discussing whether a major change in the organization of space missions and development was appropriate. Most Air Force leaders had concluded that future combat operations depended on the control and use of space, which in turn demanded normalizing space operations. Their steady efforts to do so in the wake of the *Challenger* disaster were rewarded during Operation Desert Storm, when space systems proved crucial in supporting a rapid allied victory over Iraq and Saddam Hussein.[12] This marked the first time space-based assets were used at the operational and tactical levels to enhance weapons targeting, navigation, and early warning capabilities, and they proved very effective as force multipliers.[13]

Reorganizing Military Space

After Desert Storm, the United States moved to integrate space systems into every facet of joint operations. Those systems proved vital during conflicts in Kosovo and after the attacks of September 11, 2001, on the Pentagon and the World Trade Center. U.S. forces from all branches and those of U.S. allies enjoyed unprecedented support from an enormous and regularly updated array of space-based systems that influenced every facet of warfare, including communications, early warning, position, navigation, timing, intelligence, surveillance, and reconnaissance, and weather. That dependence deepened as the twenty-first century progressed and the United States found itself in protracted conflicts in Afghanistan and the Middle East, and it mirrored a growing commercial dependence on space exemplified by the widespread use of global positioning system satellites for navigation and by a constantly expanding fleet of private satellites for delivery of internet access, television programming, and cellular telephone calls. Air Force leaders realized that in the event of a global full-spectrum conflict, those space-based assets would require protection as well, lest their military advantage dissipate and the global economy be ruined. That realization fueled an intense debate over how best to manage military space and satellite operations in the future. Some officers even wondered if the United States needed an independent space force.[14]

That question was hardly new. As early as 1981, Air Force officers Lt. Col. Dino A. Lorenzini and Maj. Charles Fox asserted the need for a U.S. space force by 2001, arguing that "in the not-too-distant future, space will be the dominant medium for the maintenance of national security."[15] They were particularly prescient when they argued that "the security and wellbeing of the United States in the next century will depend increasingly on the uninhibited use of space for the collection of environmental data, communications relay, weather monitoring, safe navigation and transportation, and the efficient use of resources and energy." They even suggested, as proponents of an independent Air Force had in 1947 when it broke away from the Army, that a separate space force would be in a better position to increase and consolidate its budget in the halls of Congress.[16]

In 1999 retired General Charles A. "Chuck" Horner, who flew combat missions in Vietnam as an F-4 Wild Weasel pilot and eventually gained fame as the commander of allied air operations during Desert Storm, added to the debate in his memoir of the Gulf War air campaign, *Every Man a Tiger*, when he wrote, "[S]pace has become too big, too important, to be treated as a subset of air operations or of the Cold War.... [Our] space force may need to become a military entity in its own right, equal and apart from our air, land, and maritime forces."[17] Horner had seen the power and potential of space-based assets to augment air and ground operations in Iraq and became an even greater proponent of spacepower after being named in 1992 as the triple-hatted commander of North American Aerospace Defense Command, U.S. Space Command, and Air Force Space Command (AFSPC). His prominent support of greater independence for space operations kept the debate going internally within the Air Force as it approached the twenty-first century.

In 2001 the idea of revising American military space operations and organization received another boost when the Commission to Assess United States National Security Space Management and Organization, better known as the space commission, warned of a possible "space Pearl Harbor" if the United States did not prioritize and improve the funding and management of military space operations and work to ensure it could defend orbital assets and defeat potential enemies in space. Critics argued the commission included overly hawkish members who were unreasonably paranoid, eager to garner more federal money for their corporations, or grandstanding for political purposes as part of what some scholars called "the military-industrial-thinktank complex," but the report received considerable national attention.[18] The space commission stopped short of calling for an independent military service, but its report found favor with those in the Air Force who believed that controlling access to the domain of space had already become a focal point of military strategy and that U.S. military organization and doctrine needed to adapt accordingly. This seemed an obvious truth to many, as Air Force space doctrine had noted the possibility of space-to-space and space-to-earth warfare as early as the 1980s. Yet there remained internal opposition to creating a separate branch among those who believed the Air Force could handle both air and space missions and others who feared another military branch would simply create more bureaucracy. As late as 2018 Secretary of Defense James Mattis opposed an independent space force for those reasons, as did Sen. Jim Inhofe, chair of the Senate Armed Services Committee, and many others in Congress.[19]

And so when Raymond became a four-star general in 2016 upon taking command of AFSPC, fifty-eight years after the United States launched its first satellite (Explorer I) in 1958, the U.S. space community was the best in the world and, at the same time, fragmented and in need of restructuring. The Air Force had become the leader and the largest player in space, but the Army and Navy also had space-based assets, as did the intelligence community, the Missile Defense Agency, the National Aeronautics and Space Administration, and many others, and they all wanted to control their assets themselves. Someone in a position of political power needed to push hard for change, and that someone was Republican congressman Mike Rogers of Alabama.

In 2017 Rogers publicly called for the creation of a separate space force within the Department of Defense (DoD). He argued forcefully that space should be a military domain equal to air, sea, land, and cyber, and that the Air Force failed to prioritize space or develop enough leaders for the space community. Rogers bemoaned interservice competition over space between the Army, Navy, and Air Force, noted the threat from potential adversaries and the successful test of a Chinese antisatellite weapon in 2007, complained that the National Reconnaissance Office had too much authority and the military too little when it came to space and national security, and noted that every Air Force argument against an independent space force had been made by the Army while fighting against losing control of its airpower mission between 1907, when the first aeronautical division emerged in the Army Signal Corps, and 1947, when Congress created the Air Force.[20]

Rogers and Democratic congressman Jim Cooper of Tennessee gained the attention of President Donald Trump, who issued an executive order in the summer of 2018 calling for the creation of a space force and requesting Marine Gen. Joseph Dunford, chairman of the joint chiefs of staff, to oversee the establishment of the new branch. However, no new branch of the armed forces could be created without congressional action, and passage of any new laws creating a space force seemed unlikely because a majority of members had indicated their opposition.[21] That left DoD in the awkward position of being asked to draw up contingency plans, which it dutifully did while being fully aware those plans might never be put into action.

At this juncture, Raymond was commander of AFSPC. He saw a need to continue normalizing space operations, a point he made in a famously misunderstood op-ed for *Defense One*, and to shift warfighting domain issues to the joint staff in the Pentagon.[22] Over the years many space-related issues had gone directly to the secretary of defense instead, and Raymond saw a need for them to be handled first by military officers, the way that air strategy and doctrine were. He was not inclined to break space operations away from the Air Force at that point. His goal was simply to make them more streamlined within the existing command structure.[23]

However, the more time Raymond spent commanding AFSPC, the more he realized that operating a major command under the auspices of the Air Force had significant limitations compared to what might be possible with a separate service. Promotion rates of space officers, for example, were sometimes 20 percent below the Air Force average, too few space officers were selected for the Air War College or the School of Advanced Air and Space Studies, and space needs often took a back seat to air needs of the Air Force in congressional budget meetings and when establishing partnerships with international allies. Raymond became convinced that to keep pace with the growing dangers in space, particularly from China, and to move more quickly and create a more agile space force, becoming an independent branch could be an advantage.[24]

And the momentum to make changes in space operations picked up noticeably after the president called for a space force. In August 2018 Raymond was discreetly told that he would soon be asked to stand up a combatant command for space.[25] The idea was not new, as U.S. Space Command (USSPACECOM) had operated from 1985 until 2002 as a unified combat command. It had been stood down to make way for U.S. Northern Command (USNORTHCOM) after the September 11 terrorist attacks because federal law limited the armed forces to a total of ten combatant commands, and in the threat environment of that period, terrorism was seen as a greater threat and USNORTHCOM deemed more vital. Combatant commands are responsible for military operations, while the military services provide forces to them, so what Raymond was being told to prepare for was significant. The Air Force had decided, either because of the growing threat in space or as a means of addressing President

Trump's desire for a space force, to support bringing USSPACECOM back, and Raymond was going to lead the way.

He started with a small planning team of five people that he took to San Antonio. They worked in isolation while he served on a promotion board and regularly checked on their progress. At the end of the first week, they had the draft of a plan that Raymond later said "was pretty much what we established. It was really close. I would say it was probably a 90-something percent solution."[26] His initial guidance had been to be ready to stand up the new command in three months, but it took a year to finalize details, complete his confirmation hearings before Congress, and finish a unified command plan, so USSPACECOM stood up in August 2019 during a ceremony in the Rose Garden at the White House.[27]

Standing Up the U.S. Space Force

During this period, the Air Force also had a "tango team" simultaneously planning for what a space force might look like in the event Congress created the new service, and by December 2019 it became apparent they were likely to do it. Language authorizing the creation of the U.S. Space Force was eventually included in the fiscal year 2020 National Defense Authorizations Act (NDAA) after a political compromise between Republicans and Democrats that saw bipartisan support for the Space Force (which even reluctant Republicans wanted because Trump did) and a guarantee that federal workers would henceforth receive twelve weeks of paid parental leave (which appealed to Democrats).[28] The law also included a crucial provision that stipulated the USSPACECOM commander could also serve as the new CSO of the Space Force for a year, and it left the decision as to whether the new commander actually would to the president. Raymond was asked by Air Force Chief of Staff Gen. David L. Goldfein and Secretary of the Air Force Barbara Barrett whether he would serve in both capacities; he said yes, but no one was certain what the president would decide until the signing ceremony at Joint Base Andrews on December 20, 2019, when Trump introduced him as the first CSO of the new branch.[29]

Raymond spent the next nine months as both CSO of the Space Force and commander of USSPACECOM, rapidly delineating the responsibilities of each organization while also dealing with the outbreak of COVID-19 and moving to Washington, DC, to be closer to his duties as the newest member of the joint chiefs of staff, another responsibility the NDAA had assigned to him as CSO. Congress had prescribed that there be no transition period—the Space Force existed from the moment President Trump signed the 2019 NDAA. This came as a surprise to the Air Force, which had planned for an eighteen-month grace period to establish the Space Force and instead got none. That fact, combined with Raymond's dual-hat command of both Space Force and USSPACECOM, allowed him to move at breathtaking speed because he could move assets and responsibilities back and forth without consulting anyone but himself.[30]

Raymond began by creating a list of "five things to scope the planning efforts—five things an independent service had to do." They included developing their own people, creating their own doctrine, controlling their own budget, designing their own force, and presenting those forces to a joint commander. Later, Raymond added a sixth requirement to "ready" the force as well, and those big-picture goals guided much of the execution and implementation that followed. Raymond also entered into discussions aimed at determining what the Air Force would look like without space integrated into it. As he put it, "I spent my whole career—largely my whole career—integrating space into the Air Force. And all of [a] sudden we're not integrated and what does that leave?"[31] Answering that question took time, and the process remained incomplete even when Raymond retired. Internally, there were those within the Air Force who opposed an independent Space Force, and within the Space Force there were others who cherished outlandish, unrealistic, largely science fiction–driven visions of what the force could and should become in the short run. Raymond had to artfully deal with both groups, and he succeeded in finding many solutions while keeping the internal squabbling out of the public eye. That by itself made his efforts more successful and smoother than the very bitter fight that occurred over separating the Air Force from the Army in 1947. The Air Force, it seems, had enough space professionals who knew the time was right for a new service, and they worked to make the transition successful. Cooperation proved vital because the Space Force was extraordinarily small and relied on the Air Force for transportation, logistics, and support personnel. It was an independent service but under the auspices of the department of the Air Force, and in that way remarkably similar to the Marine Corps, which was independent but part of the department of the Navy, and similarly dependent on the much larger Navy for some support.

Over the next two years, Raymond and his staff dedicated themselves to molding the Space Force into a very lean, efficient organization that contained all the traditional elements of an independent military branch. He maintained a desire "to flatten the bureaucracy" in order to remain responsive and retain the ability to move quickly. Private sector consultants emphasized this facet of organizations to him time and again. "You've got to be small," they said. "You can't be big, or you'll be slow." He made a dedicated effort to reduce layers between the lower ranks and himself. At the same time, he sought to "free up some colonels to go work on staffs that we didn't have space experts on," because one of the disadvantages of being small was that the Space Force was underrepresented on command staffs, joint staffs, and in the office of the secretary of defense. So he changed that. He also made many changes to mission alignment, such as combining the space-based and ground-based missile warning systems, previously divided between two Air Force wings, into a single space unit, providing "synergies" of enormous value.[32]

On the personnel front, Raymond strove to stay within his allocated resources, prioritize personnel, and emphasize processes. Space Force headquarters, for example, was allocated approximately one thousand personnel positions, and he kept that number under six hundred, freeing positions for other duties. He also immediately hired civilian personnel experts to assist with building the people part of the force and creating established norms. As he put it, "You had to be able to recruit, and assess, and develop, and promote, and have ID cards and pay [people], and all those things. And that's not easy. All that had to be built." He told his staff to build a service for "today," one that leveraged digital technology and would be innovative, and they were able to do so because they were starting from scratch.[33] The Space Force eventually revised all of the schools that educate personnel, including the airman leadership school, the noncommissioned officer academy, and the senior noncommissioned officer academy, and entered into a university partnership program with fourteen universities offering significant space science, technology, and mathematics degrees to help recruit their graduates. Over time, the Space Force also significantly increased the number of Air Force Academy graduates commissioning into the Space Force. Yet despite all this expansion, he kept the force lean. The Space Force includes only five career fields—operations, engineering, cyber, intelligence, and acquisition. Everything else they need still comes from the Air Force.[34]

Raymond also had to navigate the challenge of choosing a rank structure, uniforms, a seal, a collective name for Space Force personnel, a motto, and a service song. These ranked among his most consequential decisions because they impacted morale, public perception, and esprit de corps, and it was, therefore, important to get them right. He pushed for an Air Force–like rank structure and had to overcome a "big push in Congress to potentially take Navy ranks" and included President Trump in the decision to go with blue as a standard thread color in the uniforms and adopt the current seal, which drew inspiration from the 1982 AFSPC logo. Raymond also drew from AFSPC when he chose to call his people "guardians," for the AFSPC motto had been "Guardians of the High Frontier." For the motto he went with "Sempra Supra," which was suggested to him in an email from an airman stationed in Europe, and the motto in turn became the name of the branch song, written by Air Force veteran James Teachenor with help from Sean Nelson, the chief musician of the United States Coast Guard, and recorded for the first time by the Coast Guard band.[35]

Each of these choices, except for the motto, received a fair amount of public criticism and even derision, particularly from the science fiction community, which was (and is) too eager to see inspiration drawn from television shows like *Star Trek* and *Battlestar Galactica* that were irrelevant when it came to standing up a professional military space force. Netflix added to the negative publicity by launching a comedy series called *Space Force* that was, at best, irreverent about the armed forces in general and the Space Force in particular, and at worst disrespectful and misleading when it came to portraying why the Space Force existed, what it did, and how effective it was as a service branch. Netflix defended the show by arguing it was satire, but the series clouded public perception of the Space Force until executives canceled it after two seasons.[36]

The Space Force also struggled with being associated very closely with President Trump in the eyes of the public. Trump considered creating the force one of the crowning achievements of his administration and regularly promoted it at rallies. For his supporters, the Space Force was a good thing. For his detractors, it represented proof that it was conceived on a whim by a president who had never served in uniform and that the new branch might not have really been necessary. Much of the dissonance stemmed from simple ignorance, because few in the public or the press really understood what the Space Force did. They had not been prepared for its creation in advance and were in no position to understand its missions, because so many of them were secret and not part of the general knowledge Americans had regarding the armed forces. That challenge could be addressed over time, but in the interim Raymond admitted in interviews that even his mother did not entirely know what he did.[37] These were not problems the armed forces had back when the Air Force was created in 1947, but Raymond stoically pushed through them all, perhaps knowing that the reception of the guardians was predominantly positive and that the criticisms were largely inevitable and would fade in the long run.[38]

Raymond remained open to innovative ideas throughout this period, and the excitement of being part of creating a new military branch focused on space spread to many guardians. Initial Space Force recruiting campaigns seemed a bit futuristic and abstract, so Raymond adjusted them, and early recruiting efforts were revamped to become more selective. He organized the first doctrine summit in February 2020 and was thrilled because so many new ideas came from the bottom up from guardians who had ideas of their own and sent them upward through the chain of command for polishing.[39] Those efforts culminated in the August 2020 release of *Spacepower*, a space capstone publication that represented the "first articulation of spacepower as a separate and distinct form of military power" and became the foundation on which future evolutions of space doctrine were built. *Spacepower* reflected the reality that "adversaries have made space a warfighting domain" and fulfilled Raymond's vision of a branch able to design and implement doctrine for itself.[40]

Amid this whirlwind of activity, Raymond embraced the need for mission transfers from other services to the Space Force. AFSPC had become the core of the force the moment the latter stood up, so those assigned forces transferred relatively easily. The challenge then was to determine what else the Air Force and the other military branches did in space that properly belonged in the Space Force. Raymond started with the Air Force and, in consultations with senior leaders, identified twenty-eight squadrons that dealt with training, intelligence, research, and other space-related missions and, with support from Secretary Barrett, had them transferred to the Space Force. DoD also wanted the other services to relinquish all or most of their space missions to it, so Raymond met with his counterparts in the Army, the Navy, and the approximately sixty organizations that had a hand in space. Prior to the creation of the Space Force, the fact that, as Raymond put it, "sixty people could say no, nobody could say yes" was a significant bureaucratic hurdle. Raymond worked to consolidate space missions, bringing over an ultra–high frequency satellite system known as the Mobile User Objective System and the Ultra-High-Frequency Follow-On program from the Navy and satellite communications (SATCOM) global infrastructure known as the wideband SATCOM operations center, the regional SATCOM support center, and the joint tactical ground stations from the Army, among others. By the time he finished, the Space Force had united all responsibility for SATCOM and missile warning under a single service, which promised greater efficiency and unity of command.[41]

In August 2020 Raymond felt comfortable relinquishing command of USSPACECOM to Gen. James H. Dickinson, USA, in order to sharpen the distinction between the Space Force (a service branch) and USSPACECOM (a combatant command) and worked to clarify the roles of the National Guard and reserves when it came to space. The task was straightforward when it came to the reserves, which operate under Title 10 of the United States Code (which outlines the role of the forces), but was complicated with the National Guard, which is under Title 32 (which outlines the role of the Guard) and is partially controlled by individual states and their governors and state legislatures. Raymond's proposed solution was to combine the reserves and active duty personnel in the Space Force and then operate a separate Space National Guard, but at the time of his retirement, those suggestions required political resolution that had not taken place.[42]

Space Force Advocate
The most "consequential" work Raymond oversaw during his years with the Space Force was dedicated to, as he put it, "shifting the design of our space architecture from a system designed for a benign domain rather than a warfighting domain." He stood up a space warfighting analysis center built around civilian specialists, military engineers, and satellite operators to find ways to redesign space-based assets that are affordable and can survive in a warfighting environment. They completed force designs for missile warning, missile

tracking, offensive space capability, and ground-moving target indicators that were so well received by Congress that it made the Space Force the lead force design architect for all of DoD.[43]

Raymond also spent a significant amount of time in the public eye explaining and advocating for the Space Force. He did interviews with noted astrophysicist Neil deGrasse Tyson, among others, met with countless reporters, and appeared in numerous press conferences.[44] His themes usually centered around explaining the need for the Space Force, why it was created, and what it does and generating enthusiasm for the branch to help with recruitment. He combined thoroughly in-depth knowledge with a capacity for explaining complicated issues in ways the public could understand, a process that was necessary because the Space Force is small and therefore works with a limited natural constituency. In other words, not many people understand or work on space-related matters, either in the public or the halls of power, so Raymond strove to educate the public while promoting the Space Force at the same time.

In an interview with Mary Louise Kelly of National Public Radio, for example, he noted that space activity fell naturally into three categories: civil (science and exploration), commercial (private enterprises seeking a profit), and national security, and that all three had exponentially increased in size over the last two decades, requiring protection so that the American and global economies could function smoothly. Defense was necessary because space had transitioned from a benign domain to a warfighting domain, one in which potential adversaries (especially China and Russia) had the ability to disable, capture, or destroy U.S. satellites. Moreover, space had become inordinately congested. The Space Force tracked approximately fifty thousand objects in orbit in late 2022, including five thousand satellites. Given the dramatic decline in launch costs made possible by private companies like SpaceX, that congestion was certain to increase, and the United States thus had to operate as the world's space police, tracking objects and notifying other countries when/if collisions were imminent. And because space was congested, competitive, and contested, he pushed for the world's space powers to develop operating norms, or rules of the road like traffic laws or maritime regulations, that would mitigate or prevent conflict or accidents. Finally, he promoted the idea of integrated deterrence and partnerships with allies and promised that if deterrence failed, the United States was prepared to fight and win a war in space.[45]

Eventually, on November 2, 2022, in a ceremony at Joint Base Andrews attended by Secretary of Defense Lloyd Austin, Chairman of the Joint Chiefs of Staff Gen. Mark Milley, and Elon Musk of SpaceX, among others, Raymond closed his military career and retired. Gen. Bradley Chance Saltzman assumed leadership of the Space Force as CSO, and Jay Raymond rode, as it were, off into the sunset. He did so well before his own appointment as CSO would have naturally ended, and in the midst of such an astonishing series of achievements that it might easily have been extended. Yet Raymond's sense of duty drove him to retire because he thought it was in the best interests of the Space Force. He reasoned that because so many generals had taken their jobs when the Space Force came into being in 2019 or during the following year when other general officer positions were filled, everyone would change jobs at the same time if they all served full terms in office. That would have been severely detrimental in terms of continuity, so Raymond went to the secretary of defense and laid out a plan for rotating the senior officers and then volunteered to retire to set an example. In other words, his profound sense of duty led him to make a personal sacrifice in the service of the nation. As he put it, "I did it because I wanted to leave the service in the best position," which is perhaps the most eloquent testament to his qualities as a leader.

From his perspective, Raymond identified four major experiences that influenced his thinking prior to becoming CSO. The first came with his initial duty assignment as a missile launch officer, which he felt gave him operational focus, discipline, a commitment to checklists and crew coordination, and "all of the foundational things that are required to succeed as an operator in any job, whether it's flying airplanes or launching missiles or operating satellites." Then, as chief of commercial space launch operations, he gained an appreciation for the enormous potential of the commercial space business to lower launch costs and become a national security asset. He learned all about "commercial spaceports, payload processing centers, then commercial rockets and launch vehicles." That knowledge came a bit ahead of its time but proved invaluable when Elon Musk and SpaceX, among others, finally figured out how to make commercial space operations profitable. The third came during his time in the office of force transformation inside the office of the secretary of defense. He worked for Adm. Art Cebrowski on a project called TACSAT-1 (for Tactically Responsive Space), part of a larger effort to make space more responsive to warfighters and to accelerate development times by "doing things smaller, cheaper, faster" and using standardized equipment. His fourth critical experience came when he deployed to U.S. Central Command and worked for the combined force air component commander and saw how soldiers, sailors, airmen, and Marines used space on the battlefield. It "flipped his thinking," he said. "Rather than thinking about the satellites, I thought about the users that needed that information." Finally, he cited his tour of duty in East Asia with Fifth Air Force because it "really taught me the value of partnerships." Raymond helped lead the U.S. effort to assist Japan in the wake of the earthquake, tsunami, and disaster at the Fukushima nuclear power plant in 2011, and the power of space to assist with humanitarian efforts made an indelible impression.[46]

Raymond also benefited from serving with senior officers who mentored him throughout his career. Gen. Bob Kehler served as his commanding officer many times, consistently recommending Raymond for new and challenging assignments. Gen. Kevin P. Chilton took then-Colonel Raymond to witness Chilton's annual testimony before Congress because he thought Raymond would likely be a general one day and would benefit from the experience.[47] And there were countless others. Yet leaders are more than the sum of their experiences. Raymond combined sweeping expertise in his chosen field, an engaging, collegial personality that promoted cooperation, an openness to new ideas and a willingness to listen to subordinates of any rank, a deep understanding of organizational structure and the byzantine world of government personnel and budget processes, and a proactive approach to planning that meant he was ready when given the opportunity to make changes at operational speed.

By the end of his tenure as CSO, the Space Force had become a fully functioning service branch, with Space Force bases in Florida, Colorado, and California that were formerly Air Force bases.[48] More than eight thousand guardians worked with an approximately equal number of civilian specialists on Space Force missions around the world, partnering with government agencies and every U.S. military branch and ally while facilitating satellite launches across DoD from development to execution. Simply by being a part of that effort, Raymond would have earned a special place in Air Force and Space Force history. He was a vital conduit in what Gen. Thomas S. Moorman Jr., Air Force vice chief of staff, predicted in 2007 would be the development arc of the service when it would, he said, move through an inevitable transition from an air force to an air and space force and then to a space and air force.[49] What Moorman saw later on the space commission was the next evolution, when the Space Force broke away and became independent and the Air Force returned to being an air force. Raymond is one of the leaders who made that happen.

Conclusions

Raymond led the standup of a new military service in such a way that Secretary of Defense Lloyd Austin said he "made it look easy."[50] But he did a great deal more than that, and much that will be impossible for historians and scholars to fully appreciate in the future. Like all of us, he was a product of his experiences, each of them combined with his own attributes as a leader to prepare him for what lay ahead, and he benefited from a certain amount of luck along the way. We can only wonder, for example, what the Space Force might have looked like if Raymond had become a pilot or never been executive officer to General Coglitore, or whether it would exist at all if no president had decided to make it a priority. One presumes that a Space Force of some kind would have emerged eventually, but without Raymond it would have been different, for to lead is to make choices, and the Space Force was decisively shaped by his.

Some of his traits and skills were learned, while others were innate, and Raymond defies simple comparisons to other transformational leaders in Air Force and Space Force history. He was far more politically skilled than Billy Mitchell, far less bombastic or rigid than Curtis LeMay, and never suited by training or experience to the kind of invention required of Bernard Schriever. Perhaps the closest analog in temperament and achievement would be Henry H. "Hap" Arnold, the father of the Air Force. Theodore von Kármán described Arnold as "the greatest example of the American military man, a combination of complete logic, mingled with farsightedness and superb dedication."[51] Let that stand by Raymond's name.

Notes

1. Melissa Carl, "West Point's Legacy Families," *West Point* 2, no. 4 (Fall 2012): 41–42. Through marriage, the Raymond family can trace West Point graduates to 1825. The author would like to thank Mr. Christopher Rumley, command historian for headquarters, Space Operations Command, Dr. Gregory Ball, senior historian with the Department of the Air Force Space History Division, Dr. Rick Sturdevant at headquarters,

Space Training and Readiness Command, and Dr. Wendy Whitman Cobb at the School of Advanced Air and Space Studies for their assistance with this chapter.
2. "John Raymond," *Washington Post*, December 6, 2016, https://www.legacy.com/us/obituaries/washingtonpost/name/john-raymond-obituary?id=6075890.
3. Gen. John W. "Jay" Raymond, interview by Lance Janda, February 7, 2023, 7.
4. Cynthia Belio, "Chief of Space Operations Gen. John W. 'Jay' Raymond's Exit Interview," Defense Visual Information Distribution Service, November 1, 2022, https://www.dvidshub.net/video/862757/chief-space-operations-gen-john-w-jay-raymonds-exit-interview.
5. Clemson University, Clemson Commencement Program, May 1, 1984, https://tigerprints.clemson.edu/cgi/viewcontent.cgi?article=1201&context=comm_programs; Department of Defense, "Gen. John W. 'Jay' Raymond Biography," https://www.defense.gov/About/Biographies/Biography/Article/1996612/john-w-jay-raymond/.
6. Belio.
7. Raymond, Janda interview, 8.
8. Raymond, Janda interview, 9.
9. Interview with Gen. John W. "Jay" Raymond, USSF, Chief of Space Operations, United States Space Force, by Dr. Gregory W. Ball, Mr. Wade Scrogham, and Dr. Kathleen J. Nawn, Air Force History Office Space History Division, July 26, August 30, and September 15, 2022, 1; Department of Defense, Raymond biography.
10. Raymond, Ball, Scrogham, and Nawn interview, 3, 1.
11. Department of Defense, Raymond biography.
12. David N. Spires, *Beyond Horizons: A Half Century of Air Force Space Leadership* (Peterson AFB, CO: Air Force Space Command, 2002), chap. 7.
13. "Fireside Chats with Gen. John W. 'Jay' Raymond," Aspen Institute: Aspen Security Forum, July 19, 2022, https://archive.org/details/theaspen-Fireside_Chats_with_General_John_W._Jay_Raymond.
14. For an excellent and concise history of the USSF, see David Christopher Arnold, "The United States Space Force," in *Understanding the U.S. Military*, ed. Katherine Carroll Blue and William B. Hickman (London: Routledge, 2022), 108–23.
15. Dino A. Lorenzini and Charles L. Fox, "2001: A U.S. Space Force," *Naval War College Review* 34, no. 2 (1981): 62, https://digital-commons.usnwc.edu/nwc-review/vol34/iss2/5/.
16. Lorenzini and Fox, 49, 62.
17. Tom Clancy and Gen. Chuck Horner (Ret.), *Every Man a Tiger* (New York: G. P. Putnam's Sons, 1999), 517–18.
18. "Rumsfeld Space Commission," *Militarist Monitor* (November 1, 2007), https://militarist-monitor.org/profile/rumsfeld_space_commission/. Rumsfeld was a former secretary of defense, White House chief of staff, and three-term congressmen from Illinois. He served in a variety of other public and private sector roles and was a lifelong member of the Republican party. He chaired the space commission, as well as the Commission to Assess the Ballistic Missile Threat to the United States in 1998.

19. Katie Bo Williams, "Opposition to a Space Force Simmers in the Senate," *Defense One*, August 1, 2018, https://www.defenseone.com/policy/2018/08/opposition-space-force-simmers-senate/150227/.
20. "Remarks of Congressman Mike Rogers, Chairman, House Armed Services Strategic Forces Subcommittee," *Strategic Studies Quarterly* (Summer 2017): 1–11, https://www.airuniversity.af.edu/Portals/10/SSQ/documents/Volume-11_Issue-2/Rogers.pdf.
21. Alex Ward, "Trump Really, Really Wants Troops in Space," *Vox* (June 18, 2018), https://www.vox.com/2018/6/18/17475444/trump-space-force-speech-air-force-congress.
22. John W. "Jay" Raymond, "We Need to Focus on Space: We Don't Need a 'Space Corps,'" *Defense One*, July 12, 2017, https://www.defenseone.com/ideas/2017/07/we-need-focus-space-we-dont-need-space-corps/139360/. Despite the headline, Raymond did not argue specifically against a "Space Corps" or an independent space force. He merely pushed for more normalization of U.S. military space operations.
23. Raymond, Ball, Scrogham, and Nawn interview, 6.
24. Raymond, Ball, Scrogham, and Nawn interview, 6–7.
25. Raymond, Ball, Scrogham, and Nawn interview, 7–8.
26. Raymond, Ball, Scrogham, and Nawn interview, 8.
27. Raymond, Ball, Scrogham, and Nawn interview, 8.
28. Ellie Kaufman, Haley Byrd, and Phil Mattingly, "House Passes Defense Bill that Would Include Paid Family Leave for Federal Workers for the First Time," *CNN*, December 11, 2019, https://www.cnn.com/2019/12/11/politics/ndaa-defense-bill-passes-space-force/index.html.
29. Raymond, Ball, Scrogham, and Nawn interview, 8–9.
30. Raymond, Ball, Scrogham, and Nawn interview, 9–10, and 19–20.
31. Raymond, Ball, Scrogham, and Nawn interview, 10–11. There were still plenty of opponents of the USSF. See Robert Farley, "Space Force: Ahead of Its Time, or Dreadfully Premature?" CATO Institute Policy Analysis no. 904 (December 1, 2020), for an early example of their arguments: https://www.cato.org/policy-analysis/space-force-ahead-its-time-or-dreadfully-premature.
32. Raymond, Ball, Scrogham, and Nawn interview, 12.
33. Raymond, Ball, Scrogham, and Nawn interview, 13–14.
34. Raymond, Ball, Scrogham, and Nawn interview, 25, 29.
35. Raymond, Ball, Scrogham, and Nawn interview, 14–17; Svetlana Shkolnikova, "Space Force Unveils Official Song 'Sempra Supra,'" *Stars and Stripes*, September 20, 2022, https://www.stripes.com/branches/space_force/2022-09-20/space-force-song-guardians-7403666.html.
36. Bryan Bender, "The Netflix Show the Pentagon Can't Stop Talking About," *Politico*, May 23, 2020, https://www.politico.com/news/magazine/2020/05/23/space-force-netflix-steve-carell-274951.
37. Jacqueline Feldscher and Connor O'Brien, "Can the Space Force Shake Off Trump?" *Politico*, February 3, 2021, https://www.politico.com/news/2021/02/03/space-force-jay-raymond-white-house-465497.

38. See Martin Pengelly, "'It's Not a Banger': Response to Space Force Official Song Is Less Than Stellar," *The Guardian*, September 20, 2022, https://www.theguardian.com/us-news/2022/sep/20/us-space-force-official-song; Peter Suciu, "Space Force Introduces New Seal: Welcome to Starfleet," *ClearanceJobs*, January 27, 2020, https://news.clearancejobs.com/2020/01/27/space-force-introduces-new-seal-welcome-to-starfleet/; Toria Barnhart, "Space Force Uniforms 'Complete Rip-Off' of Sci-Fi Shows, Social Media Users Say," *Newsweek*, September 21, 2021, https://www.newsweek.com/space-force-uniforms-complete-rip-off-sci-fi-shows-social-media-users-say-1631284.
39. Raymond, Ball, Scrogham, and Nawn interview, 18.
40. Ashley M. Wright, U.S. Space Force Public Affairs, "Space Force Releases 1st Doctrine, Defines 'Spacepower' as Distinct Form of Military Power," August 10, 2020, https://www.spaceforce.mil/News/Article/2306828/space-force-releases-1st-doctrine-defines-spacepower-as-distinct-form-of-milita/.
41. Raymond, Ball, Scrogham, and Nawn interview, 18–19.
42. Raymond, Ball, Scrogham, and Nawn interview, 20–21.
43. Raymond, Ball, Scrogham, and Nawn interview, 22.
44. See "Raymond and deGrasse Tyson Talk Space," *Air and Space Forces Magazine*, April 1, 2021, https://www.airandspaceforces.com/watch-raymond-and-degrasse-tyson-talk-space/.
45. "Fireside Chats with Gen. John W. 'Jay' Raymond."
46. Raymond, Ball, Scrogham, and Nawn interview, 2–3.
47. Raymond, Janda interview, 3–4, 19.
48. The bases are Patrick Space Force Base in Florida, Buckley, Peterson, and Schriever Space Force Bases in Colorado, Los Angeles AFB (still an AFB at this writing), and Vandenberg Space Force Base in California. See spaceforce.mil for details regarding units assigned to each installation.
49. Spires, xiv.
50. Doug G. Ware, "Raymond Retires as 1st Space Force Chief, Saltzman Takes Over in Inaugural 'Change of Responsibility' Ceremony," *Stars and Stripes*, November 2, 2022, https://www.stripes.com/branches/space_force/2022-11-02/space-force-raymond-saltzman-command-7901010.html.
51. Thomas M. Coffey, *Hap: The Story of the U.S. Air Force and the Man Who Built It* (New York: Viking, 1982), 376–77.

CONCLUSION
The Space Force Needs Visionary Leaders
DAVID CHRISTOPHER ARNOLD

Insights that can be easily gleaned in all these chapters are the obvious ones like the importance of leadership or integrity, which are discussed in almost any conversation on leadership. But the concepts of strategic thinking and vision are rarely discussed in collections of biographies like this one and should be central to a book about general officers who impacted military space. If strategy is the linking of ways and means to achieve a particular end, and leadership is the ability to motivate and inspire self-sacrifice, then strategic thinking is the ability to develop strategies and implement them from a very high level—that is, with "the big picture" in mind. For the subjects of this book, what binds them all together are their abilities as strategic leaders. What does strategic leadership entail? Using the U.S. Army War College's primer in strategic leadership as a basis for discussion, we can see that each of these space leaders provided a vision, shaped a culture, led and managed change, and built and shaped relationships.[1]

Edgar Puryear starts his epilogue to *Stars in Flight* with the question, "What does command entail?" After a brief discussion, he concludes that command is a twenty-four-hours-a-day, seven-days-a-week occupation, consuming all the hours of a commander's day and indeed some of a commander's family's day as well. Puryear assesses that the reason the men he chose for his book were "great men" was because they had the highest character—they had the integrity, humility, selflessness, concern for others, respect, and self-confidence of most

American leaders. He does not assert that his subjects were perfect human beings but rather that their high character meant they could "be depended on at virtually all times in all situations and in all aspects of life."[2] Each one of the subjects of *Stars in Flight* was a combat leader who shaped the creation and early history of the United States Air Force.

The same cannot be said for the subjects of this book, *Space Force Pioneers*, in which fewer than half of the people discussed have been in a combat zone, only three have been to space, and exactly zero have space combat experience. And although they all wore the title of "commander" at some point in their career, this book does not assess that its subjects are all people of high character or that "they had the integrity, humility, selflessness, concern for others, respect, and the self-confidence of most American leaders," because even if they strove for these attributes each and every day, like many of us, they sometimes fell short. This book certainly does not assert that they were all perfect human beings and worthy of statues, even if some of them have had Space Force bases or buildings named after them. Nevertheless, this is not a book about commanding an organization but about strategic leadership for the space domain.

One of the keys to strategic leadership is having a vision of what success looks like. Each of these leaders possessed a vision, whether it was about defending the United States by deterring its enemies using space systems or creating an independent space service, and vision was key to their ability to lead others toward their desired ends. Having a vision is the ability to see what success looks like in the future. It is the ability to give a sense of identity, purpose, and direction to an organization and then to lead the organization to success, whether it was Medaris pushing the German rocket engineers toward the first successful American satellite, Helms pulling together an international team of payload specialists to achieve the scientific goals of STS-78, or Raymond willingly standing up and being counted as guardian number one on the rolls, speaking in engagement after engagement about why the need for an independent military service for space had finally arrived. Clausewitz referred to the coup d'oeil, which "merely refers to the quick recognition of a truth that the mind would ordinarily miss or would perceive only after long study and reflection," but strategic leaders must have the ability to see what the future can be and successfully achieve it.[3]

Culture is like the personality and characteristics of an organization.[4] Part of creating success is creating a culture to achieve that success; witness Hartinger's quizzes on space topics at staff meetings, Lord's driving the space professional program that ultimately included the space badge, or Hyten's creation of a warfighting focus in Air Force Space Command. These officers also had to be technically competent. They had to understand the systems, manage the relationships, and be politically adept. Space systems are and always have been highly technically complicated. Schriever was smart enough to surround himself with people who were also technically competent but would tell him when he was wrong. Although there are some examples in this book of officers with highly technical backgrounds, such as the astronauts who all graduated from flight test school, not all these leaders had engineering or science backgrounds. Power never graduated from high school. Ritland never graduated from college. Kehler had a degree in education. Moorman was a history major. But even without a scientific or technical background, they each still had to have the ability to take extremely technical ideas and translate them into easily understood concepts. Being able to understand the difference between high earth orbit and a highly elliptical orbit might seem basic to the space professional today, but being able to explain the differences and the reasons for them in noncondescending but useful ways is an important skill. The Space Force must create an organizational culture that can achieve its ends, and although that culture starts with technical competence, it ends with strategic leadership.

Further, none of that technical competence would have meant anything if these leaders had not also successfully managed change, which requires an understanding of the context of a strategic problem. A strategic leader must identify and understand the "domestic political, economic, bureaucratic, social/cultural, and technological factors that are likely to help or hinder both the strategy-making effort and [a strategy's] viability once executed."[5] One only needs to look at the Air Force's attempts to find missions for military pilots in space to understand that. Put another way, "Although technology might be a prime element in many public issues, nontechnical factors take precedence in technology-policy decisions."[6] When Lord told the Air Force Association in November 2003, "We have a separate space force. It's called the Air Force," the space context was very different from when Raymond stood in the Oval Office unfurling the flag of a new military service.[7] Yet each understood the political context of their time in leadership.

These space leaders also had to build and manage their relationships, whether within their service or agency, with other agencies in the executive branch, with the other two coequal branches of government, or even with other nations. When Power and Schriever talked about the future of space and missiles within the confines of the Air Research and Development Command, they were trying to work together to provide deterrent capabilities for the nation. Similarly, when Ritland represented the Air Force in meetings with the National Aeronautics and Space Administration in the early days of human spaceflight, he got the organization to understand that the Air Force was its partner, not its competitor. His ability to manage this particular space relationship, perhaps greater than any other reason, may have led Schriever to note that "Ossie is one of the few people that has more friends than enemies in DC."[8] Certainly Hartinger's relationship with Aerospace Defense Command commander Chappie James was important in getting the new space command implemented in 1982. Or maybe it was as simple as Bolden's command of STS-60, which included a Russian cosmonaut/mission specialist to which he was initially opposed. They all managed relationships because it is key to getting things done.

Were these people the only people responsible for the creation of the American military's newest branch? No, of course not. Another volume could and probably should include the biographies of nonmilitary people who were also present at the U.S. Space Force's creation: Congressmen Mike Rogers and Jim Cooper, any number of secretaries of defense and secretaries of the Air Force, generals and intelligence officials, program managers, and many others. This group of leaders had the ability to see the future and achieve that vision. What the U.S. Space Force needs today, and tomorrow, are strategic leaders like these who can develop strategies and implement them. Hopefully, in this volume, we have given some good examples of how to do just that.

Notes

1. Tom Galvin and Dale Watson, eds., *Strategic Leadership: Primer for Senior Leaders*, 4th ed. (Carlisle, PA: U.S. Army War College, 2019).
2. Edgar F. Puryear, *Stars in Flight: A Study in Air Force Character and Leadership* (New York: Random House, 1981), 279–83.

3. Carl von Clausewitz, *On War*, ed. and trans. Michael Howard and Peter Paret (Princeton: Princeton University Press, 1976), 102.
4. Edgar H. Schein with Peter Schein, *Organizational Culture and Leadership*, 5th ed. (Hoboken, NJ: Wiley, 2007), 127–31.
5. Steven Heffington, Adam Oler, and David Tretler, eds., *A National Security Strategy Primer* (Washington, DC: National Defense University Press, 2019), 10.
6. Melvin Kranzberg, "Technology and History: 'Kranzberg's Laws,'" *Technology and Culture* 27, no. 3 (July 1986): 544–60.
7. Gen. Lance W. Lord, presentation to the Air Force Association Space Symposium, Los Angeles, November 2003, cited in David C. Arnold, "Lt. Gen. Forrest S. McCartney, the First Space Professional," *Air Power History* 51, no. 4 (Winter 2004): 28.
8. "The Ritland Fan Club," retirement dinner, circa December 1, 1965, audio recording, track 8, 6:50–13:00, provided by Kathleen Ritland Montoya.

CONTRIBUTORS

David Christopher Arnold got the idea for this volume years ago when he read Edgar Puryear's 1981 book *Stars in Flight: A Study in Air Force Character and Leadership* and wondered why there wasn't a book about space leaders. Today Arnold is a professor of national security strategy at the National War College. He received his PhD from Auburn University in 2002 and is the author of *Spying from Space: Constructing America's Satellite Command and Control Systems* (2005). He is a career space and missile professional who retired from the U.S. Air Force in 2015 as a colonel. Dr. Arnold has taught history at 7,258 feet, 750 miles north of the Arctic Circle, and in cyberspace, and served as the editor for a decade of the award-winning journal *Quest: The History of Spaceflight Quarterly*.

Gregory W. Ball received his PhD in U.S. history from the University of North Texas in 2010 and has been a Department of the Air Force historian since 2009. Formerly the command historian for Air Force Space Command, he is now the senior historian for the U.S. Space Force, providing direct support to the space staff and ensuring policies and programs are in place to document and preserve the history of the nation's newest military service. He is the author of *They Called Them Soldier Boys: A Texas Infantry Regiment in World War I* (2013) and *Texas and World War I* (2019).

Lisa L. Beckenbaugh is the chair of the leadership and research development department at Air University's Air Command and Staff College. Dr. Beckenbaugh received bachelor's and master's degrees from St. Cloud State University and a PhD from the University of Arkansas. Dr. Beckenbaugh has taught at a variety of undergraduate and graduate civilian institutions. Her book, *The Versailles Treaty: A Documentary and Reference Guide* (ABC-CLIO), was published in 2018. Dr. Beckenbaugh also serves as the faculty advisor for the Gathering of Eagles elective and has edited six of their published books, most recently, *Why We Stay: Stories of Unity and Perseverance*. Her current research is on the 1st Mobile Army Surgical Hospital (MASH), later redesignated 8209th MASH, during the Korean War.

Contributors

Stephen J. Garber is an historian in the NASA history division. He is especially interested in applied history, making the past directly relevant for policymakers. He has edited dozens of books and written a variety of journal articles. A book he coauthored on national space policy, *Origins of 21st-Century Space Travel: A History of NASA's Decadal Planning Team and the Vision for Space Exploration, 1999–2004* (2018), won two prestigious professional awards. He has also researched and written on many other topics in national security space policy and aerospace history.

Lance Janda is professor of history and chair of the department of social sciences at Cameron University. He received his PhD from the University of Oklahoma and has taught a wide array of courses in U.S., modern European, and military history over the last two and a half decades. His published works include *Stronger than Custom: West Point and the Admission of Women* (2002), "Shutting the Gates of Mercy: The American Origins of Total War, 1860–1880," and book chapters on various topics including the Flying Tigers and the experience of the U.S. Marine Corps in the Pacific during World War II.

Col. Margaret C. Martin, USAF, is vice dean of the faculty and permanent professor of history, United States Air Force Academy. She earned her PhD from the University of North Carolina and earned her commission from the United States Air Force Academy. She is a command pilot with more than thirty-four hundred flight hours, including over three hundred combat hours in the C-17.

Jennifer Ross-Nazzal is the human spaceflight historian for NASA; for almost twenty years, she served as the Johnson Space Center historian. She earned her PhD in history from Washington State University and holds the unique distinction of being a scholar of NASA history and women's history. Ross-Nazzal is the author of two books: *Winning the West for Women: The Life of Suffragist Emma Smith DeVoe* (2011) and *Making Space for Women: Stories from Trailblazing Women of NASA's Johnson Space Center* (2022), which received the Liz Carpenter Award for Best Book on the History of Women from the Texas State Historical Association.

Col. William D. Sanders, USSF, has served twenty years in military space operations. He holds a PhD in military strategy from the School of Advanced Air and Space Studies. A 2003 graduate of the U.S. Air Force Academy, he also holds master's degrees from the Kelley School of Business at Indiana University, the Air Command and Staff College, and the National War College. His research interests include the history of military space operations and U.S. Space Force culture.

Wade A. Scrogham is a graduate of Georgia Southern University and the Air Command and Staff College. He is a former curator of the National Museum of the Mighty Eighth Air Force and joined the Department of the Air Force history and museums program in 2003. His previous assignments include chief historian of the Air Force Space and Missile Systems Center, senior historian for Space Operations Command, and lead historian for the Space Training and Readiness Command Task Force. He currently oversees historical programs at Patrick Space Force Base and Cape Canaveral Space Force Station.

Rick W. Sturdevant, PhD, joined the U.S. Air Force History and Museums Program in 1984. He served as space communications division historian (1985–91) and deputy director of history for Air Force Space Command (1991–2021) and became Space Force Training and Readiness Command historian in November 2021. He serves on the *Quest: The History of Spaceflight Quarterly* editorial board and has been the International Academy of Astronautics history series editor. His honors include the Air Force Exemplary Civilian Service Award (1995–99), the American Astronautical Society (AAS) President's Recognition Award (2005), AAS Fellow election (2007), and the AAS President's Award (2023).

John G. Terino, associate professor of military and security studies at the Air Command and Staff College, teaches military theory, airpower, contemporary warfare, leadership, joint planning, joint air planning, and an elective on the Air Force in fact, fiction, and film. Previously, as a professor at the School of Advanced Air and Space Studies, he directed courses on technology and military innovation, airpower history, and wargaming. Retired from the Air Force as a lieutenant colonel after serving for twenty-two years, he earned bachelor's, master's, and PhD degrees in the history and sociology of science from the University of Pennsylvania.

Heather P. Venable is an associate professor of military and security studies in the department of airpower and the airpower strategy and operations course director at the Air Command and Staff College. She graduated with a bachelor of arts degree in history from Texas A&M University and a master of arts degree in American history from the University of Hawaii. She received her PhD in military history from Duke University. She also is also a graduate of the space operations course. She has written *How the Few Became the Proud: Crafting the Marine Corps Mystique, 1874–1918* (Naval Institute Press, 2019).

Brent D. Ziarnick is a visiting assistant professor in the U.S. Space Force Officer Professional Military Education Program at the Johns Hopkins University School of Advanced International Studies. Dr. Ziarnick's specialty is spacepower theory and strategy. A career Air Force space operations officer who retired as a lieutenant colonel, he served in both higher headquarters strategy and operational squadron command assignments. He has written three books and received the prestigious Airpower History Book of the Year Award from the Air Force Historical Foundation.

INDEX

Advanced Extremely High-Frequency satellite, 27
Advanced Research Projects Agency (ARPA), 16, 103
aerospace concept, 4, 132–33, 139
Aerojet Corporation, 51
Aerospace Defense Command, 17, 301. *See also* Air Defense Command
Air and Space Power Journal, 262, 266. See also *Air Power Journal*
Air Combat Command, 180, 229, 234, 237, 240
Air Command and Staff College, 262–63, 280
Air Defense Command (ADCOM), 17–19, 152–54, 156, 158, 160, 174; disestablishment of, 153
Air Force, 5th, 280, 293; in World War II, 53
Air Force, 8th, 28
Air Force, 13th, 280
Air Force, 14th, 28–29, 216, 228, 233, 242, 280
Air Force, 24th, 241
Air Force, Headquarters, 17–20, 35, 124, 129, 177, 231. *See also* Air Force, Department of the; Chief of Staff of the Air Force; Space Command, Air Force
Air Force Academy, U.S., 23, 150, 159, 167, 203, 205, 211, 216–17, 241, 287. *See also* Colorado Springs
Air Force, Department of the, 26, 32–33, 37, 273, 286
Air Force Institute of Technology, 148, 159, 241
Air Force Manual 1-1, *Functions and Basic Doctrine of the United States Air Force*, 20, 156
Air Force Space Command (AFSPC or AFSPACECOM or SPACECOM). *See* Space Command, Air Force

Air Force Systems Command. *See* Systems Command
Air Mobility Command, 180
Air Power Journal, 262
Air Research and Development Command (ARDC), 15–17, 58, 120–26, 130, 134–46, 138
Air University, 231, 233
Air University Review, 17
Air War College, 152, 284
Aldridge, Edward "Pete," 22, 157, 174–76
Allard, Wayne, 237
Allen, Lew, 19–21, 23, 154, 157, 173–74
Anderson, Andrew, 19
Angel, John, 21
Annapolis, 195–96, 200
anti-satellite weapons (ASAT), 17, 160, 170–71, 182, 259–60, 262; Chinese, 238, 283
Apollo, 46, 58, 106, 111, 256, 279
Apollo 11, 210. *See also* moon landing
Armagno, Nina, 217
Armstrong, Neil, 278. *See also* moon landing
Army Air Corps, U.S., 48–49, 52–53, 120. *See also* Army Air Forces
Army Air Defense Command (ARADCOM), 152
Army Air Forces, U.S., 7–8, 13–14, 54, 97. *See also* Army, U.S.
Army Ballistic Missile Agency (ABMA), 16, 82, 85–88, 90. *See also* Redstone Arsenal; Huntsville, Alabama
Army School for Advanced Military Studies, 32
Army, U.S., 13–14, 16, 33, 47–50, 52, 55, 58, 69–71, 73–74, 76–90, 93, 100–1, 129–30, 144–46, 167, 175, 181, 184, 186, 229–30, 235–36, 259–60, 278, 281, 283, 286, 290
Army War College, 298

308 | Index

Arnold, Henry "Hap," 5, 7, 14, 48–52, 54, 59, 93, 187, 294
Aspin, Les, 183
Atlas, 7, 15, 27, 57–58, 83, 99, 101, 106–10, 121–23, 125, 127–29, 132, 138
Atomic Energy Commission, 95
Austin, Lloyd, 292, 294

Barrett, Barbara, 285, 289
Battlestar Galactica, 288
Bissell, Richard "Dick," 97, 102–3, 105
Blue Gemini, 109–11
Blue Origin, 139
blue ribbon panel, 176, 179–81, 183–84
Bolden, Charles, 195–202
Bombardment Squadron, 9th, 50
Bradburn, David, 17
Bradley, Omar, 73–76, 78–81, 210
Bureau of Aeronautics, 15
Burnt Frost, Operation, 109
Burrows, William, 98
Bush, George H. W., 186
Bush, George W., 26, 209

California Institute of Technology, 13–14, 51, 55–56. *See also* Jet Propulsion Laboratory; Guggenheim Aeronautical Laboratory at Caltech (CALCIT)
Cape Canaveral, 83, 86, 207, 215, 256
Carter, Jimmy, 9, 153, 170–72
Cebrowski, Art, 292
Central Intelligence Agency (CIA), 58, 85, 97–98, 103–4
Chaffee, Roger, 256
Chain, John, 20–23, 261
Challenger, 174, 280
Cheyenne Mountain, 155–56, 173, 263. *See also* North American Aerospace Defense Command; Air Defense Command
Chief of Space Operations (CSO), 37, 279, 285–86, 292–93. *See also* Space Force, U.S.; Raymond, John "Jay"
Chief of Staff of the Air Force, 17, 19, 21, 24, 31, 33, 109, 121, 138–39, 153–54, 157, 161, 173, 176, 180, 184, 207, 209, 228–29, 240, 262, 285; deputy for operations, 19–20, 174, 280; deputy for research and development, 55, 97, 121; assistant vice chief of staff, 18, 23, 231, 242; vice chief of staff, 167, 184, 293
Chilton, Kevin, 194, 202–10, 225–26, 237, 240, 293
China, 30, 209, 238, 244, 266, 269, 272, 283–84, 291
Cheyenne Mountain, 155–56, 173, 263. *See also* Space Command, U.S.
Clausewitz, Carl von, 2–4, 299
Clark, Gaylord "Wes," 19, 173
Clinton, William "Bill," 183, 186, 265
Coglitore, Sebastian, 279, 294
Collbohm, Franklin, 14
Collins, Eileen, 213
Collins, Lawton, 81
Colorado Springs, 17–18, 22–23, 25, 30–31, 33, 152, 154, 156–57, 159, 172–74, 233, 263
Columbia, 195
Columbia University, 203
Commission to Assess United States National Security Space Management and Organization. *See* space commission
concurrency, 55–59, 99–100, 133
Congress, U.S., 18, 22, 30–31, 33–37, 59, 70–71, 73, 81, 87, 99, 153, 160, 172, 186, 201, 209, 216, 237, 239, 243, 282–83, 285–86, 288, 291, 293. *See also* House Appropriations Committee; House Armed Services Committee; House of Representatives, U.S.; Senate, U.S.; Senate Armed Services Committee (SASC)
Consolidated Space Operations Center (CSOC), 22, 155
Constellation, 200
Continental Air Defense Command, 152. *See also* Air Defense Command (ADCOM)
Convair, 55, 106, 108, 122–24, 127, 137–38. *See also* Atlas
Cooper, Jim, 32–34, 36, 283, 301

Index

Copp, DeWitt, 5
Corona, 8, 58, 102–5, 168–69
Crosier, Clint, 35–36
Creech, Wilbur "Bill," 18
culture, military, 48, 167, 214, 235
culture, space, 2–3, 13, 33, 38, 46, 112, 139, 175, 186, 214, 226–27, 229–31, 235, 298, 300

Davis, Benjamin "Benni," 157, 160
DeBlois, Bruce, 262, 266
Defense Meteorological Satellite Program (DMSP), 24, 169, 263
Defense Satellite Communications System, 27, 169, 178
Defense Support Program, 17, 24, 27, 179, 261
DeKok, Roger, 21, 263
Department of Defense (DOD), 16, 20–22, 29–30, 33–35, 83–85, 87–89, 103, 110–11, 156–57, 159–61, 170–72, 175, 177–78, 183–84, 283, 289, 291, 293
Deptula, David, 267
Desert Shield, Operation, 25, 178, 260
Desert Storm, Operation, 25, 178–82, 224, 227–28, 231–32, 236, 244, 261–63, 280–82. *See also* Gulf War
deterrence, 3–4, 21, 28, 59–60, 94, 99, 102, 137–38, 215, 243, 262, 267, 269–73, 291, 301
Dickinson, James, 290
Dickman, Robert, 19, 171
director of space forces, 267–68, 280
disaggregation, 243
Discoverer, 8, 102–4
doctrine, Air Force. *See* Air Force Manual 1-1
Doolittle, James "Jimmy," 97, 99, 107
Douglas Aircraft Company, 14, 128, 131, 202. *See also* RAND
Dunford, Joseph, 30, 36, 272, 283
Dyna-Soar, 109, 133, 138–39

Eaker, Ira, 5, 48–49
Edwards Air Force Base, 194

Eisenhower, Dwight, 15, 57, 73, 76, 78, 81, 87–89, 101, 104, 194
Endeavour, 205
Estes, Howell, 25, 229
evolved expendable launch vehicle, 27, 183
executive agent for space, DOD, 26, 156, 172, 187
Explorer I, 7, 84, 86, 283

Faga, Martin, 184
Falcon, 27
Falcon Air Force Base/Station, 7, 156. *See also* Consolidated Space Operations Center (CSOC)
Fallujah, 268
Fighter-Bomber Squadron, 23rd, 147
Fighter-Bomber Wing, 428th, 147
Fighter-Interceptor Squadron, 331st, 148
first information war, 227
first space war, 25, 178–79
Foale, Michael, 198
Fogleman, Ronald, 184–85
Ford, Gerald, 9, 170
Foulois, Benjamin, 6
Fox, Charles, 281
Frisbee, John, 5

Gabriel, Charles, 24, 161
Gambit, 169
Gardner, Trevor, 15, 57, 98, 121–23, 125–26, 132; and scientific advisory strategic missiles evaluation committee, 15; and ICBM scientific advisory committee, 125–26. *See also* Teapot Committee
Gemini, 106, 109–11, 113. *See also* Blue Gemini
Gemini B. *See* Blue Gemini
General Accounting Office, 157
General Atomics, 137
George, Harold, 6
Geosynchronous Space Situational Awareness Program, 27, 244
Gibson, Robert "Hoot," 198
Gilruth, Robert, 109

310 | Index

Glenn, John, 109, 203
Glennan, Keith, 87–88
Global Positioning System (GPS), 24–25, 27, 158, 160, 178–79, 239, 242, 281
Global Strike Command, Air Force, 240
Goldfein, David, 31, 33, 285
Greenleaf, Abbot, 18
Grissom, Virgil "Gus," 203, 256
Guardian Challenge, 264
guardians, 235, 288–89, 293
Guggenheim Aeronautical Laboratory at Caltech (CALCIT), 13, 51. *See also* Jet Propulsion Laboratory
Gulf War, 59, 167, 178–80, 184, 187, 261–62, 282. *See also* Desert Storm, Operation
Gunter Air Force Base, 257–58

Habednank, Otto "Ken," 203
Haeussermann, Walter, 86
Hall, Ed, 58
Hall, Harvey, 14–15
Hallion, Richard, 182
Hamel, Michael, 176
Hansen, James, 94
Hart, Gary, 173
Hartinger, James, 22–24, 144–62, 173–74, 300–1
Helms, Susan, 210–17, 299
Henry, Richard, 22
Henry Award, 264
Herres, Robert, 18, 25
Hexagon, 105, 169
High Frontier, 232–33, 235, 236, 242
Hill, James, 18–19, 154
Hill, Jimmie, 183
Hillard, Victor, 183
Hiroshima, 95, 119
Horner, Charles "Chuck," 282
House Appropriations Committee, 18, 36, 181
House Armed Services Committee, 30, 32–33
House of Representatives, U.S., 22, 30, 36–37, 87, 185
Hoyer, Steny, 167, 185
Hughes Aircraft, 56–57

Hughes, Thomas Parke, 3, 94
Hungerford, John, 21
Huntsville, Alabama, 7, 16, 58, 82–84, 86–88, 255–56, 259–60
Hutto, James "Cal," 235
Hyten, John, 29, 35–36, 181, 187–88, 242, 245, 254–74, 300

Industrial College of the Armed Forces, 96, 149
Inhofe, Jim, 283
intelligence community, 28–29, 35, 168, 207, 283, 301
intercontinental ballistic missile (ICBM), 1, 7, 14, 54–55, 83, 99, 107, 109–10, 119, 121–34, 138–39, 151, 179, 207, 225, 227, 229–32, 234–35, 238–40, 261, 268, 279. *See also* Atlas; Minuteman; Polaris; Titan
intermediate range ballistic missile (IRBM), 57, 82, 100, 102, 125–26, 129, 132–34. *See also* Jupiter; Thor
International Geophysical Year, 16
International Space Station (ISS), 200–2, 206–7, 214, 217

James, Daniel "Chappie," 155, 301
Jet Propulsion Laboratory, 14, 55, 86–87
Johnson, Kelly, 97, 105
Johnson, Lyndon, 87, 195–96
Johnson, Nicholas, 209
Johnson, Roy, 16
Johnson Space Center, 21–22, 200–1, 241; as Manned Spacecraft Center, 109
Joiner, William Howard, 13–14
joint chiefs of staff (JCS), 10, 20, 24–25, 30, 37, 152, 156, 161, 183, 187–88, 208–9, 254, 272, 283, 286, 292
joint force air component commander, 267
joint force component commander for space (JFCC Space), 29
joint force space component commander (JFSCC), 29, 271
Joint Force Quarterly, 233
joint functional component command for space and global strike, 28

Index | 311

joint interagency combined space operations center (JICSpOC), 28–29
joint space operations center (JSpOC), 28–29
Joint Task Force Space Defense, 32
Jones, Carl, 227
Jones, David, 17–18, 153–54, 156, 205
Jumper, John, 207–9
Jupiter, 82–86, 88, 100–1

Kehler, Robert, 238–42, 244, 293, 300
Kenney, George, 6, 53
Kirtland Air Force Base, 58, 95–96
Klotz, Frank, 225, 237
Kramer, Ken, 22–23, 157, 173
Kregel, Kevin, 206, 210–11
Krikalev, Sergei, 198–99
Kutyna, Donald, 177–78, 180, 182

Langley Air Force Base, 234
Langley Test Center, 94
Leaf, Dan, 235
LeMay, Curtis, 8, 14–15, 54, 109, 119, 121–22, 125, 139, 271, 294
Lipp, James, 15
Lord, Lance, 59, 186, 225, 230–39, 241–42, 244, 300
Lorenzini, Dino, 281
Los Angeles Air Force Base, 20, 22, 58, 98–99, 102, 109, 123, 172, 259–60
Los Angeles, city of, 196, 202

major force program for space, 26
Manned Orbiting Laboratory (MOL), 111
March Field, 50–52
Marine Corps, U.S., 68, 196, 198–200, 202, 268, 286
Mark, Hans, 18, 20, 172–73, 175
Marshall, George, 79, 210
Marshall Space Flight Center, 88
Massachusetts Institute of Technology, 50, 53, 257
Mattis, James, 31, 33, 283
Maxwell Air Force Base, 152, 170, 180, 262, 280. *See also* Air Command and Staff College

McCartney, Forrest, 58
McCaskill, Claire, 216
McElroy, Neil, 16, 84–85
McLucas, John, 19
McNamara, Robert, 16, 112, 139
McNair, Ronald, 197
McPeak, Merrill "Tony," 180–82, 184, 228–29
Medaris, Bruce, 16, 67–90, 100, 299
Mercury, 106–8, 110, 112–13, 194, 202
Military Academy, U.S. *See* West Point
Military Airlift Command, 23
Milley, Mark, 292
Milstar, 27, 158
Minuteman, 57–58, 83, 138, 230, 239, 279
Miramar Naval Air Station, 34
Missile Defense Agency, 283
Mitchell, William "Billy," 48, 187, 241, 294
Mobile User Objective System, 27, 290
moon landing, 7, 58–59, 106, 110, 241, 255, 278
Moorman Space Education and Training Center, 187
Moorman, Thomas, 19, 166–88, 229, 254, 262, 293–94, 300
Mosely, Michael, 240
Musk, Elon, 215, 292
Myers, Richard, 207–8

Nagasaki, 119
National Advisory Committee for Aeronautics, 87
National Aeronautics and Space Administration (NASA). *See* NASA
National Defense Authorization Act (NDAA), 36; Fiscal Year 2018, 30, 33–34; Fiscal Year 2019, 30–31, 286; Fiscal Year 2020, 36–37, 285–86
National Reconnaissance Office (NRO), 16, 21, 28, 104, 168–72, 173–74, 183, 209, 269–70, 283
NASA, 2, 7, 9, 16, 21, 46, 56, 58–59, 67, 87–88, 90, 94, 106–13, 175, 183, 185–86, 194, 197–202, 207, 210, 214, 217, 255–56. *See also* Office of Manned Space Flight

312 | Index

National Public Radio, 291
National Space Act, 16, 87
National Space Defense Center, 29, 270
National Security Space Institute, 187
national security space integration office, 226
National Space Council, 30, 33–34
National Space Defense Center, 29, 270
National War College, 55, 172, 228
Naval Academy, U.S. *See* Annapolis
Naval War College, 280
Navy, Department of the, 33, 287
Navy Space Command, 161
Navy, U.S., 4, 15, 24, 33, 58, 82–83, 85, 88, 90, 100, 104, 145, 161, 175, 181, 184, 186, 195–96, 229–30, 236, 283, 287–88, 290
North American Aerospace Defense Command (NORAD), 8–9, 18, 22, 25, 27, 145, 151–57, 160–61, 173, 180, 186
Northern Command, U.S. (USNORTHCOM), 25, 284

O'Malley, Jerome, 20–23, 156, 169, 171, 174
Obama, Barack, 200, 216
Office of Management and Budget (OMB), U.S., 200
Office of Manned Space Flight, 110–11
Officer, Lyn, 112
Offutt Air Force Base, 168, 263, 265
operationally responsive space office, 33
Operations Group, 30th, 280
Operations Group, 50th, 242
Orion nuclear space propulsion, 139–40
Orion multipurpose space vehicle, 200, 202
Orr, Verne, 22, 161, 172

Pacific Air Forces, 149, 267
Pagano, Vito, 21
Parker, Geoffrey, 4
Patrick Air Force Base, 215–16
Patterson, Charles, 74
Patterson, Robert, 14
Patton, George, 73–74, 204
Patuxent River, Naval Air Station, 195, 197

Pearson, Doug, 259
Pence, Mike, 34
Perry, Robert, 104, 124–26
Peterson Air Force Base, 22, 25, 35, 159, 174, 241, 280
Phillips, Samuel, 58–59
Pilot Training Squadron, 3526th, 147
Pioneer, 88
Polaris, 58, 83, 100
Powell, Colin, 263
Power, Thomas, 118–40, 300–1
Preliminary Design of a World-Circling Spaceship, 15, 131. *See also* RAND
Presidential Directive 37, 20, 153, 171–72
Presidential Directive 42, 153
Program 437, 17
Project Feedback, 15, 124
Puryear, Edgar, 5, 298
Putt, Donald, 97, 99, 121, 129

Quesada, Elwood, 6

Ramo, Simon, 56–57
Ramo-Wooldridge Corporation (R-W), 57, 122–24, 126, 129
RAND, 14–15, 101–2, 124, 131, 135
Raymond, John "Jay," 29, 31, 37, 188, 268, 270–72, 278–94, 299–300
Reagan, Ronald, 24, 161, 279
Reconnaissance Technical Squadron, 432nd, 169
reconnaissance satellite. *See* Corona; Gambit; Hexagon; Weapon System 117L
Redstone Arsenal, 7, 16, 82–83, 129–30. *See also* Army Ballistic Missile Agency
Reed, Jack, 34
Reed, Lisa, 199
Research and Development Command. *See* Air Research and Development Command
Reserve Officer Training Corps (ROTC), 49, 68–69, 168, 257, 279
revolution, 23, 118, 133, 139–40,
revolution in military affairs, 4, 255
Ride, Sally, 212

# Index | 313

Ritland, Osmond, 58, 93–113, 300–1
Roberts, Michael, 4
Rogers, Mike, 32–34, 36, 283, 301
Rumsfeld commission. *See* space commission
Rumsfeld, Donald, 25–26, 33, 185–86, 229–30, 265–67. *See also* space commission
Russia, 30, 99, 198–99, 202, 206–7, 214, 244, 269, 291, 301. *See also* Soviet Union

Salter, Robert, 15
Saltzman, Bradley Chance, 292
Sanborn, Morgan, 17
Sanders, William, 3
satellite communications (SATCOM), 4, 25, 27, 290
satellite communications system, Air Force, 24
Saturn, 87–88, 255
Savage, Bill, 21
Schein, Edgar, 2, 226–27, 235
Schriever Air Force Base, 267. *See also* Falcon Air Force Base
Schriever, Bernard, 1, 3, 5–8, 15–16, 29, 51, 53–54, 82, 93–95, 98–102, 104, 110, 113, 122–34, 138–40, 172, 187, 235–36, 254, 262, 294, 300–1
School of Advanced Air and Space Studies, 284
School of Advanced Military Studies, 32
School of Aerospace Medicine, 112
Scientific Advisory Board, 54–59, 155
Scientific Advisory Group, 14–15, 54. *See also* Scientific Advisory Board
scientific satellite, 84–85, 125, 129–30
Scud, 179, 260–61
Senate, U.S., 30–31, 33, 37, 87, 272
Senate Armed Services Committee (SASC), 22, 269, 272, 283
September 11th attacks, 10, 25, 208, 225, 281, 284
Shannahan, Patrick, 35–36, 272
Shaw, John, 35
Shelton, William, 188, 225–26, 241–45
Shepard, Alan, 203
Soviet Union (USSR), 15–16, 55, 57, 78, 84, 98, 101, 104–5, 112, 125, 134, 138, 148, 151, 168–70, 175, 179, 261, 269. *See also* Russia
Spaatz, Carl "Tooey," 5, 51, 93
Space and Missile Defense Command, Army, U.S., 28
Space and Missile Systems Center (SMC), 26, 38, 229, 237
Space and Missile Systems Organization (SAMSO), 21. *See also* Space Division
space badge, 10, 235–36, 244, 300
Space-Based Infrared System, 27
Space-Based Space Surveillance satellite, 27
Space Command, Air Force (AFSPC or AFSPACECOM or SPACECOM), establishment of, 17–24, 156–61
Space Command, U.S. (USSPACECOM), establishment of, 24–26; reestablishment of, 27–32, 161–62
space commission, 26–27, 34, 185–88, 226, 229–32, 236, 265–67, 282, 287, 294
Space Corps, 26, 33–34, 36–37, 186
space defense, 153–54, 160, 175, 259–60. *See also* Joint Task Force Space Defense
Space Division, of NORAD, 173; of the Air Staff, 20–21, 157, 174; of Air Force Materiel Command's Space and Missile Systems Center, 22; at the Air Force Weapons School, 181
space enterprise vision, 269–70
Space Force, Netflix television series, 288
Space Force, U.S. (USSF), 5, 9–11, 13, 30–38, 46, 59, 118, 139–40, 187, 226, 235, 237, 254, 270, 300–1; advocacy for, 290–91; CSOC as nucleus of, 157; debate over, 225, 240, 272–73, 281–86, 293–94, 300; establishment of, 32–37, 59, 187–188, 245, 266, 285–90; planning task force, 31, 35–36. *See also* Chief of Space Operations (CSO)
Space Foundation, The, 33, 185
Space Group, 595th, 267
space mission force, 269–70
Space Mission Organization Planning Study, 19, 21, 171, 173

Space Operations Command, 38
Space Operations Squadron, 2nd, 242
Space Operations Squadron, 6th (6 SOPS), 263
Space Policy Directive 4, "Establishment of the United States Space Force," 35
space Pearl Harbor, 26, 34, 266, 282
space rapid capabilities office, 33
Space Systems Command, 38
Space Surveillance Squadron, 5th, 280
Space Training and Readiness Command, 38
space warfighting construct, 269
Space Warfare Center, 181, 228
Space Wing, 1st, 174
Space Wing, 21st, 280
Space Wing, 30th, 239, 280
Space Wing, 45th, 215
Space Wing, 50th, 263, 267
Space X, 139, 215, 291–92
Spires, David, 134, 172, 175, 182–83, 186, 229
Sputnik, 16, 84, 102, 236
Sputnik 2, 84
spy satellite. *See* Corona; Gambit; Hexagon; Weapon System 117L
Squadron Officer School (SOS), 148, 205, 231
Stanford University, 53, 146–47, 169, 212
Star Trek, 288
Star Wars. *See* Strategic Defense Initiative
Staver, Robert, 14
Stetson, John, 18
Strategic Aerospace Division, 1st, 279
Strategic Arms Limitation Treaty, 151
Strategic Air Command (SAC), 8, 18–19, 23–24, 85–86, 95–96, 119–22, 125, 132, 138–39, 153–55, 157–58, 160, 168–69, 228, 230, 260–61
Strategic Defense Command, 260
Strategic Defense Initiative (SDI), 24, 161, 175, 259–60
STS-31, 202
STS-45, 198
STS-49, 205
STS-54, 213

STS-60, 198, 301
STS-61C, 197
STS-64, 213
STS-76, 206
STS-78, 213, 299
STS-101
study system requirement program, 135–38
Stuhlinger, Ernst, 86
Sullivan, Kathryn, 198, 200
Swegel, Jeffrey, 32
Systems Command, Air Force (AFSC), 1, 17, 21, 27, 55, 58, 110–11, 133, 155, 157, 173–74, 177, 259

TACSAT-1, 292
Tactical Air Command (TAC), 18, 23, 150–52, 154
Talbott, Harold, 15
Teapot Committee, 15, 121–22
Test Group (Atomic), 4925th, 93, 95
Thomas, Shirley, 5
Thompson, David "D.T.," 187–88
Thor, 57, 82–85, 99–102, 125–26, 129, 132
Titan, 7, 57, 83, 99, 110–11, 129, 138, 175, 231, 239
Titov, Vladimir, 199
Todd, Harold "Pete," 18
Toward New Horizons, 14
Transportation Command, U.S., 180,
Trump, Donald, 30–31, 34–35, 37, 283, 285–86, 288
Twining, Nathan, 5–6, 121
Tyson, Neil de Grasse, 291

U-2, 58, 93, 97–98, 100–2, 104–5, 208
Ultra-High-Frequency Follow-On, 27, 290
Unified Assistance, Operation, 267
Urgent Fury, Operation, 228
USA-193, 209
USSR. *See* Soviet Union; Russia
Uy, Robert, 267

V-1 missile, 77
V-2 missile, 13, 55, 77, 82

Vandenberg Air Force Base, 28–29, 239, 241, 279
Vandenberg, Hoyt, 5–6
Vanguard, 16, 84, 86
Van Inwegen, Earl, 21–22,
Vautrinot, Suzanne, 214
Vietnam, 4, 8–9, 144, 149–50, 169, 196, 228, 257, 282
Von Braun, Wernher, 5, 7, 14, 16, 55, 58, 77, 82–84, 86–88, 90, 255–56
Von Kármán, Theodore, 5, 13–14, 51, 54, 294
Von Neumann, John, 15, 123, 125, 132, 134

Weapon System 117L (WS-117L), 16, 125–30, 132–34, 139
Webb, James, 59, 112
Weinberger, Caspar, 24
Welch, Larry, 176–77
West Point, 22, 146–48, 157, 278–79
Western Development Division (WDD), 15–16, 57, 98, 121–26, 129–34, 139
White, Ed, 256
White House, 30, 32, 34–35, 37, 87, 285

White Sands Missile Range, 82, 86
White Space Trophy, 161, 179
White, Thomas, 5, 132, 138–39
Wideband Global SATCOM, 27
wideband SATCOM operations center, 290
Widnall, Sheila, 182, 184
Wilcutt, Terrence, 199, 214
Wilson, Charles, 83–84
Wilson, Heather, 31, 33, 35–36, 83, 272
World War I, 1, 45, 48–49, 53
World War II, 5, 7–8, 13–14, 45, 50, 52–55, 80–81, 93–95, 100, 119–20, 134, 138, 144, 167, 204, 207, 235
Wright Air Development Center, 16, 125, 133
Wright Field, 7, 51, 93, 95, 101–2
Wright, Orville and Wilbur, 50
Wynne, Michael, 239

X-20. *See* Dyna-Soar
X-37B, 27

Zuckert, Eugene, 110, 139

The **Naval Institute Press** is the book-publishing arm of the U.S. Naval Institute, a private, nonprofit, membership society for sea service professionals and others who share an interest in naval and maritime affairs. Established in 1873 at the U.S. Naval Academy in Annapolis, Maryland, where its offices remain today, the Naval Institute has members worldwide.

Members of the Naval Institute support the education programs of the society and receive the influential monthly magazine *Proceedings* or the colorful bimonthly magazine *Naval History* and discounts on fine nautical prints and on ship and aircraft photos. They also have access to the transcripts of the Institute's Oral History Program and get discounted admission to any of the Institute-sponsored seminars offered around the country.

The Naval Institute's book-publishing program, begun in 1898 with basic guides to naval practices, has broadened its scope to include books of more general interest. Now the Naval Institute Press publishes about seventy titles each year, ranging from how-to books on boating and navigation to battle histories, biographies, ship and aircraft guides, and novels. Institute members receive significant discounts on the Press' more than eight hundred books in print.

Full-time students are eligible for special half-price membership rates. Life memberships are also available.

For more information about Naval Institute Press books that are currently available, visit www.usni.org/press/books. To learn about joining the U.S. Naval Institute, please write to:

Member Services
U.S. Naval Institute
291 Wood Road
Annapolis, MD 21402-5034
Telephone: (800) 233-8764
Fax: (410) 571-1703
Web address: www.usni.org